Lipid Oxidation in Food

ACS SYMPOSIUM SERIES **500**

Lipid Oxidation in Food

Allen J. St. Angelo, EDITOR
Agricultural Research Service
U. S. Department of Agriculture

Developed from a symposium sponsored
by the Division of Agricultural and Food Chemistry
of the American Chemical Society
at the Fourth Chemical Congress of North America
(202nd National Meeting of the American Chemical Society),
New York, New York,
August 25–30, 1991

American Chemical Society, Washington, DC 1992

Library of Congress Cataloging-in-Publication Data

Lipid oxidation in food / Allen J. St. Angelo, editor.

p. cm.—(ACS symposium series, ISSN 0097–6156; 500)

"Developed from a symposium sponsored by the Division of Agricultural and Food Chemistry of the American Chemical Society at the Fourth Chemical Congress of North America (202nd National Meeting of the American Chemical Society), New York, New York, August 25–30, 1991."

Includes bibliographical references and indexes.

ISBN 0–8412–2461–7

1. Lipids—Oxidation—Congresses. 2. Food spoilage—Congresses.

I. St. Angelo, Allen J. II. American Chemical Society. Division of Agricultural and Food Chemistry. III. American Chemical Society. Meeting (202nd: 1991: New York, N.Y.) IV. Series.

TX533.L5L56 1992
664'.028—dc20
92–19970
CIP

The paper used in this publication meets the minimum requirements of American National Standard for Information Sciences—Permanence of Paper for Printed Library Materials, ANSI Z39.48–1984. ∞

Copyright © 1992

American Chemical Society

All Rights Reserved. The appearance of the code at the bottom of the first page of each chapter in this volume indicates the copyright owner's consent that reprographic copies of the chapter may be made for personal or internal use or for the personal or internal use of specific clients. This consent is given on the condition, however, that the copier pay the stated per-copy fee through the Copyright Clearance Center, Inc., 27 Congress Street, Salem, MA 01970, for copying beyond that permitted by Sections 107 or 108 of the U.S. Copyright Law. This consent does not extend to copying or transmission by any means—graphic or electronic—for any other purpose, such as for general distribution, for advertising or promotional purposes, for creating a new collective work, for resale, or for information storage and retrieval systems. The copying fee for each chapter is indicated in the code at the bottom of the first page of the chapter.

The citation of trade names and/or names of manufacturers in this publication is not to be construed as an endorsement or as approval by ACS of the commercial products or services referenced herein; nor should the mere reference herein to any drawing, specification, chemical process, or other data be regarded as a license or as a conveyance of any right or permission to the holder, reader, or any other person or corporation, to manufacture, reproduce, use, or sell any patented invention or copyrighted work that may in any way be related thereto. Registered names, trademarks, etc., used in this publication, even without specific indication thereof, are not to be considered unprotected by law.

PRINTED IN THE UNITED STATES OF AMERICA

1992 Advisory Board

ACS Symposium Series

M. Joan Comstock, *Series Editor*

V. Dean Adams
Tennessee Technological
  University

Mark Arnold
University of Iowa

David Baker
University of Tennessee

Alexis T. Bell
University of California—Berkeley

Arindam Bose
Pfizer Central Research

Robert F. Brady, Jr.
Naval Research Laboratory

Margaret A. Cavanaugh
National Science Foundation

Dennis W. Hess
Lehigh University

Hiroshi Ito
IBM Almaden Research Center

Madeleine M. Joullie
University of Pennsylvania

Mary A. Kaiser
E. I. du Pont de Nemours and
  Company

Gretchen S. Kohl
Dow-Corning Corporation

Bonnie Lawlor
Institute for Scientific Information

John L. Massingill
Dow Chemical Company

Robert McGorrin
Kraft General Foods

Julius J. Menn
Plant Sciences Institute,
  U.S. Department of Agriculture

Vincent Pecoraro
University of Michigan

Marshall Phillips
Delmont Laboratories

A. Truman Schwartz
Macalaster College

John R. Shapley
University of Illinois
  at Urbana–Champaign

Stephen A. Szabo
Conoco Inc.

Robert A. Weiss
University of Connecticut

Peter Willett
University of Sheffield (England)

Foreword

THE ACS SYMPOSIUM SERIES was first published in 1974 to provide a mechanism for publishing symposia quickly in book form. The purpose of this series is to publish comprehensive books developed from symposia, which are usually "snapshots in time" of the current research being done on a topic, plus some review material on the topic. For this reason, it is necessary that the papers be published as quickly as possible.

Before a symposium-based book is put under contract, the proposed table of contents is reviewed for appropriateness to the topic and for comprehensiveness of the collection. Some papers are excluded at this point, and others are added to round out the scope of the volume. In addition, a draft of each paper is peer-reviewed prior to final acceptance or rejection. This anonymous review process is supervised by the organizer(s) of the symposium, who become the editor(s) of the book. The authors then revise their papers according to the recommendations of both the reviewers and the editors, prepare camera-ready copy, and submit the final papers to the editors, who check that all necessary revisions have been made.

As a rule, only original research papers and original review papers are included in the volumes. Verbatim reproductions of previously published papers are not accepted.

M. Joan Comstock
Series Editor

Contents

Preface xi

1. **Lipid Oxidation in Foods: An Overview** 1
   J. R. Vercellotti, Allen J. St. Angelo, and
   Arthur M. Spanier

   ## MECHANISMS OF FREE-RADICAL PROCESSES

2. **Mechanisms of Lipid Oxidative Processes and Their Inhibition** 14
   Michael G. Simic, Slobodan V. Jovanovic, and Etsuo Niki

3. **Lipid Oxidation in Muscle Foods via Redox Iron** 33
   Eric A. Decker and Herbert O. Hultin

4. **Mechanism of Nonenzymic Lipid Peroxidation in Muscle Foods** 55
   Joseph Kanner

5. **Role of Lipoxygenases in Lipid Oxidation in Foods** 74
   J. Bruce German, Hongjian Zhang, and Ralf Berger

6. **Relationship Between Water and Lipid Oxidation Rates: Water Activity and Glass Transition Theory** 93
   Katherine A. Nelson and Theodore P. Labuza

7. **Lipid Oxidation: Effect on Meat Proteins** 104
   Arthur M. Spanier, James A. Miller, and John M. Bland

   ## PREVENTION OF LIPID OXIDATION

8. **Maillard Reaction Products and Lipid Oxidation** 122
   Milton E. Bailey and Ki Won Um

9. Chemical and Sensory Evaluation of Flavor in Untreated and Antioxidant-Treated Meat 140
   Allen J. St. Angelo, Arthur M. Spanier, and Karen L. Bett

10. Prevention of Lipid Oxidation in Muscle Foods by Nitrite and Nitrite-Free Compositions 161
    Fereidoon Shahidi

11. Lipid Oxidation of Seafood During Storage 183
    George J. Flick, Jr., Gi-Pyo Hong, and Geoffrey M. Knobl

12. Seafoods and Fishery Byproducts: Natural and Unnatural Environments for Longer Chain Omega-3 Fatty Acids 208
    R. G. Ackman and H. Gunnlaugsdottir

METHODOLOGIES FOR ASSESSING LIPID OXIDATION PRODUCTS

13. Gas Chromatographic Analyses of Lipid Oxidation Volatiles in Foods 232
    J. R. Vercellotti, O. E. Mills, Karen L. Bett, and D. L. Sullen

14. Characterization of Off-Flavors by Aroma Extract Dilution Analysis 266
    Werner Grosch, Ute Christine Konopka, and Helmut Guth

15. Sensory Evaluation of Lipid Oxidation in Foods 279
    G. V. Civille and C. A. Dus

PROCESSING EFFECTS ON LIPID OXIDATION

16. Influence of Food Processing on Lipid Oxidation and Flavor Stability 292
    Hans Lingnert

17. Factors Affecting Lipid Autoxidation of a Spray-Dried Milk Base for Baby Food 302
    J. P. Roozen and J. P. H. Linssen

18. Effect of Lipid Oxidation on Oil and Food Quality in Deep Frying 310
    Edward G. Perkins

19. **Effect of Storage on Roasted Peanut Quality: Descriptive Sensory Analysis and Gas Chromatographic Techniques** 322
    Karen L. Bett and T. D. Boylston

20. **Changes in Lipid Oxidation During Cooking of Refrigerated Minced Channel Catfish Muscle** 344
    M. C. Erickson

Author Index 351

Affiliation Index 351

Subject Index 352

To Althea
for Always Being There

Preface

PROBLEMS RELATED TO LIPID OXIDATION IN FOODS are still with us three decades after the first symposium on foods was held at Oregon State University. The mechanisms have been upgraded from the days of mere "autoxidation or metal catalysis" to catalytic systems that include free radicals, singlet oxygen, superoxide radicals, and Haber–Weiss–Fenton reactions. Far-reaching consequences of such oxygen chemistry on food quality and human health have also been discovered. However, what was needed 30 years ago is still needed today: conferences to bring together those scientists who conduct basic studies and those who apply the technologies to solve practical problems.

The first symposium was so successful that a second biennial symposium was held in September 1961; the proceedings were subsequently published (*Symposium on Foods: Lipids and Their Oxidation*; Schultz, H. W.; Day, E. A.; Sinnhuber, R. O., Eds.; Avi Publishing: Westport, Conn., 1962). In the preface, Schultz stated that the purpose of the symposium was "to bring together scientists who concern themselves with research not directly related to commercial foods and scientists who are attempting to find practical solutions to problems pertinent to the commercial food supply".

The mission of the symposium on which this book is based was to bring together many scientific leaders in the field to present their latest basic findings on lipid oxidation in foods and to stimulate their fellow scientists to participate in discussions on the up-to-date mechanisms and methodologies presented. We are all aware of the essential roles that lipids play in the metabolism of cells in regard to providing a source of high energy and reserve storage. However, lipid oxidation can lead to primary and secondary products that affect flavor, aroma, taste, nutritional value, and in general, overall quality. Overall quality formed the subject of this symposium.

The book is organized into four sections. Following an overview chapter, the first section covers mechanisms and describes the latest theories on how lipids oxidize, the different catalysts that promote oxidation, including free-radical chemistry, iron-redox systems, and enzymic and nonenzymic catalysts. The next section discusses prevention of lipid oxidation through the use of natural and synthetic antioxidant systems. In the third section, methodologies are presented for assessing lipid oxidation products and for identifying compounds that contribute to specific flavor and aroma characteristics. Finally, the last section discusses particular commodities, the influence of processing on lipid oxidation, and chemical changes that take place in specific foods as they relate to lipid oxidation. A significant point is made in several chapters, particularly in the last one, that free radicals not only generate off-flavors by oxidative breakdown of polyunsaturated fats but also actively contribute to destroying molecules that make positive flavor attributes.

This book contains new and pertinent information on lipid oxidation in muscle foods such as beef, pork, marine and freshwater fish, and in other commodities such as baby foods, vegetable oils, and peanuts. It is my hope that this book will inspire food scientists and others engaged in related areas of research to produce higher quality and healthier foods to benefit the consumer.

Finally, I am indebted to my coworkers, John R. Vercellotti and Arthur M. Spanier, for serving as session chairmen and for their assistance in writing the introductory chapter. I also extend my sincere gratitude to the many scientists who reviewed the chapters found herein.

ALLEN J. ST. ANGELO
Agricultural Research Service
U.S. Department of Agriculture
New Orleans, LA 70124

October 8, 1991

Chapter 1

Lipid Oxidation in Foods
An Overview

J. R. Vercellotti, Allen J. St. Angelo, and Arthur M. Spanier

Agricultural Research Service, U.S. Department of Agriculture, Southern Regional Research Center, 1100 Robert E. Lee Boulevard, New Orleans, LA 70124

>Very real and important problems remain in food quality preservation from free radical oxidative deterioration. The chemistry of polyunsaturated fatty acids and their reactivity to oxygen and carbon to carbon bond scissions or rearrangements in foods is an active, exciting area. Each paper in this *Symposium on Lipid Oxidation in Foods* posed unique challenges. Solutions involve understanding mechanisms of free radical oxidation and their control through critical chemical interactions. The future food supply as well as nutritional health of the public and institutions supporting highest quality of products for the consumer will benefit from knowledge shared on this occasion.

Lipids play a vital role in the metabolism of cells by providing a source of energy and reserve storage materials. As lipids oxidize, they form hydroperoxides, which are susceptible to further oxidation or decomposition to secondary reaction products such as aldehydes, ketones, acids and alcohols. In many cases, these compounds adversely affect flavor, aroma, taste, nutritional value and overall quality. It is the latter overall food quality that formed the basis of this symposium. Therefore, the mission of this symposium was to serve as a forum for scientific leaders in the field of lipid oxidation to present their results and to stimulate fellow scientists to participate in discussions on the latest mechanisms and methodologies for the purpose of producing food products of highest quality.

Mechanisms of Free Radical Processes Involving the Deterioration of Food Lipids.

There are many catalytic systems that can oxidize lipids. Among these are light, temperature, enzymes, metals, metalloproteins and microorganisms. Most of these reactions involve some type of free radical and/or oxygen species, the chemistry of which will be described in this first section. One of the purposes of this symposium is to cover the overall phenomenon of oxygen free radical modification of food lipids.

Introductory remarks in Chapters 2 through 7 combined the long term experience of researchers to study mechanisms involving free radicals as they affect lipid oxidative processes in foods *(1)*. The oxidation of lipids occurs to a much fuller extent when the polyunsaturated glycerides are in bilayers such as those that occur in cellular membranes. Inhibitors such as BHA, BHT, and propyl gallate decrease in redox potential according to Hammett sigma plot substituent effects. Key questions still remain about the energetics of propagation of these radical species and their reactivity with molecules other than allylic sites on polyunsaturated fatty acids.

In lipid oxidation by way of redox iron, not all oxygen radical types are generated by the same forms of iron *(2)*. In various organelles such as mitochondria, Fe(III)-ADP and NADH are required for enzyme catalyzed lipid oxidation by oxygen free radical generation. In the sarcoplasmic reticulum of muscle foods these cofactors are also involved in the generation of oxygen radicals. Chelators such as pyrophosphate, citrate, EDTA, etc., prevent these reactions by occupying all reaction sites and by disrupting redox cycles. Ceruloplasm (a copper containing plasma protein which functions as a peroxidase) is also an effective inhibitor of free radical oxidations as are superoxide dismutase and catalase found in many food systems.

The actual sources and distribution of iron catalyzing oxygen activation in foods was described in Chapter 3. It was pointed out that >95% of the iron in muscle is protein bound as ferritin, myo- and hemoglobin, etc. The iron from ferritin is released and non-enzymic oxygen activation occurs via Haber-Weiss reactions. Superoxide oxygen radical-anion can be a product of leucoflavins and xanthine oxidase and results from these processes involving iron, ascorbic acid, NAD(P)H (as reducing agent), cysteine, and hydrogen peroxide *(3)*. Ascorbic acid is present at 3.5-15 micromoles/100 g in typical muscles (turkey or mackerel) and so is an abundant source of redox prooxidant at higher concentrations. The decay curve of ascorbic acid is such that it disappears about seven days post mortem. Halogen anions are also prooxidants in muscle: fluoride<chloride<bromide< iodide. The chloride operates through Fe(III). In the sarcoplasmic reticulum, the reaction ADP-Fe(II)<-----NADH ---->ADP-Fe(III) occurs to regenerate the reduced species. This same chemistry also applies in the mitochondrion in other animal and plant tissue. Although chelators are promising for control of iron activity in muscle, they must occupy all the available reaction sites for coordination. The redox cycles that regenerate the active iron species [Fe(II)] must be blocked to prevent more Fe(III) from getting to the reaction site. Pyrophosphates, ceruloplasm, phytate, citrate, etc., tend to decrease available iron, but still require the elimination of oxygen. Reducing agents tend to create serious color problems in meats. Microsomal enzyme activity needs to be controlled also because of release of iron as well as hydrolyses of pyrophosphates.

Alternate pathways of non-enzymic lipid peroxidation as they occur in foods after cellular injury are described in Chapter 4. These processes are active in muscle tissue pre- and postslaughter, and are dependent on balances of pH, availability of metalloenzymes, or free metal ion and its complexes *(4)*. The ferryl group (Fe=O) is formed by cytosolic hydrogen peroxide *in vivo* from peroxisomes. From the ferryl activated species superoxide radical anion and hydroxyl groups then propagate *(5)*. In fish, meat, dairy products, and oilseed fractions, these highly reactive groups cause destruction of polyunsaturated fatty acids. Kanner *(4, 5)* showed that introduction of synthetic chelator (EDTA), phytate, and desferryl mesylate causes a significant decrease in oxidation occurs. Other free radical scavengers such as mannitol, methyl sulfoxide, glucose, or formate are ineffective. Ascorbate oxidase is the chief enzyme to prevent propagation of free radicals by reversing the process of Fe(III)----> Fe(II), and oxidized ascorbic acid is probably the chief protectant in muscle foods.

The role of lipoxygenases in lipid oxidation of muscle foods such as meat and fish as well as in oilseed products is shown in Chapter 5. The volatiles made by lipoxygenases differed in their mode of attack on polyunsaturated fatty acids from autoxidations by light or transition metal catalysis. Retro-aldol reactions of these conjugated hydroperoxides contributed important fish flavor attributes *(6)* through formation of volatiles such as hexanal, 1-penten-3-ol, 1-octen-3-ol, 1,5-octadien-3-ol, 2-octen-1-ol, and 2,5-octadien-1-ol. Molecules such as TBHQ stabilize the enzyme and 12 HETE has a $t_{1/2}$ of 1 to 2 hours, whereas 15 HETE is much more stable. The transition metal catalyzed reactions yield a racemic mixture and proliferation of many carbonyl, alkane, or alcohol products. In either case, the vague quality of "freshness" is lost, i.e., the expected and anticipated sensory determinants of the food, as interpreted by the brain, are diminished. Freshness also implies contributions of essential volatile oxidation products from oxidative processes and not just off-flavors resulting from lipid oxidations. Utilization of fish gill lipoxygenases in generation of fish flavor oxidation products from lipids has potential for biotechnological production of flavoring agents for processed seafood *(6)*.

The relationship between rates of lipid oxidation and moisture content is complex *(7)* and the moisture composition, water activity, and state of water in the food must be considered in predicting stability *(8)*. Water activity of a food represents a physical state of the matrix which protects lipids from oxidation by glass transitions of the protein and polysaccharide structures, giving a type of microencapsulation effect *(9)*. In Chapter 6 the collapse of the glassy encapsulating matrix of fatty components is considered and, although it is not certain whether it will always protect flavor, it is certain that the control of moisture is important for the control of lipid oxidation in many foods.

Polypeptides are also susceptible to attack by oxygen free radicals with degradation to products influencing meat flavors *(10)*. Chapter 7 presents a novel approach to the study of the interactions of several classes of flavor molecules in food. It shows that some proteins of meat are directly affected by the free radicals and lipid oxidation while others are affected only secondarily by these reactions. Free radical oxidations, as propagated through lipid hydroperoxides and other catalysts initiating these reactions *(11)*, are responsible for the changes in the functionality of several proteins in muscle foods and for changes observed in important meat peptides with a meaty taste as well as amino acid and peptide Maillard flavor precursors *(12)*. The reactive sites of free radical degradations and the flavor products they generate are very poorly understood for proteins and peptides when compared to what is known about lipids, lipid oxidation, and the off-flavors produced *(13)*. The susceptibility, disappearance, activation, or inhibition of key meaty flavor peptides and meat enzyme/enzyme-systems by the free radical reactions and cascade of lipid oxidation have been followed by capillary electrophoresis, HPLC, and analysis of enzyme activity along with changes in the lipid system *(12)*.

This work should stimulate a significant new research towards understanding the action and interaction of food lipids with other food components.

Autoxidation of Foods and Its Prevention.

Preservation of meat flavor by inhibition of lipid oxidation with free radical scavengers generated from natural meat extract as well as synthetic precursors has been a worthwhile goal of meat flavor research *(14)*. The use of alternate antioxidant systems that affect lipid oxidation in foods is discussed in Chapters 8 through 12. The first chapter on this topic (Chapter 8) describes the role played by Maillard reaction reductones and heterocyclic products as natural flavoring agents that inhibit flavor deterioration of foods. The authors show that cooked meat exudate or synthetic mixtures of its components as occur in gravies, sauces, or syrups and used to accentuate flavors, inhibit free radical oxidation. A well correlated reduction of the off-flavors by such synthetic inhibitors of free radical reactions was previously shown *(15)*. The inhibition of free radical processes is attributed to reductones in the synthetic mixtures or the cooked extracts of meat. As electron donors, these naturally formed reductones quench propagation of free radical chain reactions.

Synergistic interaction of antioxidants and the effects of diminished oxygen concentration on antioxidant efficiency has long been a question of food manufacturing practice *(16)*. The chemical and sensory evaluation of antioxidant effects on meat flavor and its preservation *(17)* was studied with compounds that function as chelators and those that are free radical scavenging agents. Correlation of elements contributing to meat flavor quality have been examined with respect to microtemperature environments that lead to flavor-zones. Structurally dependent barriers to oxygen in meat were examined in the presence and absence of oxygen (Spanier, A.M., unpublished data) as they influence sensory attributes, which are correlated to instrumental and chemical characteristics. As demonstrated further in Chapter 9, vacuum exerts a very strong enhancing effect on antioxidants of all kinds, extending shelf life and, overall, making a product of much greater flavor value.

Although a matter of controversy in health safety, nitrite remains a key inhibitor of oxidation in preservation of processed muscle foods *(18)*. The differences between phenolic or semiquinone organic free radical inhibitors (e.g., propyl gallate or TBHQ) and nitrite in meat curing and flavor or color preservation had been earlier shown through correlations between TBARS, sensory scores, and key lipid oxidation products such as hexanal in meat flavor deterioration reactions *(19)*. Nitrite is probably acting in several ways to bring about its antioxidative effects which range from formation of nitric oxide after nitrite complexes with heme pigments to the formation of antioxidant species from its reaction with polar lipid compounds. Nitric oxide is a highly efficient radical chain terminator, and nitrite effectively inhibits lipid oxidation in foods at concentrations of less than 50 ppm. In Chapter 10, flavonoids and phenolic benzoic acids are shown to inhibit free radicals as do phenolic (cinnamic type) acids. Herbs or oleoresinous extracts *(20)* inhibit lipid oxidation in meat with an additive protective effect from trisodium polyphosphate. On the other hand, the additive should not be permitted to alter flavor significantly even though it does inhibit lipid oxidation.

Postharvest preservation of fish to prevent flavor degradation caused by lipid oxidation is particularly difficult for the marine fisheries industry *(7)*. Lipid oxidation of fish during storage-after-catch is a widespread problem and one that varies with species enzyme, degree of hydration of the flesh, climate during the time after catch, transportation, temperature, microbial population, microconstituent concentrations such as mineral salts, and quality of cooking *(21)*. Peroxide values are lower in salted sardines, and the lower the salt the faster the hydrolysis to free fatty acids *(7)*. Immediate postharvest preservation is necessary to preservation of fish value according to the report in Chapter 11, and questions are raised that cover fish production from catch to distribution and final cooking.

Peroxidation of fish fat after the catch or during storage and its preservation through naturally occurring inhibitors are both interesting and important questions in food lipid oxidation research. The peroxidation of fish skin glyceride lipids produces characteristic fish flavor attributes *(21)*. There seems to be present an endogenous antioxidant which regulates production of these fish flavor molecules under appropriate physiological conditions. Natural antioxidants such as tocopherol are found seasonally in fish. They come from marine plants or algae and drop to near zero in July and August. Therefore, fish like sole caught in those months contain no natural antioxidants *(21)*. Fish meal fed to farm animals or aquaculture species must be clear of oxidation products so that they do not taint the fatty meat animal muscle. In Chapter 12, BHA and BHT are shown to function well to prevent this oxidation in fish meal. Ethoxyquin or santoquin are better in keeping polyunsaturated fish oil chains from oxidizing and preventing free fatty acids from building up in the feed. Fish meal has a very high conversion ratio as a feed with 1.1 lbs converted to 1.0 lb of muscle food when used as a nutrient. Prevention of lipid oxidation in this important feedstuff will be a continuing challenge to the industry.

Methodology for Assessing Lipid Oxidation Products as Affecting Food Quality.

Determining the extent of lipid oxidation by chemical means, i.e., peroxide values, TBA numbers, etc., or by instrumental techniques, i.e., gas chromatography, are important measurements of food composition. However, when analyzing a food for quality, it is very important to know how the food tastes, smells, etc. Therefore, Chapters 13 through 15 introduce sensory aspects of measuring food quality.

An efficient estimation of lipid oxidation volatiles in foods requires a combined GC/sniffer port profiling of flavor quality. As an added perspective on delineating the effects of lipid oxidation in food flavor, Chapter 13 provides a brief history of the question and new approaches being developed. Improved concentration of volatiles with an external closed inlet device (ECID) *(22)* as well as better capillary separations are shown. Moisture in samples not only freezes at cryogenic trapping temperature but also lowers the efficiency of the bonded phase of the capillary column. Volatiles trapping on solid phase in high yield along with high resolution gas chromatography of the separated components must ideally be resolved to permit dividing the GC effluent peaks for FID and sniffer port analysis by human bioassay. Few economically important foods have been characterized completely for flavor intensity contributions of component volatiles. An understanding of how different compounds and their mixtures elicit sensations of taste and smell and subsequently interact to generate flavor experiences requires that volatile compounds in a food be separated and evaluated for olfactory contribution to flavor quality. Most highly significant flavor components are very low threshold aroma compounds (ppb or much lower concentrations) that form only small peaks in the capillary GC baseline. With the advent of improved gas chromatographic and computerized olfactory methodology (23), the complex stimuli in aromas can now be sorted out for their individual contributions to flavor and their locations in the GC profile of the food's aromatics.

Aroma extract dilution analysis for comparison of olfactory intensity with GC quantitation of lipid oxidation products presents a powerful tool for characterizing food flavor problems (24) in Chapter 14. The perceptions of food flavors depend on the size of food sample, the amount of fat present, dilution of the odor in the vapor phase, and aroma concentration *vs* its threshold of perception (25). Guth and Grosch (26) reported that it is equally important to consider generation of lipid oxidation products with simultaneous destruction of positive flavor attributes to define deterioration of flavor. A stable isotope dilution assay was developed for quantification of important lipid oxidation volatiles by gas chromatography/mass spectroscopy: *n*-hexanal, 1-octen-3-one, (Z)-2-octenal, (Z)-2-nonenal, (*E,E*)-2,4-nonadienal, (*E,E*)-2,4-decadienal, *trans*-4,5-epoxy-(*E*)-2-decenal, and 3-methyl-2,4-nonanedione (26). With a stable isotope method and GC/MS, it was possible to analyze quite accurately parts per billion quantities of off-flavors in oils.

Earlier Grosch et al. (25) had defined the differences between samples with flavor dilution-factors (FD) as the ratio of the concentration of the compound in the initial extract to the concentration of the compound after serial dilution to the most diluted form in which the odor was still detected by GC/O (O = olfactory or sniffing analysis). On storage of cooked beef at refrigerated temperature, the FD-factors of significant lipid oxidation products (26) increased from the freshly cooked to the meat flavor deteriorated form during the storage period. In terms of comparison to volatiles from fats (26), five compounds were delineated by flavor units (FU values obtained by dividing concentration of a flavor compound by its olfactory sensory threshold) as the lipid oxidation components of meat storage flavor deterioration: *n*-hexanal, 1-octen-3-one, (*E,E*)-2,4-nonadienal, (*E,E*)-2,4-decadienal, and *trans*-4,5-epoxy-(*E*)-2-decenal. This work supports the role of *n*-hexanal as a key contributor of green, tallowy notes to the GC profile of off-flavored lipid oxidation products in meat flavor deterioration or the so called "warmed-over flavor". *trans*-4,5-Epoxy-(*E*)-2-decenal is probably also a principal contributor to the olfactory response in "warmed-over flavor" of meat according to their GC/O analyses, and it causes a metallic off-flavor in the stored cooked meat. A question of some interest concerns long chain furan fatty acids in glycerides that are precursors for many off-flavors because the furan ring is opened by peroxidation to phenolic cyclization products (26). In Chapter 14 these furan acids in vegetable oils, fish, meat, and other foods such as nuts, are shown to be a potential source of serious off-flavors. The plants probably make the furan carboxylic acids as natural products, which are eaten by domestic animals or fish. They are then assimilated into fatty tissue in meat or passed into milk and dairy products. For instance, a furan carboxylic acid can be oxidized to 2-methoxyphenol, which has a burnt meaty odor, or to 2,5-dimethoxyphenol, which has a smokey, burnt odor. This area of furan carboxylic acids in fats as off-flavor precursors needs more research but is an area of concern.

Descriptive sensory analysis as an objective measurement of flavor quality as well as discrimination of attributes has developed into a scientifically sophisticated tool. There is a great emphasis today on accurately measured sensory evaluation of foods because consumer acceptance depends upon high impact olfactory and taste stimulation that is reproducible throughout the marketing chain *(27)*. To carry out these food quality and value processes successfully, descriptive measurements must involve evoking psychophysical responses from a trained sensory panelist evaluating a food and measuring flavor quality by calibrated intensity scales *(28)*. These panelists should be capable of accuracy not only in their qualitative identification of attributes with defined descriptors but also in conditioned reflex for estimation of that attribute's intensity. The professional food scientist must analyze the data generated by the panels using modern computerized statistical methods and compare these values for validity with analytical standards generated from expected attributes in consumer-acceptable foods. A recurring phenomenon in foods as they age is the loss of defining positive flavor attributes and the simultaneous generation of off-flavors. Accurate measurement by olfactory methods is necessary if profiles of change are to be plotted or comparisons of sensory intensities made with other chemical and physical changes occurring in the food. In Chapter 15, a leading expert and consultant to the food, cosmetics, and personal care products industry details examples of modern industrial application of descriptive sensory analysis: meat flavor deterioration (including poultry) *(29)*; the dynamics of roasted peanut flavor *(30)*; generation and application of a lexicon of descriptors for pond raised catfish *(31)*; and vegetable oil flavor scoring.

Food Processing as it Affects Lipid Oxidation.

Postharvest and storage effects promote gradual lipid oxidation with decreases in shelf life stability due to continuing organic chemical processes set up by the initial peroxidations *(32)*. The effects of postharvest storage and processing of the raw commodities on lipid oxidation formed the theme of the last five chapters. Appropriate steps must be taken all through the life of a food to insure initial high quality as well as a long term stability, according to the authors of Chapter 16. From harvest, postharvest raw storage, processing, and packaging through distribution to the retailer and shelf life periods cumulative damage can result in an unacceptable product. Thus, in potato products, lipid oxidation is occurring at the surface as well as inside. In meat that is extruded, comminuted, flaked-and-formed, chopped, or mixed with other ingredients that expose membrane surface to oxygen, much processing damage is done leading to permanent off-flavors. Lipases, lipoxygenases, and metals can catalyze destruction of foods through modification of lipids *(33)*. However, Maillard reaction products are excellent oxidation inhibitors. Synthetic browning products, e.g., histidine/glucose, can also be used to protect against lipid oxidation. Carbon dioxide is more soluble in fatty foods than is nitrogen as an inert gas, but neither are as good as a vacuum. Fermentations such as aging of sausage also produce antioxidants. A caution that referenced A. L. Tappel's earlier work reminds the audience that lipid oxidation products are not only contributors to poor flavor quality or physical properties, but are also potential health hazards that promote physiological deterioration of muscle function by cross linking proteins or initiating tumors.

A particularly challenging problem in food processing is the preservation of milk-based dried powdered infant formulations from lipid oxidation. Processed dairy products are difficult to protect from lipid oxidation effects. In earlier publications cited in Chapter 17 spray dried baby foods with high levels of polyunsaturated fatty acids present are very susceptible to lipid oxidation *(34)*. Enrichment with minerals and vitamins also contribute to enhancing off-flavors by promoting off-flavor reactions. By careful control of water activity over a period of 2 to 13 weeks, reductones determined by *p*-anisidine values *(35)* showed that more lipid oxidation damage is done at 37° C and a_w of 0.11 and 0.34, respectively, than at 20° C and a_w 0.24. The formation of browning spots in the dried product is least under this latter condition. In order to control the water activity more closely at 0.24, sodium acetate was incorporated into the sample. Chloride salts of minerals contribute less to oxidation than carbonates.

Effects of lipid oxidation degradation products on vegetable oils used in frying foods has great significance for many food processing industries of the world as addressed in Chapter 18. The "fast food" segment is especially sensitive to these changes from lipid oxidation off-flavors that are carried over into the eaten food product *(9)*. Over a period of only a few hours of use, high concentrations of non-saponifiable, unoxidized, low molecular weight acids, cyclic monomers of fatty acids, dimers or trimers of fatty acids, hydroxylated or oxo-forms of the unsaturated fatty acids, and polar polymers of acids or glycerides develop *(36)*. Consumer tests show that after 3 to 5 days of use, the deterioration of flavor quality of the fried food (from the oil) is best correlated with the percent of polar acids or oxygenated polymers present in the frying oil.

Autoxidation as it affects both positive *and* negative flavor attributes of roasted peanuts during storage is an example of lipid oxidation, free radical chemistry, and olfactory change that is quite complex. It has been found by descriptive sensory analysis *(27,28,30)* that when roasted peanuts undergo autoxidation, positive attributes such as roasted peanutty decrease while negative character notes such as cardboard and painty increase. Using a well tested headspace GC method on peanut volatiles (e.g., quantitation as well as internal/external standards for recovery estimation), the authors of Chapter 19 were able to designate a large number of reliable markers for statistical comparison with descriptive sensory data (Boylston, T.D., Washington State University, personal communication). A number of questions arise concerning the relationships of sensory changes measured and the physical as well as perceptual loss of pyrazines and other high impact heterocycles during the oxidative storage conditions of the roasted peanuts. For instance, one suggestion is made that the roasted peanutty flavor principles were not really being destroyed by free radical reactions but rather were being adsorbed into the hydrocolloid matrix consisting of protein, hydratable polysaccharides or cell wall material, and lipid. Such encapsulations are common in foods and can be freed by suspension in aqueous salt medium. The other aggravating factor to roasted peanutty flavor is the high concentrations of lipid oxidation products *(ppm)* which tend to occupy many olfactory epithelium sites and literally mask out the heterocycles there in much lower *(ppb)* concentrations. The data reported in this chapter are both intriguing and original and hopefully will bring answers to these questions.

Cumulative hydroperoxidation in fish muscle during frozen storage and additive effects to oxidation during cooking is potentially a very important deterioration of the final product *(3)*. The differences in lipid oxidation reactions of stored minced *vs* whole catfish muscle on cooking are discussed in Chapter 20. This question is also important because state of subdivision of muscle foods and exposure of membrane components to oxygen are largely rate determining for formation of off-flavor components. The TBARS of the raw fish is, e.g., 0.5 mg/kg while after cooking this value increases to 2.91 mg/kg *(7)*. In addition to polyunsaturated fatty acids being oxidized, cholesterol is also modified to oxides and hydroxy derivatives. Tocopherol is a potent antioxidant in such systems. Tocopherols are lost on cooking, depending on their structures *(32)*. Other effects which change the generation of off-flavor in stored muscle foods are microbial metabolite buildup and time of frozen storage during which not only the hydroperoxides are formed but also the degradation to volatile off-flavor products. Lipid oxidation in fish products must be avoided at all costs as the author of Chapter 20 points out.

Concluding Remarks

In closing, very real and important problems remain in food quality preservation from free radical oxidative deterioration. Making the most of agricultural materials and foodstuffs in the coming decades with overwhelming demographic projections to tax the food system will require effective solutions to the questions asked in this *Symposium*. The field of chemistry relating to polyunsaturated fatty acids and their reactivity to oxygen and carbon to carbon bond scissions or rearrangements in food, medicine, and industrial materials applications is very active and exciting *(37)*. Each paper in this *Symposium on Lipid Oxidation in Foods* poses unique challenges. Yet, each set of challenges, hypotheses, approaches and their significance to the biochemical science of lipid oxidation in foods signal that important breakthroughs are soon coming to the entire food and nutritional health system *(37)*. These solutions will involve understanding mechanisms of free radical oxidation and their control through critical chemical interactions. The future food supply as well as nutritional health of the public and its institutions supporting highest quality of products for the consumer will benefit from solving the problems framed by each speaker in this 1991 ACS Division of Agricultural and Food Chemistry *Symposium on Lipid Oxidation in Foods*.

Literature Cited

1. Simic, M. G.; Taylor, K. A. In *Warmed-Over Flavor of Meat*; St. Angelo, A. J.; Bailey, M. E., Eds.; Academic Press, Inc.: Orlando, FL, 1987; pp 69-117.
2. Decker, E. A.; Hultin, H. O. *J. Food Sci.* 1990, 55, 947-953.
3. Erickson, M. C.; Hultin, H. O. In *Oxygen Radicals in Biology and Medicine*; Simic, M. G.; Taylor, K. A.; Ward, J. F.; Sonntag, C., Eds.; Plenum Press, Inc.: New York, NY, 1988; pp 307-312.

4. Kanner, J.; Harel, S. *Arch. Biochem. Biophys.* 1985, *237*, 314-321.
5. Kanner, J.; German, J. B.; Kinsella, J. E. *CRC Crit. Rev. Food Sci. Nutrition.* 1987, *25*, 317-364.
6. Josephson, D. B.; Lindsay, R. C.; and Stuiber, D. A. *J. Agric. Food Chem.* 1984, *32*, 1347-1350.
7. Hsieh, R.J.; Kinsella, J.E. *Adv.in Food and Nutrition Research.* 1989, *33*, 233-341.
8. Labuza, T.P. *CRC Crit. Rev. in Food Technology.* 1971, *2*, 355-405.
9. Frankel, E.N. *J. Sci. Food Agric.* 1991, *54*, 495-511.
10. Davies, K.J.A. *J. Biol. Chem.* 1987, *262*, 9895-9901.
11. Davies, K.J.A.; Delsignore, M.E.; Lin, S.W. *J. Biol. Chem.* 1987, *262*, 9902-9907.
12. Spanier, A.M.; McMillin, K.W.; Miller, J.A. *J. Food Sci.* 1990, *55*, 318-322 and 326.
13. Davies, K.J.A.; Delsignore, M.E. *J. Biol. Chem.* 1987, *262*, 9908-9913.
14. Bailey, M. E. In *The Maillard Reaction in Foods and Nutrition*; Waller, G. R.; Feather, M.S., Eds.; ACS Symposium Series No. 215; American Chemical Society: Washington, D.C., 1983; pp 169-184.
15. Bailey, M. E.; Shin-Lee, S. Y.; Dupuy, H. P.; St. Angelo, A. J.; Vercellotti, J. R. In *Warmed-Over Flavor of Meat*; St. Angelo, A.J.; Bailey, M.E., Eds.; Academic Press, Inc.: Orlando, FL, 1987; pp 237-266.
16. Bentley, D.S.; Reagan, J.O.; Miller, M.F. *J. Food Sci.* 1989, *54*, 284-286.
17. St. Angelo, A. J.; Crippen, K. L.; Dupuy, H. P.; James, C., Jr. *J. Food Sci.* 1990, *55*, 1501-1505.
18. Asghar, A.; Gray, J.I.; Buckley, D.J.; Pearson, A.M.; Booren, A.M. *Food Technology* 1988, *42*, 102-108.
19. Shahidi, F.; Yun, J.; Rubin, L. J.; Diosady, L. C. *Can. Inst. Food Sci. Technol. J.* 1987, *20*, 104-106.
20. Stoick, S. M.; Gray, J. I.; Booren, A. M.; Buckley, D. J. *J. Food Sci.* 1991, *56*, 597-600.
21. Ackman, R. G. *Prog. Food Nutr. Sci.* 1989, *13*, 161-241.
22. Legendre, M. G.; Fisher, G. S.; Schuller, W. H.; Dupuy, H. P.; Rayner, E. T. *J. Am. Oil Chem. Soc.* 1979, *56*, 552-555.
23. Acree,T.E.; Barnard,J.; Cunningham,D.G. *Food Chemistry* **1984**, *14*, 273-286.
24. Ullrich, F; Grosch, W. *Z. Lebensm. Unters. Forsch.* 1987, *184*, 277-282.
25. Blank, I.; Fischer, K.-H.; Grosch, W. *Z. Lebensm. Unters. Forsch.* 1989, *189*, 426-433.
26. Guth, H.; Grosch, W. *Lebensm.-Wiss. und -Technol.* 1990, *23*, 513-522.
27. Meilgaard, M.; Civille, G. V.; Carr, B. T. *Sensory Evaluation Techniques*; CRC Press: Boca Raton, FL, 1991; pp 187-199.
28. Meilgaard, M.; Civille, G. V.; Carr, B. T. *Sensory Evaluation Techniques*; CRC Press, Inc.: Boca Raton, FL, 1987; Vol. 2, pp 1-23.
29. Johnsen, P. B.; Civille, G. V. *J. Sensory Studies.* 1986, *1*, 99-104.
30. Johnsen, P. B.; Civille, G. V.; Vercellotti, J. R.; Sanders, T. H.; Dus, C. A. *J. Sensory Studies.* 1988, *3*, 9-18.

31. Johnsen, P. B.; Civille, G. V.; Vercellotti, J. R. *J. Sensory Studies.* **1987**, *2*, 85-92.
32. Heath, H. B.; Reineccius, G. A. *Flavor Chemistry and Technology;* AVI Publishing Co., Inc.: Westport, CT, **1986**; pp 112-141.
33. Lilja Hallberg, M.; Lingnert, H. *J. Am. Oil Chem. Soc.* **1991**, *68*, 167-170.
34. Karel, M. In *Concentration and Drying of Foods;* MacCarthy, D., Ed.; Elsevier Applied Science Publishers: London, UK, **1986**; pp 37-51.
35. Paquot, C. *Standard Methods for the Analysis of Oils, Fats, and Derivatives;* 6th Edition, IUPAC; Pergamon Press: Oxford, UK, **1979;** pp 143-144.
36. Perkins, E. G.; Qian, C.; Caldwell, J. D.; Yates, R. A. *J. Am. Oil Chem. Soc.* **1989**, *66*, 483-489.
37. Johnsen, P. B.; Vercellotti, J. R. *Off-Flavors in Foods,* In *The Encyclopedia of Food Science and Technology;* Hui, Y. H., Ed.; J. Wiley and Sons: New York, NY; **1991**; pp 1895-1900.

RECEIVED February 19, 1992

Mechanisms of Free-Radical Processes

Chapter 2

Mechanisms of Lipid Oxidative Processes and Their Inhibition

Michael G. Simic[1], Slobodan V. Jovanovic[2], and Etsuo Niki[3]

[1]Department of Pharmacology and Toxicology, University of Maryland, Baltimore, MD 21201
[2]Radiation Laboratory, Institute for Nuclear Science Vinca, Belgrade, Yugoslavia
[3]Research Center for Advanced Science and Technology, University of Tokyo, Komaba, Meguro-ku, Tokyo, Japan

A concise overview of the kinetics, energetics, and mechanisms of oxidative processes involving fatty acids in simple model systems relevant to autoxidation of foods is presented. Endogenous and exogenous factors that initiate oxidative processes in biochemical systems are reviewed and elements of and conditions for propagation of chain peroxidation processes are defined. Mechanistic aspects of chain-breaking antioxidants, including redox potentials, are presented. Biomarkers of fatty acid peroxidation are briefly reviewed and their relevance and specificity assessed.

Food quality, excluding bacterial and enzymatic spoilage, deteriorates on standing due to oxidative processes induced and propagated by atmospheric oxygen (*1*). The appearance, texture, flavor, and odor of foods are affected by oxidative processes. The extent of these changes depends on the type of food and conditions. The rancidity of fats, which is unpleasant, can be reduced by storing fats at lower temperatures. Although low-temperature storage efficiently preserves saturated and lower unsaturated fats, reduced temperature alone is not sufficient to protect polyunsaturated fish oils against oxidation. For example, refrigeration of ham at 4 °C preserves its quality for days or even weeks. On the other hand, refrigerated cooked mackerel is inedible after only two days.

In addition to the undesirable sensory characteristics of autoxidized foods, questions have been raised regarding the safety of oxidation products in foods (*2*). Can hydroperoxides, epoxides, and their decomposition products (diverse carbonyl compounds) contribute to carcinogenesis through the stages of initiation, promotion, or progression? Hydroperoxides are known to damage DNA. Carbonyl compounds, although unreactive with DNA, may affect cellular signal transduction.

Foods are complex systems with numerous components. Hence, oxidative mechanisms are often difficult, if not impossible, to study directly. The very short lifetime of free radicals further complicates mechanistic studies. To fully understand the underlying mechanisms and promote phenomenological predictability, comprehensive studies of model systems are invaluable (*3*). However, the need to correlate the findings of mechanistic studies in model systems with studies of oxidative processes in foods through molecular biomarkers must be emphasized.

This chapter outlines free radical processes pertinent to oxidation of food and discusses the energetics, kinetics, and mechanisms of isolated reactions. Since fatty acid derivatives are found in homogeneous phases (fat and oil) as well as in heterogeneous systems such as micelles, liposomes, and membranes, the effects of structural features of aggregates on oxidation of fatty acids are presented. Secondary reactions initiated by the decomposition of hydrogen peroxide and, in particular, hydroperoxides are reviewed. Inhibition of free radical reactions through inactivation of free radicals by antioxidants is discussed and the guiding principles of antioxidant efficacy are outlined. Mechanisms of formation of specific products that may serve as biomarkers of oxidative processes are considered.

Initiation

Oxidative degradation of lipids in foods may be initiated by endogenous species (H_2O_2, ROOH) and radicals ($\cdot O_2^-$, ROO$\cdot$, $\cdot$OH, GS$\cdot$) or by exogenous species ($^1O_2$, O_3), radicals (NO_x, $SO_3^{-}\cdot$), and agents (UV, ionizing radiation, heat) (*2, 3*).

The targets of attack of these agents are diverse and are specific to each agent and the conditions.

Endogenous Agents. One of the simplest free radical-generating reactions is

$$O_2 + Fe(II) \rightarrow \cdot O_2^- + Fe(III) \quad (1)$$

Many complexes of iron with a low redox potential, such as heme, generate superoxide radical (*2, 3*),

$$O_2 + Heme(II) \rightarrow \cdot O_2^- + Heme(III) \quad (2)$$

Reducing agents, such as ascorbate, contribute to redox cycling via the following reaction,

$$Fe(III) + AH^- \rightarrow Fe(II) + \cdot A^- + H^+ \quad (3)$$

In this reaction ascorbate, a well-known antioxidant, acts as a pro-oxidant. Under hypoxic (ischemic) conditions mitochondria may generate superoxide radical via ubiquinone semiquinone radical,

$$\cdot UQ^- + O_2 \rightarrow UQ + \cdot O_2^- \quad (4)$$

Meats rich in myoglobin (Mb) may generate superoxide radical either by detachment of heme from Mb or from damaged Mb molecules in which heme is not in an autoxidation-resistant configuration (4),

$$Mb(II)\ O_2 \rightarrow Mb'(II)\ O_2 \rightarrow Mb'(III) + \cdot O_2^- \qquad (5)$$

$$Mb'(III) + AH^- \rightarrow Mb'(II) + \cdot A^- + H^+ \qquad (6)$$

$$Mb'(II) + O_2 \rightarrow Mb'(III) + \cdot O_2^- \qquad (7)$$

In addition to H_2O_2 certain enzymes, such as xanthine oxidase (Xox), generate superoxide radical (5),

$$X + O_2 \xrightarrow{Xox} Urate + n\ \cdot O_2^- + (n-1)\ H_2O_2 \qquad (8)$$

The above reaction is used to generate $\cdot O_2^-$ for measuring reaction rate constants of superoxide radicals via chemiluminescence methodologies (6, 7).

Superoxide radicals may react with diverse molecules despite their fairly low reactivities (8). If they react via their conjugate acid, $\cdot O_2H$ (pK_a = 4.8), the oxidative reactivities are greatly enhanced.

Superoxide radicals recombine with or without the assistance of superoxide dismutase (SOD) (9),

$$2\ \cdot O_2^- + 2H^+ \rightarrow H_2O_2 + O_2 \qquad (9)$$

Hydrogen peroxide generated via reaction 9 or directly by other metabolic processes, such as the P450 system, is a potential source of other radicals, provided it survives degradation by catalases, peroxidases, and certain thiols. Hydrogen peroxide and some metal (Fe, Cu) ions, as well as some of their complexes, are capable of generating highly reactive hydroxyl radicals via the well-known Haber-Weiss reaction (2, 3),

$$H_2O_2 + Fe(II) \rightarrow \cdot OH + OH^- + Fe(III) \qquad (10)$$

Alternatively, reaction 10 may generate very reactive peroxo species.

Hydroxyl radicals are a source of diverse C-centered, S-centered, and heterocyclic radicals, which are formed via three basic reactions (10): abstraction of H from C-H and S-H bonds,

$$\text{>C-H} + \cdot OH \rightarrow \text{>C}\cdot + H_2O \qquad (11)$$

$$-S-H + \cdot OH \rightarrow -S\cdot + H_2O \qquad (12)$$

addition to double bonds,

$$>C=C< + \cdot OH \rightarrow >\dot{C}-C\triangleleft OH \qquad (13)$$

$$-N=C< + \cdot OH \rightarrow -N-\dot{C}\triangleleft OH \qquad (14)$$

and addition to aromatic rings,

$$-C_6H_5 + \cdot OH \rightarrow -\dot{C}_6H_5-OH \qquad (15)$$

The rate constants of the reactions of $\cdot OH$ and other free radicals with diverse substrates are shown in Table I.

Table I. Reaction Rate Constants for Oxyl Radicals with Bio-Related Substrates in Solutions at Room Temperature[a]

Substrate, S	$k(R\cdot + S)$, $M^{-1} s^{-1}$				
	ROO·	$\cdot O_2H$	O_2^-	RO·	·OH
Stearic acid	$10^{-3} - 10^{-4}$	low	low	2.3×10^6	$\sim 10^9$
Oleic acid	0.1 - 1	low	low	3.3×10^6	$\sim 10^9$
Linoleic acid	~ 60	1.2×10^3	low	8.8×10^6	9.0×10^9
Linolenic acid	~ 120	1.7×10^3	low	1.3×10^7	7.3×10^9
Arachidonic acid	~ 180	3.0×10^3	low	2.0×10^7	$\sim 10^{10}$
Aldehyde	2.7×10^3	50	n.m.[b]	n.m.	$\sim 10^9$
GSH	$< 10^6$	1.8×10^5	< 15	n.m.	1.4×10^{10}
BHT	10^4	2.4×10^3	n.m.	4×10^7	$\sim 10^{10}$
BHA	2.6×10^6	n.m.	n.m.	n.m.	$\sim 10^{10}$
QH_2	1.2×10^5	$10^4 - 10^5$	n.m.	n.m.	$\sim 10^{10}$
α-Tocopherol	5.7×10^6	2.6×10^5	6	n.m.	$\sim 10^{10}$
Ascorbate (AH⁻)	2.2×10^6	–	5.0×10^4	n.m.	1×10^{10}

[a] Adapted from ref. 4.
[b] n.m., not measured.

Exogenous Agents. Singlet oxygen may be generated exogenously or endogenously by oxyl radical recombination. One of the main reactions of singlet oxygen yields a hydroperoxide (e.g., 5-α-cholesterol hydroperoxide),

$$R-H + {}^1O_2 \rightarrow ROOH \qquad (16)$$

which leads indirectly to free radical formation,

$$ROOH + Fe(II) \rightarrow RO\cdot + OH^- + Fe(III) \tag{17}$$

Alkoxyl radicals, RO·, are in many respects similar to ·OH radicals, although they are not as reactive (Table I).

UV light generates radicals via photosensitization (*11*),

$$S + h\nu \rightarrow {}^3S \tag{18}$$

$$RH + {}^3S \rightarrow \cdot R + \cdot SH \tag{19}$$

Ionizing radiations (X- and γ-rays, high-energy electrons and particles) generate free radicals via ionization of molecules (*10, 11*). Unlike UV, distribution of ionizing radiation interaction within a system is proportional to the electron density fraction of constituent molecules. Because many foods contain a high water concentration, the major reaction is,

$$H_2O \rightsquigarrow H_2O^+ + e^- \tag{20}$$

followed rapidly (< ps) by,

$$H_2O^+ + H_2O \rightarrow \cdot OH + H_3O^+ \tag{21}$$

Hence, ionizing radiations may mimic the Haber-Weiss reaction, and they are used for mechanistic studies.

High temperature, like ionizing radiation, may split food molecules into free radicals. Even at 37°C some azo derivatives such as AAPH and AMVN (Figure 1) may generate radicals at sufficiently high rates for mechanistic studies (*12*),

$$RN = NR \rightarrow 2 \cdot R + N_2 \tag{22}$$

followed by,

$$\cdot R + O_2 \rightarrow ROO\cdot \tag{23}$$

Because AAPH is water soluble and AMVN is lipid soluble, these molecules provide an ideal system to generate peroxyl radicals in either membranes or cytosol (*2*).

The reaction rate constants of some exogenously generated species with fatty acids or fatty acid moieties are shown in Table II.

Table II. Initiation Rate Constants of Fatty Acid Oxidation by Singlet Oxygen, Ozone, Sulfite, and Thiyl Radicals

Substrate, S	k(X + S), $M^{-1} s^{-1}$			
	$^1O_2$ [a]	O_3 [b]	$SO_3^{-\cdot}$ [c]	GS· [d]
Oleic	0.74×10^5	9.5×10^5		$< 2 \times 10^6$
Linoleic	1.3×10^5	1.1×10^6	1.8×10^6	8×10^6
Linolenic	1.9×10^5		2.8×10^6	1.9×10^7
Arachidonic	2.4×10^5		3.9×10^6	3.1×10^7

[a] In pyridine (13)
[b] In aqueous 140-160 mM SDS solutions at 25°C (14)
[c] In aqueous solution at pH 11.5 (15)
[d] In ethanol/water 1/1 at pH 5.1 (16)

Propagation

Elements of Propagation. Peroxidation of fatty acids is predominantly a chain reaction because of the high reactivity of peroxyl radicals with the weak allylic and bisallylic C-H bonds and the high reactivity of resulting fatty acid radicals with oxygen. We shall briefly present the elements that govern these complex reactions. One should note the mechanistic similarity of these reactions in foods and *in vivo*.

Reaction Kinetics of Free Radicals with Oxygen. Most C-centered free radicals react with oxygen at close to diffusion-controlled rates,

$$\geqslant C\cdot\ +\ O_2\ \rightarrow\ \geqslant COO\cdot \tag{24}$$

$$k \sim 1 - 4 \times 10^9\ M^{-1}\ s^{-1}$$

Such high rates are expected for saturated fatty acid radicals.

Because of the resonant character of unsaturated and polyunsaturated fatty acid (PUFA) radicals, the rate constants for reactions with oxygen are lower (17). For example, the bisallylic radical (HL·) of linoleic acid (H_2L), reacts with oxygen an order of magnitude slower,

$$HL\cdot\ +\ O_2\ \rightarrow\ HLOO\cdot \tag{25}$$

$$k = 1.8 \times 10^8\ M^{-1}\ s^{-1}$$

Energetics of Peroxidation. The reason for the formation of resonant allylic and bisallylic radicals can be found in the vast differences of C-H bond strengths (BS) in monounsaturated fatty acids and PUFAs.

The weakest C-H bond in PUFAs is 75 kcal/mol (Figure 2) in the bisallylic position. Consequently, the bisallylic C-H bond is the most reactive site for H-atom abstraction by free radicals, although other C-H bonds may also be involved in abstraction. Their involvement, however, depends on the relative reaction rate constants, which can be derived from measurements of specific products.

Comparison of the following bond strengths (in units of kcal/mol) is instructive: HO-H, 119; RO-H, 104-105; ROO-H, ~90; Ar-H, 112: ArO-H, 85; NH_2-H, 107; RS-H, ~90; and ArS-H, ~84 (18). In SI units bond energies are expressed in kJ/mol (1 cal = 4.18 J).

Resonance. The initially generated unpaired electron of a free radical may interact with other atoms or groups within the radical. As a result, the unpaired electron becomes delocalized and the electron density at the original site drops.

The bisallylic radical in linoleic acid shown in Figure 3 does not react with oxygen due to a considerably reduced electron density at C11 (17). The mesomeric forms with an unpaired electron at C9 and C13, however, react with oxygen but at the reduced rate indicated in equation 25.

Resonance also contributes to the reversibility of reaction 25 and its corollary, the exchange of dioxygen between oxygen in the atmosphere and peroxyl radicals.

Kinetics and Mechanisms of Propagation. Other unsaturated fatty acids would be expected to have a reaction mechanism similar to that of linoleic acid, since the predominant sites of attack are the weakest C-H bonds. In oleic acid (*cis*-9-octadecenoic) these are the allylic positions, C8 and C11. In linoleic acid (*cis,cis*-9,12-octadecadienoic), as previously discussed, the most reactive site is the double allylic position, C11. Linolenic acid (all-*cis*-9,12,15-octadecatrienoic) has two reactive bisallylic sites, C11 and C14. Arachidonic acid (all-*cis*-5,8,11,14-eicosatetraenoic) has three bisallylic sites, C7, C10, and C13.

The kinetics of the propagation reaction (k_p) for different fatty acids (H_2Fa),

$$ROO\cdot + H_2Fa \xrightarrow{k_p} ROOH + HFa\cdot \quad (26)$$

are given in Table I. The rate constants are much lower for fatty acids lacking bisallylic sites. In PUFAs, k_p is proportional to the number of bisallylic sites.

It should be noted that k_p values do not vary much for different ROO· radicals unless the peroxyl radicals have a halogen substituent in the α position.

The positions of oxygen addition to free radicals of unsaturated fatty acids and the corresponding hydroperoxides are shown in Table III. It is evident that oxygen does not add to bisallylic radicals sites because no corresponding hydroperoxides are found.

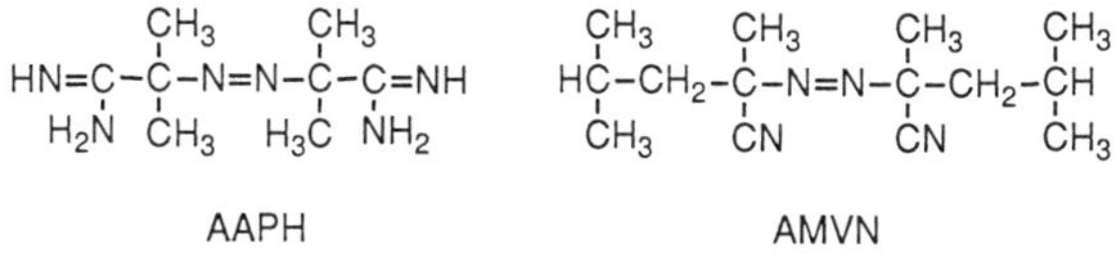

AAPH structure: HN=C(NH₂)-C(CH₃)₂-N=N-C(CH₃)₂-C(NH₂)=NH

AMVN structure: H₃C-CH(CH₃)-CH₂-C(CH₃)(CN)-N=N-C(CH₃)(CN)-CH₂-CH(CH₃)-CH₃

Figure 1. Water-soluble AAPH and lipid-soluble AMVN as radical initiators.

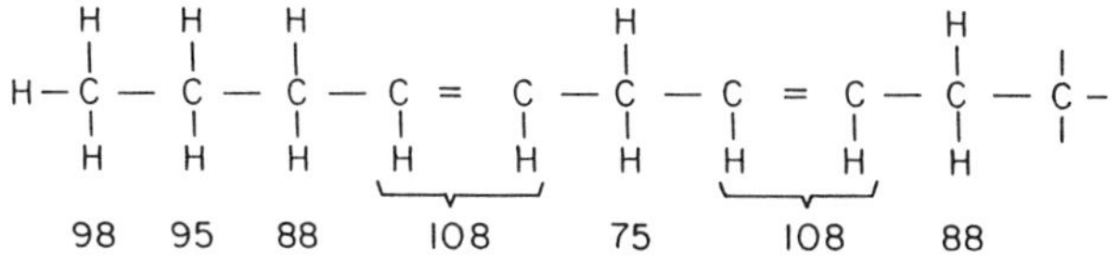

Bond energies (kcal/mol): 98, 95, 88, 108, 75, 108, 88

Figure 2. Bond energies in kcal/mol of C-H bonds in polyunsaturated fatty acids. (Adapted from ref. *19*)

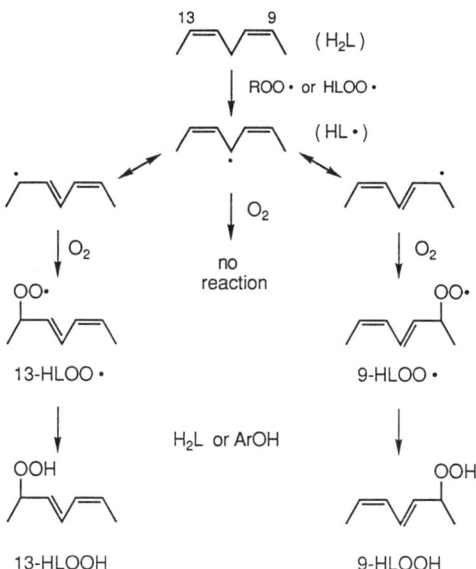

Figure 3. Peroxidation mechanism of linoleic acid (H_2L). (Adapted from ref. *17*)

Table III. Location of H Atom Abstraction (→) and Resulting Hydroperoxides, with Yields Indicated (%), in Peroxidation of Fatty Acids[a]

C Location	Oleic	Linoleic	Linolenic	Arachidonic
5	-	-	-	27
6	-	-	-	-
7	-	-	-	→ -
8	→ 27	-	-	7
9	23	50	32	9
10	→ 23	-	-	→ -
11	27	→ -	→ -	11
12	-	-	11	6
13	-	50	11	→ -
14	-	-	→ -	-
15	-	-	-	40
16	-	-	46	-

[a]Adapted from ref. 19.

Propagation Conditions. Peroxidation of fatty acids is a chain reaction, and the chain length would be infinite if it were not for the reaction between peroxyl radicals or peroxyl radicals and antioxidants. Radical-radical reactions are quite fast (20),

$$2ROO\cdot \rightarrow \text{Intermediate(s)} \rightarrow \text{Products} \quad (27)$$

$$2k_t \sim 10^6 - 10^8 \text{ M}^{-1}\text{ s}^{-1}$$

The intermediate(s) is in many cases a tetroxide, which decomposes via the Russell mechanism into -OH and >CO derivatives (10). Conditions that determine the chain-propagation length (CPL) are considered below (21).

Initiation Rate. The chain length in model systems is usually determined from the rate of oxygen consumption (21),

$$\frac{-d[O_2]}{dt} = \frac{k_p[H_2L]\, R_i^{1/2}}{(2k_t)^{1/2}} \quad (28)$$

where k_p is the propagation rate constant, $2k_t$ is the termination rate constant, and R_i is the rate of generation of initiating radicals (e.g., R·).

In ionizing radiation experiments (21, 22), oxidation of Phe, HCOO⁻, and H_2L was initiated by ·OH radicals (reactions 11-15), which generated the initial radical, ·R. The rate of initiation, R_i, is determined by the radiation dose-rate (D_r, Gy s^{-1}). The yield of oxygen uptake is given as G(-O$_2$), and the initial free radical yield under controlled conditions is $G(\cdot R) = 6.4$. Hence, a ratio of $G(-O_2)/G(\cdot R)$ gives the number of oxygen molecules consumed per initial radical $\cdot R$ (i.e., $\cdot OH$). The experimental data are shown in Figure 4.

The CPL for Phe and formate was 0 because oxidation of either substrate is not a function of dose-rate. Hence, neither reaction is a chain reaction. For linoleic acid, in contrast, the CPL increased markedly for lower dose-rates. Figure 4 illustrates a reliable test for chain peroxidation processes.

Structure of Aggregates. At higher concentrations fatty acids aggregate into micelles. The effect of concentration on oxygen uptake is shown in Figure 5. Above critical micellar concentration (cmc = 2.3 mM for linoleic acid) spherical micelles are formed, a process that is completed at 6 mM. Above 12 mM cylindrical micelles are formed. It is obvious from Figure 5 that the CPL increases with increasing structure of the aggregates.

Temperature. Temperature affects not only the CPL but also product distribution. Product distribution may be quite different at 50-70°C, the temperature range in which accelerated autoxidation experiments are conducted, than at room temperature due to different activation energies for different reactions involving different C-H bonds. Similarly, temperature affects the stereochemical product distribution, e.g., the $c,t/t,t$ ratios (23).

The effect of temperature on oxygen uptake is shown in Figure 6 for linoleic acid below cmc. For H_2L at 0°C the CPL is 0, whereas at 48°C the CPL is 7.2 under these experimental conditions. It is interesting to note that at those temperatures, the CPL in cylindrical micelles is 7 and 23, respectively.

Chain-Branching. In the presence of metal ions or certain complexes, fatty acid hydroperoxides generate the corresponding alkoxyl radicals,

$$HLOOH + Fe(II) \rightarrow HLO\cdot + OH^- + Fe(III) \quad (29)$$

followed by initiation of a new chain (chain branching).

$$HLO\cdot + H_2L \rightarrow HLOH + HL\cdot \quad (30)$$

Chain branching may be prevented by compounds that chelate metals and prevent reaction 29 from occurring.

Antioxidants

Antioxidants are a class of compounds capable of inhibiting chain peroxidation reactions ($2, 3$). Natural plant antioxidants and physiological antioxidants in animals sustain life by inhibiting deleterious oxidative processes. Since foods are derived from many different living organisms, a variety of antioxidants is present in foods.

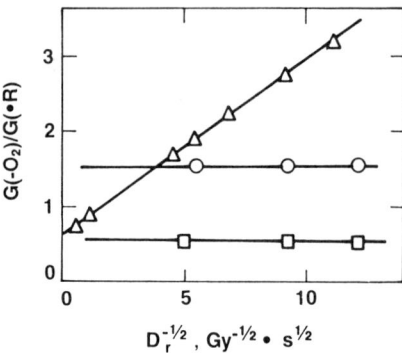

Figure 4. Oxygen uptake per initial radical ·R as a function of dose-rate$^{-1/2}$ in aqueous solutions of (△) 0.1 mM linoleic acid, pH 8.4; (○) 1 mM phenylalanine, pH 6.7; (□) 1 mM sodium formate, pH 7. All solutions were saturated with N_2O/O_2 (5:1) at 20 °C. (Adapted from ref. 22.)

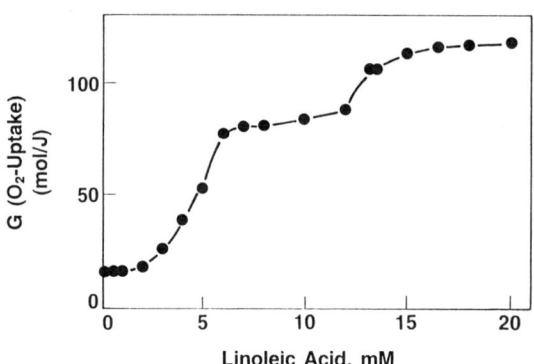

Figure 5. Effect of linoleic acid concentration on $G(-O_2)$ in γ-radiolysis of N_2/O_2 (4:1) aqueous solutions of linoleic acid. Dose rate = 1.2×10^{-2} Gy s^{-1}, pH = 9. (Reproduced from ref. 21. Copyright 1988 American Chemical Society.)

Mechanisms. The basic mechanism of the interaction of antioxidants with peroxyl and alkoxyl radicals has not yet been fully resolved. Oxygen radicals, in principle, could abstract an H atom from ArO-H bonds,

$$ROO\cdot + ArOH \rightarrow ROOH + ArO\cdot \tag{31}$$

Arguments based on bond strength considerations that favor reaction 31 should be viewed with caution because many bond strength values are derived from kinetic data. There is growing evidence, however, that the interaction of peroxyl radicals and antioxidants is a redox process, which takes place in two steps (24),

$$ROO\cdot + ArOH \rightarrow ROO^- + ArOH^{+}\cdot \tag{32}$$

followed by rapid deprotonation of the charged antioxidant intermediate,

$$ArOH^{+}\cdot + H_2O \rightarrow ArO\cdot + H_3O^+ \tag{33}$$

ArO· radicals have mesomeric forms with specific configurations, depending on the substituents. Resonant forms of the simplest phenoxyl radical are shown in Figure 7.

Antioxidant phenoxyl radicals have a stronger resonant character than bisallylic radicals, and the reaction with O_2 does not occur (25),

$$ArO\cdot + O_2 \rightarrow \text{not observed} \tag{34}$$

$$k < 10^5 \, M^{-1} \, s^{-1}$$

Kinetics. In general, the rate constant for the reaction of a peroxyl radical and an antioxidant,

$$XROO\cdot + ZAH \rightarrow XROO\cdot + ZAH^{+}\cdot \tag{35}$$

is a function of the electrophilic character of substituents X and Z (26). Hence, an electron-withdrawing substituent, X (e.g., $-Cl$), and electron-donating substituent, Z (e.g., $-OCH_3$), increase the redox potential difference, ΔE (i.e., ΔG), and the reaction rate of reaction 35. For example, the reaction rate constant for trichloro-substituted methylperoxyl radical with α-tocopherol is about 30 times greater than that for a nonsubstituted radical (27).

α-Tocopherol (α-Toc), the major constituent of vitamin E, is one of the antioxidants most reactive with peroxyl radicals (Table I). Prevention by α-tocopherol of hydroperoxide formation via chain propagation is shown in Figure 8.

Redox Potential. In the past, redox potentials, E, of compounds were usually determined by electrochemical methods or by calculations. More recently, the

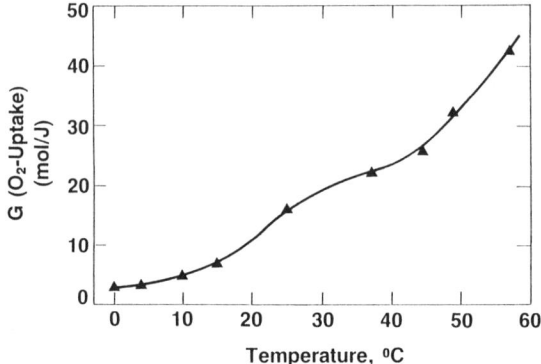

Figure 6. Effect of temperature on $G(-O_2)$ in aqueous solutions of linoleic acid, N_2O/O_2 (4:1). Dose rate = 1.1×10^{-2} Gy s^{-1}, pH = 9.4, 0.1 mM H_2L. (Reproduced from ref. 21. Copyright 1988 American Chemical Society.)

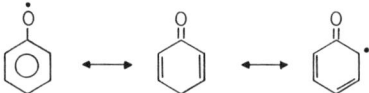

Figure 7. Mesomeric forms of the phenoxyl radical.

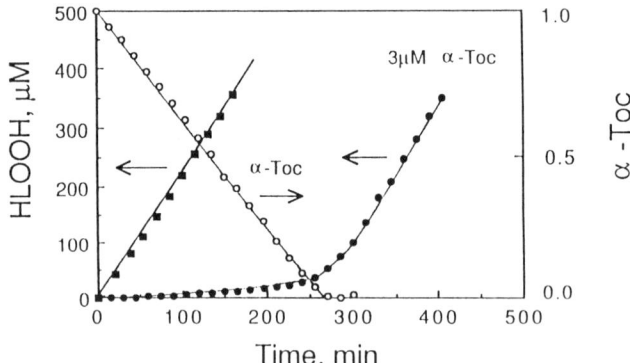

Figure 8. Oxidation of methyl linoleate (453 mM) in hexane induced by AMVN (0.20 mM) in the absence (■) and presence (•) of α-tocopherol (3 μM) at 37°C in air, as well as consumption of α-Toc (o).

E-value for a one-electron redox couple can be determined under homogeneous conditions using pulse radiolysis or laser photolysis (28). These time-resolved techniques are necessary because one-electron oxidation of a compound yields a free radical, which usually has a very short lifetime.

The one-electron redox potentials are derived from the equilibrium between a known redox standard, S^- (or SH), an antioxidant A^- (or AH), and their free radical intermediates (26, 28),

$$\cdot A + S^- \underset{k_r}{\overset{k_f}{\rightleftarrows}} A^- + \cdot S \qquad (36)$$

and from either k_f and k_r, or from absorptions (concentrations) of intermediates at the equilibrium,

$$K = \frac{k_f}{k_r} = \frac{[A^-][\cdot S]}{[\cdot A][S^-]} \qquad (37)$$

The following equation is used to calculate ΔE at 20°C,

$$\Delta E = E(\cdot A/A^-) - E(\cdot S/S^-) = 0.059 \log K \qquad (38)$$

The oxidation potentials of some well-known antioxidants are shown in Table IV.

Table IV. One-Electron Oxidation Potentials (vs. NHE) of Selected Compounds in Aqueous Solutions at pH 7 and 20 °C[a]

Compound	E_7, V	Compound	E_7, V
4-Nitrophenol	1.23	Uric acid	0.59
4-Hydroxybenzoate	1.04	2,6-Dimethoxyphenol	0.58
Phenol	0.97	α-Tocopherol	0.48
Tyrosine	0.89	Hydroquinone	0.46
4-Methylphenol	0.87	DOPA	0.44
4-Methoxyphenol	0.73	4-Aminophenol	0.41
Serotonin	0.64	4-Dimethylaminophenol	0.36
Sesamol	0.62	Ascorbate	0.28

[a]Adapted from ref. 4.

When the radical is reduced by the transfer of one electron from a suitable donor, we express this as the reduction potential of the antioxidant radical, $E(\cdot A/A^-)$ or $E(\cdot A/AH)$. By convention the oxidation potential of the parent compound is equal to the reduction potential of the corresponding free radical. It is important, however, to indicate the redox couple to which the potential refers because the parent compound, in addition to an oxidation potential, may also have a reduction potential, $E(AH/\cdot AH^-)$ or $E(AH/\cdot AH_2)$.

Because the one-electron oxidation potential of ascorbate (vitamin C) is lower than that of α-tocopherol (Table IV), ascorbate repairs vitamin E chromanoxyl radicals (29),

$$\text{Chr}-\text{O}\cdot \ + \ \text{AH}^- \ \rightarrow \ \text{Chr}-\text{OH} \ + \ \cdot\text{A}^- \quad (39)$$

$$k = 5.7 \times 10^6 \text{ M}^{-1} \text{ s}^{-1}$$

This regeneration process is demonstrated in Figure 9.

One-electron redox potentials can be calculated from the Hammett correlation using Brown σ^+ substituent constants (30). At pH 7,

$$E_7 = 0.95 + 0.31 \Sigma \sigma^+ \quad (40)$$

Hence, polysubstituted phenols, which include some plant compounds, may have a much lower E_7 value than, for example, a monosubstituted phenol such as tyrosine. Epigallocatechin gallate (EGCG), a major constituent of green tea, is a polysubstituted phenol that was shown to be an outstanding scavenger of superoxide radicals (Niki and Simic, unpublished data). Similarly, 2-hydroxyestradiol has a much lower E_7 than estradiol (Figure 10) and, consequently, a higher and more versatile protective capacity (31, 32).

Biomarkers of Oxidation

There are numerous products of oxidation of fatty acids. None of them, however, provides a measure of the extent of oxidation. The only quantitative indicator of peroxidation is oxygen consumption, which is invaluable in model studies under controlled conditions but cannot be measured in foods or biological systems after the fact. We shall very briefly review some of the products currently used in the assessment of oxidative processes.

Hydroperoxides. Hydroperoxides, the main products of oxidative processes, can be measured by HPLC, GC, GC/MS, or via their reaction with iodide (3). While simple in model systems (Figure 8), chromatographic separation in foods is considerably more complex. Furthermore, hydroperoxides are labile and decompose readily, e.g., via reaction 17 or on heating.

Alkanes. One of the products of hydroperoxide decomposition is an unstable alkyl radical, ·R, which reacts with surrounding fatty acids, H_2Fa, to generate an alkane (*33*),

$$\cdot R + H_2Fa \rightarrow HR + HFa\cdot \qquad (41)$$

Reaction 41 is relatively slow, $k < 1\ M^{-1}\ s^{-1}$, and the rate depends on the fatty acid composition and microenvironment. It competes with reaction 24.

Alkyl radicals are generated via formation of alkoxyl radicals and β-scission on reduction of hydroperoxides. The reaction is shown in Figure 11.

The n-6 fatty acids generate pentyl radical and pentane, as shown in Figure 11, whereas n-3 fatty acids generate ethyl radical and ethane. Pentane (*34*) and ethane (*35*) in breath have been used for the assessment of oxidative processes in living organisms (*34-36*).

TBA Test. Thiobarbituric acid (TBA) reacts with carbonyls to give a colored product (*37, 3*). One of the products of endoperoxide of arachidonic acid is malondialdehyde (MDA), which binds with two molecules of TBA. The product can be assessed either by measuring absorbance (523 nm) or fluorescence (excitation at 515 nm and emission at 553 nm). The fluorescence test is considerably more sensitive and allows determination of TBA-reactive substances (TBARS) at picomolar concentrations levels in blood plasma (*37*).

It is well documented in the literature that the TBA test is not specific for MDA. However, the usefulness of the TBA test in practice (*37*) greatly outweighs its nonspecificity.

4-Hydroxy-2-nonenal (HNE). HNE produced *in vitro* and *in vivo* from decomposition of hydroperoxides of n-6 PUFAs (linoleic and arachidonic) has been demonstrated (*38, 39*). A complex mechanism of its generation was suggested recently (*40*).

HNE oxidatively generated from fatty acids can be determined by HPLC techniques at nanogram levels in biosystems. Recent GC/MS measurements used negative ion chemical ionization (NICI) and exploited the high volatility of the bifunctional pentafluorobenzyl (PFB) and trimethylsilyl (TMS) derivatives of HNE (*41*). Formation of HNE in retina, blood plasma, and various organs was demonstrated to occur *in vivo* as a result of oxidative stress (*41*).

Conclusions

Despite major advances in our understanding of oxidative mechanisms in the last 15 years, gaps in our knowledge remain. We have relatively little data about food hydroperoxides, their decomposition products in foods, or their potential carcinogenic effects in humans. Mechanistic studies of heterogeneous systems such as foods and their protection from oxidative spoilage by antioxidants as well as the anticarcinogenic character of antioxidants are areas for further investigation.

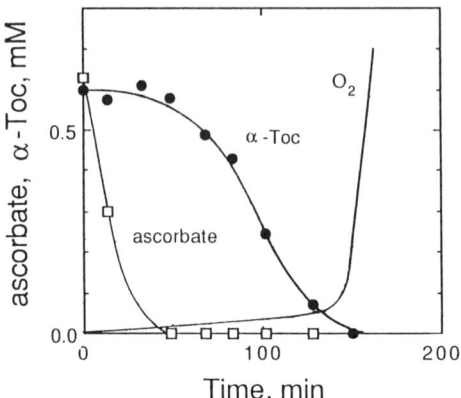

Figure 9. Disappearance of α-tocopherol (●), ascorbate (□), and O_2 consumption (−) in the induced oxidation of methyl linoleate by AMVN at 37 °C in *tert*-butyl alcohol/methanol (3:1 by volume). [Methyl linoleate] = 0.60 M, [AMVN] = 10 mM, [α-Toc] = 0.959 mM, [ascorbate] = 0.620 mM. (Adapted from ref. 29.)

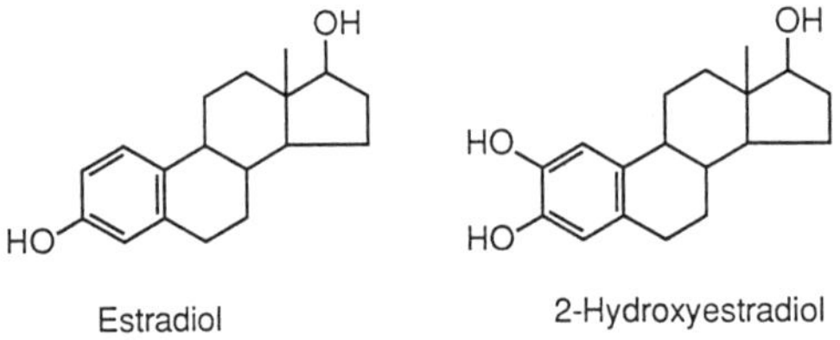

Figure 10. Estrogen antioxidants.

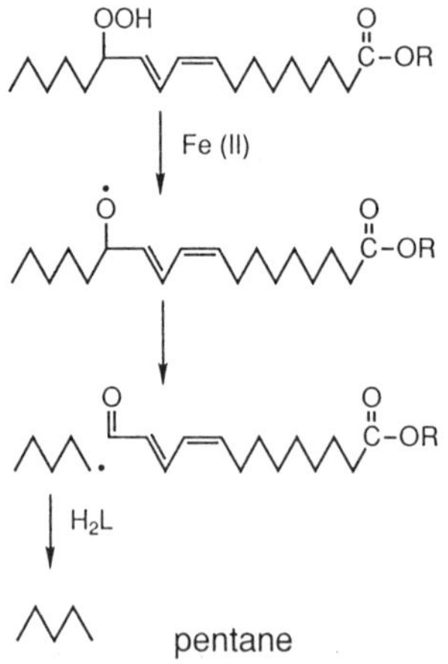

Figure 11. Formation of pentane from 13-hydroperoxide of linoleic acid. (Adapted from ref. *19*)

Progress in these areas may ultimately lead to development of safer products with a longer shelf life and cancer-preventing characteristics.

Acknowledgments

Special thanks are due to Karen A. Taylor for numerous suggestions and discussions and to Gloria Wiersma for technical assistance.

Literature Cited

1. *Flavor Quality of Fresh Meat: Warmed-Over Flavor*; St. Angelo, A. J.; Bailey, M. E., Eds.; Academic Press: New York, NY, 1987.
2. *Oxygen Radicals in Biology and Medicine*; Simic, M. G., Taylor, K. A., Ward, J. F., and von Sonntag, C., Eds.; Plenum Press: New York, NY, 1988.
3. *Oxygen Radicals in Biological Systems*; Packer, L.; Glazer, A. N., Eds.; Methods Enzymol., 1990, Vol. 186.
4. Simic, M. G. *J. Environ. Sci. Health* **1991**, *C9*, 113-153.
5. McCord, J. M.; Fridovich, I. *J. Biol. Chem.* **1968**, *243*, 5753-5759.
6. Gotoh, N.; Niki, E. *Chem. Lett.* **1990** (8), 1475-1478.
7. *Lipid Peroxides in Biology and Medicine*; Yagi, K., Ed.; Academic Press: New York, NY, 1982.
8. Buxton, G. V.; Greenstock, C. L.; Helman, W. P.; Ross, A. B. *Phys. Chem. Ref. Data* **1988**, *17*, 513.
9. McCord, J. M.; Fridovich, I. *J. Biol. Chem.* **1969**, *244*, 6049.
10. von Sonntag, C. *The Chemical Basis of Radiation Biology*; Taylor and Francis: New York, NY, 1987.
11. Bensasson, R. V.; Land, E. J.; Truscott, T. G. *Flash Photolysis and Pulse Radiolysis*; Pergamon Press: New York, NY, 1983.
12. Niki, E. *Methods Enzymol.* **1990**, *186*, 100-108.
13. Doleiden, F. H.; Fahrenholtz, S. R.; Lamola A. A.; Trozzolo, A. M. *Photochem. Photobiol.* **1974**, *20*, 519-521.
14. Giamalva, D. H.; Church, D. F.; Pryor, W. A. *J. Am. Chem. Soc.* **1986**, *108*, 6646-6651.
15. Erben-Russ, M.; Michael, C.; Bors, W.; Saran, M. *Radiat. Environ. Biophys.* **1987**, *26*, 289-294.
16. Schöneich, C.; Asmus, K.-D. *Radiat. Environ. Biophys.* **1990**, *29*, 263-271.
17. Simic, M. G.; Jovanovic, S. V.; Al-Sheikhly, M. *Free Radical Res. Comm.* **1989**, *6*, 113-115.
18. Egger, K. W.; Cocks, A. T. *Helv. Chim. Acta* **1973**, *56*, 1516-1536.
19. Simic, M. G.; Taylor, K. A. In *Flavor Quality of Fresh Meat: Warmed-Over Flavor*; St. Angelo, A. J.; Bailey, M. E., Eds.; Academic Press: New York, NY, 1987; p. 69-117.
20. Neta, P.; Huie, R. E.; Ross, A. B. *J. Phys. Chem. Ref. Data* **1990**, *19*, 413-513.
21. Al-Sheikhly, M.; Simic, M. G. *J. Phys. Chem.* **1988**, *93*, 3103-3106.

22. Al-Sheikhly, M.; Simic, M. G. In *Free Radicals, Methodology, and Concepts*; Rice-Evans, C.; Halliwell, B., Eds.; Richelieu Press: London, 1988, pp. 481-497.
23. Porter, N. A. In *Membrane Lipid Oxidation*; Vigo-Pelfry, C., Ed.; CRC Press: Boca Raton, FL, 1990, Vol. 1; pp. 33-62.
24. Jovanovic, S. V.; Simic, M. G. In *Oxygen Radicals in Biology and Medicine*; Simic, M. G.; Taylor, K. A.; Ward, J. F.; von Sonntag, C., Eds.; Plenum Press: New York, NY, 1988, pp. 115-122.
25. Hunter, E. P. L.; Desrosiers, M. F.; Simic, M. G. *Free Radical Biol. Med.*, **1989**, *6*, 581-585.
26. Simic, M. G.; Jovanovic, S. V. *J. Am. Chem. Soc.* **1989**, *111*, 5778-5782.
27. Simic, M. G. *Mutation Res.* **1988**, *202*, 377-386.
28. Jovanovic, S. V.; Steenken, S.; Simic, M. G. *J. Phys. Chem.* **1990**, *94*, 3583-3588.
29. Niki, E. *Chem. Phys. Lipids* **1987**, *44*, 227-253.
30. Jovanovic, S. V.; Tosic, M.; Simic, M. G. *J. Phys. Chem.* **1991**, *95*, 10824-10827.
31. Yagi, K.; Komura, S. *Biochem. Internat.* **1986**, *13*, 1051-1055.
32. Niki, E.; Nakano, M. *Methods Enzymol.* **1990**, *186*, 330-333.
33. Cohen, G. In *Lipid Peroxides in Biology and Medicine*; Yagi, K., Ed.; Academic Press: New York, NY, 1982, pp. 199-211.
34. Dillard, C. J.; Downey, J. E.; Tappel, A. L. *Lipids* **1984**, *19*, 127-133.
35. Sword, J. T.; Pope, A. L.; Hoekstra, W. G.; *J. Nutr.* **1991**, *121*, 251-257.
36. Jeejeebhoy, K. N. *Free Radical Biol. Med.* **1991**, *10*, 191-194.
37. Yagi, K. In *Lipid Peroxides in Biology and Medicine*; Yagi, K., Ed.; Academic Press: New York, NY, 1982, pp. 223-242.
38. Esterbauer, H.; Cheeseman, K. H. *Methods Enzymol.* **1990**, *186*, 407-421.
39. Esterbauer, H.; Schaur, R. J.; Zollner, H. *Free Radical Biol. Med.* **1991**, *11*, 81-128.
40. Pryor, W. A.; Porter, N. A. *Free Radical Biol. Med.* **1990**, *8*, 541-543.
41. van Kuijk, F. J. G. M.; Thomas, D. W.; Stephens, R. J.; Dratz, E. A. *Methods Enzymol.* **1990**, *186*, 399-407.

RECEIVED February 19, 1992

Chapter 3

Lipid Oxidation in Muscle Foods via Redox Iron

Eric A. Decker[1] and Herbert O. Hultin[2]

[1]Food Science Section, Department of Animal Sciences, University of Kentucky, Lexington, KY 40546-0215
[2]Massachusetts Agricultural Experiment Station, Department of Food Science, University of Massachusetts—Amherst, Marine Foods Laboratory, Marine Station, Gloucester, MA 01930

> A small portion of the iron in biological tissues is present as low molecular weight complexes probably with amino acids and/or nucleotides. This low molecular weight (LMW) Fe can activate O_2 and initiate lipid oxidation. LMW Fe may increase during post harvest storage, perhaps by release of Fe from ferritin. Reducing components of the tissue such as ascorbate, NAD(P)H or superoxide ion may reduce ferric to active ferrous. Membrane systems can enzymatically transfer electrons from NAD(P)H to ferric ion. The concentration of these reducing agents or their effectiveness change with post harvest storage. The particular Fe complex formed is crucial in determining its ability to catalyze oxidation. Chloride has mixed effect, stimulating some systems and inhibiting others. Chelation of LMW Fe to form non-active complexes and reduction of exposure to O_2 are promising techniques to minimize lipid oxidation in post harvest biological tissue.

Iron is essential for life because it is required for oxygen transport, respiration and the activity of many enzymes. However, iron is an extremely reactive metal and if its activity is not controlled, iron will catalyze oxidative changes in lipids, proteins and other cellular components. In foods these changes are often associated with changes in sensory parameters such as texture, flavor and color and changes in nutritional value. Muscle foods contain high amounts of iron which can lead to rapid iron-catalyzed oxidative changes. Therefore, iron-catalyzed lipid oxidation has been extensively studied in meat systems and will therefore be the primary source of data for this review.

Catalysis of lipid oxidation by ionic iron requires ferrous iron and the subsequent production of oxygen radicals. An example of this reaction is the Fenton reaction in which ferrous ions catalyze the decomposition of hydrogen peroxide to hydroxyl anion and hydroxyl radical with the production of the ferric ion. This reaction results in the oxidation of iron; therefore for the reaction to be continuous, reducing agents must be present to promote the production of ferrous ions. Haber and Weiss (*1*) described an example of this reaction in which superoxide anion acts as the reducing agent. Other reducing systems besides superoxide can also be involved in the production of ferrous ions and hydroxyl radicals through Haber-Weiss-type reactions. The hydroxy radical is believed to initiate lipid oxidation. In muscle foods the rate of iron-catalyzed lipid oxidation is dependent on both the concentration and activity of iron and reducing systems which can promote the formation of ferrous ions.

Iron Distribution in Tissue

Low Molecular Weight Iron in Muscle. Iron is basically found in two states in biological tissue, free and protein-bound. Free iron is believed to be primarily responsible for catalysis of lipid oxidation in biological tissue (*2*). Free iron is not necessarily in a free or uncomplexed state since the solubility of the ferric ion is low at pH 7.0 (*3*). Free iron might better be described as low molecular weight iron since free iron is most likely complexed with organic phosphates (e.g. NAD(P)H, AMP, ADP and ATP), inorganic phosphates, amino acids and carboxylic acids (*4*). Bakkeren and coworkers (*5*) have found iron bound to glycine, cysteine and citrate in a low molecular weight fraction of rat reticulocyte cytosol. Several researchers have reported that complexes between iron and nucleotides are capable of stimulating lipid oxidation (*6-8*). Kinetic evidence has been presented for complex formation between iron, ADP (or other nucleotides) and histidine (*9,10*)

Low molecular weight (LMW) or catalytic iron concentrations vary depending on species and type of muscle tissue. Hazel (*11*) reported LMW (<12 kilodaltons) iron concentrations to be between 3.5-5.9% of the total soluble iron and 2.4-3.9% of total muscle iron in beef, lamb, pork and chicken (Table I). Decker and coworkers (*12*) and Decker and Hultin (*13*) found LMW (<10,000 daltons) iron concentrations to range from 6.7-13.9% of the total iron concentration of a press juice isolated from flounder and mackerel muscle (Table I). Kanner and coworkers (*14*) determined the concentration of "catalytic" iron in chicken and turkey muscle by measuring an iron-bleomycin complex which was capable of catalyzing the degradation of DNA. Using this technique, "catalytic" iron concentrations ranged from 0.24 to 2.5 μg/g muscle (Table I). Several researchers have also measured nonheme iron in muscle foods in an attempt to predict the amount of free or reactive iron. While this method seems to correlate well with the development of lipid oxidation in cooked muscle, it does not necessarily measure free iron since the method involves acid digestion (*15*) or EDTA chelation followed by precipitation of proteins with trichloroacetic acid (*16*). This method only differentiates between heme and nonheme iron and not between free and protein-bound iron. Therefore these techniques generally give higher iron concentrations than low molecular weight or "catalytic" iron values. The problem arises in the fact that nonheme iron chelated to proteins such as ferritin and transferrin does not catalyze lipid oxidation (*2*). However, nonheme iron measurements have been

Table I. Low molecular weight or "catalytic" iron concentrations in muscle foods

MUSCLE	µg Fe/g MUSCLE	SOURCE
BEEF	0.7	1
LAMB	0.7	1
PORK	0.2	1
CHICKEN (DARK)	0.2	1
FLOUNDER	0.03	2
MACKEREL(LIGHT)	0.04	3
MACKEREL (DARK)	0.20	3
TURKEY (LIGHT)	0.92	4
TURKEY (DARK)	2.5	4
CHICKEN (LIGHT)	0.24	4
CHICKEN (DARK)	0.55	4
LAMB	6.7	5
BEEF	8.4	5
PORK	4.2	5

1. Hazell, 1982 (*11*); LMW Fe
2. Decker et al., 1989 (*12*); LMW Fe
3. Decker and Hultin, 1990 (*13*); LMW Fe
4. Kanner et al., 1988 (*14*); "Catalytic Fe"
5. Schricker et. al., 1982 (*15*); Nonheme Fe

found to be useful in monitoring changes in iron bound to hemoglobin and myoglobin especially during cooking (see below).

The distribution of iron in muscle foods is affected by storage and processing conditions. Storage of unfrozen mackerel ordinary muscle at 4°C for 7 days resulted in a 1.4-fold increase in low molecular weight iron from 0.16 to 0.23 µg Fe/g muscle (*13*). Frozen-thawed muscle exhibited a greater change in low molecular weight iron than unfrozen muscle during storage at 4°C for 7 days (1.9 fold increase; 0.11 to 0.2 µg Fe/g muscle). Kanner and coworkers (*14*) have reported 2-3 fold increases in "catalytic" iron in chicken and turkey muscle stored at 4°C for 7 days.

Cooking muscle also causes changes in iron distribution in muscle and these changes are thought to be partially responsible for the development of warmed-over flavor. Igene et al. (*16*) studied the effect of heating on water extracts of beef muscle and found that heating increased the nonheme iron from 1.80 to 4.18 µg/g of meat. Schricker et al. (*15*) and Schricker and Miller (*17*) reported increases in nonheme iron in cooked beef, pork and lamb. They found that increasing cooking time increased nonheme iron and baking released more nonheme iron than microwave cooking. The amount of nonheme iron released during cooking correlates well with the development of oxidative rancidity.

The source of protein-bound iron which is responsible for the observed increases in LMW and catalytic iron during storage and cooking is not completely understood. However, several sources of protein-bound iron exist in biological tissue including myoglobin, hemoglobin, ferritin, transferrin and hemosiderin.

Protein-bound Iron Sources in Muscle

Soluble Proteins. The amount of soluble iron is 20.0, 12.3, 3.6 and 3.4 µg/g muscle in beef, lamb, pork and chicken muscle, respectively (Table II; *11*). A soluble fraction (press juice) isolated from flounder, mackerel ordinary and mackerel dark muscle contained 0.3, 0.6 and 2.6 µg Fe/g muscle, respectively (Table II; *12,13*). The amount of low molecular weight iron in these muscle foods range from 0.03 to 0.7 µg Fe/g muscle (Table I) indicating that over 90% of the soluble iron in muscle is bound to proteins. The major iron containing proteins in foods include myoglobin, hemoglobin and ferritin.

Table II. Concentration of soluble iron in muscle foods

MUSCLE	TOTAL SOLUBLE Fe (µg Fe/g MUSCLE)	SOURCE
BEEF	20.0	1
LAMB	12.3	1
PORK	3.6	1
CHICKEN (DARK)	3.4	1
FLOUNDER	0.3	2
MACKEREL (LIGHT)	0.6	3
MACKEREL (DARK)	2.6	3

1. Hazell, 1982 (*11*)
2. Decker et al., 1989 (*12*)
3. Decker and Hultin, 1990 (*13*)

Hemoglobin and Myoglobin. Hemoglobin (Hb) and myoglobin (Mb) both contain iron in the form of hematin for the purpose of oxygen transport and storage. Both contain 1 heme per polypeptide but hemoglobin is a tetramer and myoglobin is a monomer. The relative concentration of hemoglobin and myoglobin in muscle foods depends on species and muscle type. Beef, lamb and pork generally contain more myoglobin than hemoglobin while chicken contains a greater amount of hemoglobin (*11*). Hemoglobin and myoglobin contribute 77.9, 56.3, 44.9 and 22.0% of the total iron and 95.0, 87.0, 86.1 and 67.6% of the soluble iron in beef, lamb, pork and chicken, respectively (11). Schricker and coworkers (*15*) reported heme iron concentrations in beef, lamb and pork to be 62, 57 and 49% of the total iron, respectively. Igene et al. (*16*) found that bound heme iron contributed to over 90% of the total iron in a beef muscle extract.

Release of iron from Hb and Mb has been studied in meat extracts (*16,18*) and intact muscle (*15,17*) by measuring changes in heme iron concentrations. Igene and coworkers (*16*) reported that cooking or addition of hydrogen peroxide to a water extract of beef caused a decrease in the concentration of heme iron suggesting that iron was being released from heme proteins. Schricker and Miller (*17*) found similar results in intact muscle with both cooking and hydrogen peroxide increasing the concentration of nonheme iron in ground beef. Hydrogen peroxide has been shown to release iron from both myoglobin (*19*) and hemoglobin (*20*).

Ferritin. Ferritin is a water soluble iron storage protein which has a molecular weight of 450,000 daltons and contains 4,500 iron atoms when fully loaded (*21*). Ferric ions are bound to the core of the ferritin protein at a concentration which is comparable to iron ore (*4*). Ferritin releases ferrous ions in the presence of reducing agents such as superoxide anion, ascorbate and thiols (*22*). Both ascorbate and cysteine are capable of releasing iron from ferritin at temperatures (2-37°C) and pH values (5.5-6.9) common to many foods during handling and processing (*23*).

Hazell (*11*) reported that the concentration of iron in a water-soluble high molecular weight fraction (>130,000) was 0.3, 0.9, 0.3 and 0.9 μg Fe/g muscle in beef, lamb, pork and chicken, respectively which represents 1.2 to 11.1% of the total iron in these muscles. The iron in this fraction was suggested to be ferritin due to similarities in its absorbance spectrum with purified ferritin. Decker and Welch (*23*) found beef psoas major muscle contained between 1.1 to 2.8 μg ferritin-bound Fe/mg muscle as determined by precipitation of ferritin from a muscle extract by ferritin serum antibodies. The amount of ferritin-bound iron in beef psoas major muscle decreased 62% after 11 days of storage at 4°C suggesting that ferritin-bound iron is release *in situ*. Kanner and Doll (*24*) reported ferritin-bound iron concentrations of 0.3 and 1.0 μg/g in turkey light and dark muscle, respectively. Kanner and Doll (*24*) also found that ferritin-bound iron concentrations decreased in turkey muscle during storage.

A high molecular weight fraction (>150,000) of mackerel ordinary muscle press juice which would be expected to contain ferritin was found to be an active catalyst of lipid oxidation in the presence of 100 μM ascorbate (*25*). Catalysis of lipid oxidation by this fraction was inhibited by ceruloplasmin suggesting that iron was the primary catalyst in this fraction. Ferritin isolated from a turkey muscle extract ($\approx$5.0 μg Fe/mL) was capable of initiating lipid oxidation in the presence of ascorbate (*24*). A ferritin-containing extract of beef diaphragm muscle catalyzed the oxidation of phosphatidylcholine liposomes in the presence of ascorbate and cysteine (10-500 μM; *26*). Heating ferritin does not result in the direct release of iron but does increase the ability of ascorbate to release ferritin-bound iron and subsequently catalyze lipid oxidation (*23,27*).

Transferrin, Ovotransferrin and Lactotransferrin and other Miscellaneous Iron-containing Proteins. There are three types of transferrin: serum transferrin, which is responsible for iron transport in vertebrates, ovotransferrin, which is found in avian eggs and lactotransferrin, which is found in milk. The transferrins can bind up to 2 ferric ions/molecule. Lactotransferrin and transferrin do not seem to be active catalysts of lipid oxidation (28,29) except in the presence of EDTA (28). Other iron-containing proteins in food include cytochrome c, iron-sulfur proteins of the electron transport chain, peroxidases, and lipoxygenase. It is not well understood if conditions in foods result in the release of the iron from these proteins.

Insoluble Proteins. The amount of insoluble iron in beef, lamb, pork and chicken is 5.9, 5.9, 3.0 and 4.7 μg/g muscle, respectively (11). This represents from 22.8 to 58.0% of the total muscle iron. While the amount of insoluble iron in muscle is relatively high, very little is known about its role in lipid oxidation in foods.

Hemosiderin. Hemosiderin is an insoluble complex of iron, other metals and proteins. It is thought to be a ferritin degradation or polymerization product (30). Iron bound to hemosiderin is released in the presence of reducing agents such as ascorbate, dithionite and superoxide anion resulting in the production of hydroxyl radicals (31-33). The amount of hemosiderin in foods is not known.

Importance of Chelators and Ceruloplasmin in Establishing the Role of Ionic Iron

Several methods have been described which attempt to determine the concentration of iron involved in lipid oxidation reactions in foods. These methods involve physical separation of soluble extracts of muscle into low molecular weight fractions (11-13) or measurement of iron which can form complexes with bleomycin to catalyze DNA degradation (14). The disadvantage of these methods is that they do not directly measure the ability of iron to catalyze lipid oxidation. An alternative method to measure the ability of iron to catalyze lipid oxidation is by the use of chelating agents such as EDTA and desferroxamine. However, caution must be used in the interpretation of results when using chelators because both EDTA and desferroxamine can nonspecifically inhibit lipid oxidation by acting as hydrogen donors at high concentrations (34,35). EDTA can also act to accelerate iron-catalyzed lipid oxidation when EDTA:Fe ratios are low (36). An additional disadvantage of chelators is that they also inhibit catalysis of lipid oxidation by other metals such as copper. Ceruloplasmin can be used to elucidate the role of iron in lipid oxidation since it maintains iron in the inactive ferric ion state (25,26,37). An advantage of using ceruloplasmin is that it actively inhibits iron-catalyzed lipid oxidation at low concentrations and has high specificity for iron.

Nonenzymic Lipid Oxidation via Redox Iron

As mentioned earlier, iron can catalyze lipid oxidation via a Haber-Weiss like reaction. The Haber-Weiss reaction mechanism involves the reduction of ferric ions by superoxide anion; however, other reducing agents in biological systems are also active in this pathway including ascorbate, cysteine and NAD(P)H.

Superoxide anion by itself can not initiate lipid oxidation but its conjugated acid (perhydroxyl radical) has been shown to initiate the oxidation of linoleic and arachidonic acids (38). The pK_a of the perhydroxyl radical is 4.8 suggesting that only about 1% would exist under physiological conditions. The primary role of superoxide in the catalysis of lipid oxidation appears to be its ability to reduce iron to the ferrous state and to dismutate to produce hydrogen peroxide either spontaneously or via superoxide dismutase.

Superoxide anion is produced in biological systems by a number of mechanisms. The most widely studied superoxide generating system is xanthine oxidase. Xanthine oxidase catalyzes the production of superoxide and hydrogen peroxide in the presence of hypoxanthine and oxygen. Kuppusamy and Zweier (39) have suggested that this reaction also generates hydroxyl radicals but production of hydroxyl radicals by xanthine oxidase has been suggested to be due to iron contamination in the enzyme preparations (40,41). Superoxide can also be produced in foods by autoxidation of leucoflavins, hydroquinones and other small molecules; enzymes such as peroxidases and flavoprotein dehydrogenases; cell organelles such as chloroplasts and mitochondria; and cells such as leukocytes (42) as well as through oxidation of ascorbate by copper (43) or iron (22) and autoxidation of reduced hemoglobin and myoglobin (44). Little work has been done to quantitate and study the effect of processing and storage on superoxide concentrations in foods. However, superoxide dismutase, which converts superoxide to hydrogen peroxide and oxygen, is widely distributed in foods (42) suggesting that superoxide is generated in these systems.

Ascorbate has also been shown to be active in reducing iron to promote lipid oxidation (45-47). Ascorbate is thought to be a more important reductant *in vivo* than superoxide since it is generally found in higher concentrations (46). Ascorbate concentrations in muscle range from 5.1 μmoles/100 g of mackerel ordinary muscle press juice (13) to 13.7 μmoles/100 g turkey dark muscle (48). Storage (4°C) of mackerel ordinary muscle for 7 days resulted in a decrease in ascorbate concentration from 7.8 to 0.2 μmoles/100 g muscle.

Ascorbate can both promote and inhibit lipid oxidation depending on the concentration of ascorbate present. Low concentrations of ascorbate generally promote lipid oxidation by causing the reduction of metal ions while high concentrations inhibit lipid oxidation by the donation of hydrogen resulting in the inactivation of free radicals (45). The concentrations of ascorbate which are prooxidant or antioxidant are dependent on iron concentration. Maximal catalysis of phosphatidylcholine liposome oxidation was observed at 50 μM ascorbate in the presence of 50 and 100 ppb iron and >100 μM ascorbate in the presence of 500 ppb iron (Figure 1a; 49). Inhibition of lipid oxidation

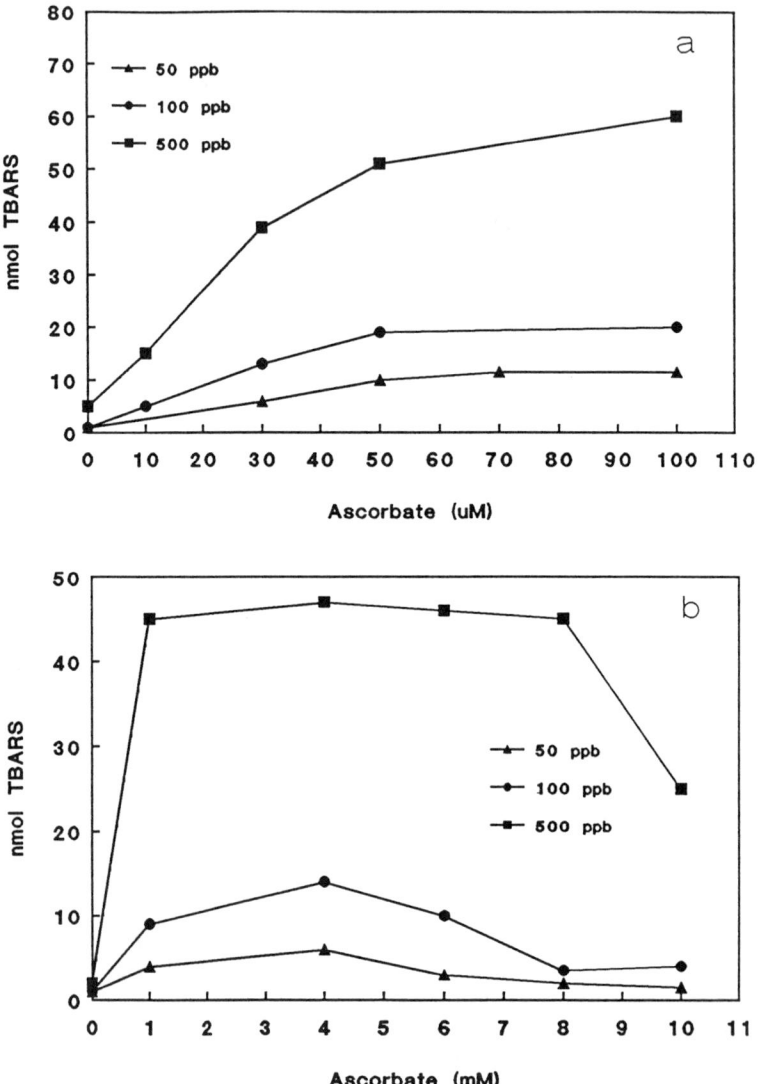

Figure 1a and b. Production of thiobarbituric acid reactive substances (TBARS) by various concentrations of ascorbate (0-10 mM) and $FeCl_3$ (50-500 ppb). All assays (5 mL) contained 0.2 mg phosphatidylcholine liposomes per mL 0.12 M KCl, 5 mM histidine, pH 6.8 and were incubated for 2.5 hr at 6°C.

occurred between 4-6 mM ascorbate for 50 ppb Fe, between 4-8 mM for 100 ppb Fe and at >8 mM for 500 ppb Fe (Figure 1b; *49*). At concentrations of ascorbate and free iron in animal tissue, ascorbate is most likely in the prooxidant range. For example, at the concentration of LMW iron (103 ppb) found in the press juice of mackerel ordinary muscle the rate of lipid oxidation was nearly maximal at 50 μM ascorbate which is approximately equal to the average ascorbate concentration of the press juice (51.0 μM; *13*).

The involvement of ascorbate in the catalysis of lipid oxidation was shown by decreases in the ability of iron to stimulate lipid oxidation in the presence of a mackerel ordinary muscle press juice in which the ascorbic acid had been oxidized with ascorbic acid oxidase (*25*). The ability of EDTA to inhibit catalysis of lipid oxidation by mackerel ordinary muscle press juice decreased as mackerel fillets were stored on ice (*25*). EDTA became less effective within one day of storage as inhibition decreased from 51% to 8%. The decrease in the effectiveness of EDTA corresponds to the post-mortem decrease in ascorbate in the press juice from aged muscle (*25*). Decreased ascorbate concentrations and loss of antioxidant activity by EDTA during storage suggests that ascorbate promoted iron-catalyzed lipid oxidation was primarily active shortly after the death of the animal.

Both NADH and NADPH can promote iron-catalyzed hydroxyl radical production; however, the activity is much lower than with ascorbate or superoxide (*46*). Glutathione is not capable of stimulating iron-catalyzed hydroxyl radical production (*46*). Cysteine will stimulate iron-catalyzed lipid oxidation and like ascorbate, will act as a prooxidant at low concentrations and an antioxidant at high concentrations (*47*). Cysteine is generally less active than ascorbate at promoting lipid oxidation (*47*). Murata and Sakaguchi (*50, 51*) reported free cysteine concentrations in yellowtail tuna ranged from 80-400 μmoles/100 g of raw or boiled muscle and the concentrations changed very little during storage.

Effect of sodium chloride on nonenzymic lipid oxidation. The effects of added sodium chloride on lipid oxidation in stored muscle foods is variable, although most workers report an accelerating effect, e.g. in beef, pork, chicken and fish (*52-57*). However, sodium chloride has also been observed to inhibit lipid oxidation (*58,59*). Rhee et al. (*60*) found that salt inhibited lipid oxidation in ground pork when it was added at concentrations greater than 2% but accelerated lipid oxidation when added at lower concentrations. Sodium chloride increased the rate of oxidation of phosphatidylcholine liposomes by a press juice prepared from mackerel muscle (*61*). Inhibition by EDTA and ceruloplasmin indicated the reaction was mediated through ionic iron. The chloride ion was responsible for this stimulation of oxidation; no differences were observed among K^+, Na^+, or Li^+. At least a portion of the chloride stimulation of the press juice involved a component of the soluble fraction of molecular weight greater than 5 kilodalton. Chloride improved the effectiveness of ascorbate in activating ferritin to catalyze this oxidation. Results indicated that the effect of chloride iron was not mediated through a chlorine radical formed by a peroxidase enzyme (62). Kanner and coworkers (63) showed that the pro-oxidant effect of NaCl in minced turkey muscle could be inhibited by EDTA and ceruloplasmin again implicating ionic iron in the effect of NaCl. These workers made the observation that NaCl appeared to displace ionic iron from sites on macromolecules.

Enzymes Involved in Lipid Oxidation via Redox Iron

Lipid oxidation was first described in rat liver microsomes by Hochstein and Ernster (64). The liver microsomal enzyme system requires iron, NAD(P)H and ADP or other pyrophosphates for activity. Bidlack and co-workers (65) suggested that cytochrome b_5 reductase is responsible for the reduction of ferric irons in the presence of NAD(P)H. Although this activity was at one time doubted to exist in muscle microsomes because of the very low cytochrome content of those membranes (66), a system similar to that of the liver was identified in the chicken muscle microsomal fraction (67). This work was the outgrowth of studies that had showed that phospholipase added to ground beef inhibited lipid oxidation (68); it was proposed that an enzyme system dependent on phospholipid was required for lipid oxidation in the ground beef. Identification of a similar system in fish muscle followed (69). Two important differences were observed between the fish system and that of the chicken. First, whereas the chicken microsomes utilized NADPH as an electron donor more efficiently then NADH, the fish system responded very weakly to NADPH but utilized NADH well. Secondly, the fish system was shown to have high activity at temperatures of refrigerated storage. The chicken microsomal system had very low activities at these temperatures. Similar systems were later identified in beef (70), pork, (71) and several species of fish (72-75). An interesting point about the lipid peroxidation systems in the several species of fish studied is that microsomes isolated from the fatty pelagic species have higher rates of lipid oxidation than do those of white fleshed fish (72,74).

All evidence indicates that the function of the NADH-driven enzymic system in fish is to reduce ferric iron to ferrous iron. Varying the concentrations and ratios of Fe^{+3} and Fe^{+2} in the absence of reducing agent (NADH) showed that it is the reduced form of iron that is responsible for lipid oxidation (76). Kanner et al. (37) demonstrated that turkey microsomal lipid peroxidation was inhibited by ceruloplasmin, indicating that the reaction was catalyzed by ferrous iron. Ferrous iron reduced by the microsomal fraction could initiate lipid peroxidation via a Fenton-type reaction with hydrogen peroxide and could act as a propagating agent via its ability to decompose lipid hydroperoxides (77). Increasing concentrations of ferrous iron caused an increase in the lag phase of lipid oxidation catalyzed by fish muscle sarcoplasmic reticulum (a further purified preparation of the microsomal fraction, 76). The lag phase is also decreased by an increasing concentration of sarcoplasmic reticulum. This indicates that ferrous iron is interacting directly with some antioxidant component in the membrane, possibly tocopherol.

Most studies of membrane lipid oxidation utilize iron concentrations much higher than found in physiological conditions. However, it has been demonstrated that iron can catalyze lipid oxidation in fish muscle sarcoplasmic reticulum at concentrations found in the low molecular weight fraction of muscle tissue (12); longer reaction times are required under these conditions. Copper which shows little pro-oxidative activity in the enzyme-catalyzed membrane system at the high concentrations usually used *in vitro*, has considerable activity at the low levels actually found in the muscle tissue compared to iron.

There is an absolute requirement for NAD(P)H in enzyme-catalyzed lipid oxidation by the microsomal or sarcoplasmic reticular fraction of muscle tissue. Indeed, along with heat inactivation, the enzymic nature of the reaction is defined based on the requirement for NAD(P)H and the specificity for either NADPH or NADH depending on whether the system is taken from the muscle of warm-blooded animals or fish. The sensitivity of the flounder microsomal fraction for NADH is extremely high. The system has a Km for NADH in the presence of ADP of approximately 1 uM and shows an optimal activity at approximately 10 uM under the same conditions (77). The concentrations of NADH which are found in post mortem fish muscle after several days are higher than this (78). The significance of this is that it is unlikely that the NADH concentration would decrease sufficiently during normal storage to prevent this membrane fraction from serving as a potential reducer of ferric to ferrous iron *in situ*. It also implies that the initial physiological concentration of NADH would be inhibitory to the system; this inhibitory effect would decrease with time post-mortem as the NADH concentration approached closer to that which would give maximal stimulation of lipid oxidation.

The kinetic effects of a number of other nucleotides is complex (10). ATP, ADP, and AMP produce the same results as NADH; at low concentrations they slightly inhibit the reaction, but at higher concentrations they stimulate the lipid oxidative activity up to a concentration where they are inhibitory. Since this effect is seen with both AMP and IMP the pyrophosphate moiety is not involved. In the system of fish sarcoplasmic reticulum, $NADP^+$ gives the same results as the other nucleotides, but NAD^+ is inhibitory at all levels studied. This latter is probably a reflection of product inhibition since NADH was used as the electron donor in these experiments. It seems likely that the function of these nucleotides is in binding the iron; they may either maintain it in soluble form and/or change its redox potential, making it easier to be reduced. The higher activity observed with no ADP is possibly due to the complexing ability of the NADH itself (9).

Histidine and histidine dipeptides occur at very high levels in some species of fish, particularly pelagic, dark-fleshed species (79); the concentration of histidine may be more than 7.0 mmoles/100 g muscle. Histidine is highly stimulative to lipid oxidation in the sarcoplasmic reticular fraction of winter flounder. Although many other amino acids are capable of binding iron, none exhibited the high stimulatory activity of histidine. D- and L-histidine stimulated lipid oxidation to the same extent, but a number of histidine analogs were not stimulatory (9). Using these analogs, it was concluded that the alpha carboxylate, the alpha amino, and the N-1 (and to a lesser extent the N-3) nitrogen were required for the stimulatory effect. These results implied that histidine functioned by forming a complex with ferric iron. The necessity of having one coordination bond on the iron free or coordinated to a very weakly held molecule such as water and optimal stimulation by ADP at a 1:1 ratio to iron, suggests that an active complex is formed in a 1:1:1 ratio of iron, ADP, and histidine. The stimulatory activity of histidine is complex and appears to have more than one phase. In any case, it is stimulatory to at least a concentration of 100 mM *in vitro*.

Ferric chloride in the presence of equimolar ADP, 1 M dimethylsulfoxide and a system capable of generating superoxide and hydrogen peroxide, i.e., hypoxanthine and xanthine oxidase, produces hydroxyl free radical. Huang (76) observed that a greater amount of hydroxyl radical was produced when histidine was used as the buffer as compared to several of its analogs; these were the same analogs that did not stimulate lipid oxidation (Table III). A

Table III. Methane sulfinic acid (MSA) generated by Fe-ADP and histidine analog buffers in a hypoxanthine/xanthine oxidase system. All treatments contained 25 μM FeCl$_3$, 25 μM ADP, 1.0 M dimethyl sulfoxide, 0.2 mM hypoxanthine, 0.1 unit xanthine oxidase, and 5 mM buffers, pH 6.8 and were performed at 23°C for 20 min. Data are results of duplicate experiments

Buffer	MSA (μM)
Histidine	47.5[a]
Imidazole	26.4[b]
1-methylhistidine	29.0[b]
3-methylhistidine	29.0[b]
Urocanic acid	26.4[b]
Histidinol	27.7[b]

[ab] Data with same superscript are not significantly different (P>0.05).
Adapted from C.-H. Huang(76).

maximal level of hydroxyl radical production occurred in this system at a histidine concentration of only 1 mM, and no further increase was seen up to 100 mM (Table IV). Thus, activation by histidine at high concentrations does not appear to be related to its ability to stimulate hydroxyl radical production in the presence of iron. It is possible that histidine serves as a hydroxyl radical scavenger (80). The free radical produced from histidine by reaction with hydroxyl would have a lower energy level than the hydroxyl radical and a greater ability to diffuse to the lipid sites on the membrane; thus, its ability to stimulate lipid oxidation at high concentrations might be related to its ability to serve as a secondary free radical.

Table IV. Effect of histidine concentration on methane sulfinic acid (MSA) production in hypoxanthine/xanthine oxidase system. All treatments contained 25 μM $FeCl_3$, 25 μM ADP, 1.0 M dimethyl sulfoxide, 0.2 mM hypoxanthine, 0.1 unit xanthine oxidase, and histidine, pH 6.8 and were performed at 23°C for 20 min. Data are results of quadruplicate samples

Histidine (mM)	MSA (μM)
0.025	23.4 ± 1.7[a]
0.1	25.6 ± 4.8[a]
0.5	33.4 ± 6.9[ab]
1	42.0 ± 3.8[bc]
5	45.5 ± 2.4[c]
50	44.6 ± 3.6[c]
100	46.9 ± 4.1[c]

[abc] Data with same superscript are not significantly different ($P > 0.05$). Adapted from C.-H. Huang (76).

Compounds that chelate iron (EDTA, sodium tripolyphosphate) or remove the active reduction products of oxygen from the medium inhibit enzymic catalyzed lipid oxidation by the sarcoplasmic reticulum. The latter include superoxide dismutase, catalase and peroxidases for the removal of hydrogen peroxide, and hydroxyl free radical scavengers. Lipid oxidation of this membrane fraction is also inhibited by phenolic antioxidants like propyl gallate or t-butylhydroquinone.

The original observation that led to the investigations of the lipid peroxidative system in muscle membranes was that phospholipase A_2 added to ground muscle mince prevented lipid oxidation (68). Earlier investigators had observed similar results (reviewed by Shewfelt; 81). Phospholipase activity was identified in the microsomal fraction of winter flounder muscle (73). Hydrolysis of the phospholipid fraction of the membrane *in vitro* by exogenously added phospholipase A_2 inhibited lipid oxidation in a manner that was dependent on free fatty acid production by the enzyme (82). Phospholipase perferentially released eicosapentaenoic acid from the phospholipid fraction. Phospholipase inhibited both enzymic (ferric iron + NADH)- and non-enzymic (ferrous iron)-catalyzed lipid oxidation. The fatty acids hydrolyzed remained for the most part with the membrane. On their removal from the membrane by BSA, activity was restored. Phospholipase C also inhibited lipid peroxidative activity. Hydrolysis of the phospholipid by phospholipase may inhibit lipid oxidation by modifying membrane structures, or the released free fatty acids may form an inactive complex with the iron (83). The free fatty acids could also inhibit the enzyme(s) involved in electron transfer from NADH to iron. Microsomal fractions from both fish and chicken are inhibited by KCl and NaCl (*69,84*); with fish microsomes the inhibitory effect of NaCl is due *in vitro* to an increased lag period (*76*).

A number of inhibitors of membrane lipid oxidation are found in the soluble fraction of the muscle tissue. Most of these inhibitors are unknown; they may function in different ways. Low molecular weight and high molecular weight compounds, thermolabile and thermostable components, and those that act immediately or require pre-incubation with the membrane have been identified (*10,75,84-86*). It is interesting that in the case of fish muscle, the nature of the inhibitors (as well as activators) varies between the muscle tissues of different species. In particular, there appears to be a significant difference between white and dark muscles (*85*).

Lipid oxidation in muscle foods is primarily a problem during frozen storage when spoilage by microorganisms is controlled. Fish muscle microsomes *in vitro* catalyze a rapid oxidation of their lipids during frozen storage (*84,87*). The extent of oxidation is higher at higher storage temperatures. When the microsomal fraction was removed from the muscle tissue and stored frozen in the presence of peroxidizing components, the rate of oxidation was very fast, being essentially over in two days; a large part of the maximal value had already been achieved in twelve hr. These results give some sense of the extent to which the antioxidant systems of muscle tissue can protect against ionic iron-stimulated oxidation. Cytosolic inhibitors were found to be effective in the frozen state, as well as the unfrozen state (*84*).

Lipid peroxidative activity of fish muscle sarcoplasmic reticulum is sensitive to the post mortem age of the membrane whether aged *in situ* or *in vitro*. Aged flounder sarcoplasmic reticulum becomes more sensitive to a soluble high molecular weight thermolabile activator in the soluble fraction of the muscle with increased post mortem age of the membrane (*86*). Non-enzymic catalyzed (ferrous iron) oxidation is also sensitive to post mortem age of the sarcoplasmic reticulum with the lag phase becoming shorter with time post mortem (*76*). There is a loss of membrane tocopherol in the sarcoplasmic reticulum with time post mortem. It is possible that the greater susceptibility of the membrane lipids to oxidation is related to the loss of this lipid-soluble antioxidant.

Lipid oxidation is understood on a molecular level far better than it is in a food. Model systems studies have helped to define some of the problems and indicate some of the specific pathways, but they usually do not tell very much about what actually occurs or can occur in a stored food. There is some indication that the membrane lipids may be the first lipids in stored fish that are oxidized. Halpin (*88*) demonstrated that the highly unsaturated fatty acids of membranes oxidized more rapidly than those of the neutral storage lipids in minced herring muscle. This was done by measuring the loss of docosahexaenoic acid (DHA) in the total lipid fraction of minced herring muscle stored at -20°C and comparing it to the loss of DHA in the sarcoplasmic reticulum fraction of the same muscle.

While there was no loss of DHA (as measured by the ratio of DHA/palmitic acid) in the total lipid fraction, there was a substantial decrease of the DHA/palmitic acid ratio in the sarcoplasmic reticulum over a storage period of eight months (Figure 2). That is, no change could be detected in the total DHA of the tissue even though the DHA of the membrane fraction of the herring muscle decreased from a DHA/palmitic acid ratio of about 3 to about 1 which represents a loss of 2/3 of the DHA. The loss in the membrane fraction was not detected in the total lipid fraction because the lipid of the SR represents a relatively small fraction of the total lipid. There was a roughly linear increase in the production of TBA-reactive substances during this period of storage.

Studies similar to these were carried out with a lean fish, winter flounder (Figure 3). Since in flounder the total tissue lipids are comprised mainly of membrane lipids, one would expect to see similar results comparing the sarcoplasmic reticulum and the total lipids. There were measurable decreases in DHA (again expressed as the ratio of DHA to palmitic acid) during 18 months of storage at -20°C in both the total lipids and the lipids of the SR. In fact, it appears as if there may have been a greater decrease in the DHA of the total tissue lipids than in the SR. This could be due to greater oxidation taking place in other membrane phospholipids compared to those of the sarcoplasmic reticulum, e.g., mitochondria. Whether peroxidizing membrane lipids can initiate lipid oxidation in neutral stored lipids is not known. In a model system it was demonstrated that peroxidizing sarcoplasmic reticulum of herring muscle could initiate lipid oxidation in an emulsified herring lipid fraction (*89*); oxidation of the emulsified lipid was completely dependent on oxidation of the lipid in the microsomal fraction.

The question of the relationship between pigment oxidation and lipid oxidation in stored muscle tissue has been of interest. It seems likely that once the oxidation starts in either fraction, it could stimulate oxidation in the other. In a model system it was demonstrated that the peroxidizing microsomal fraction from chicken muscle could stimulate the oxidation of oxymyoglobin to metmyoglobin (*90*). Oxidation of the iron myoglobin from the +2 to the +3 state could be inhibited by glutathione peroxidase-glutathione, a system which removes lipid hydroperoxides and/or hydrogen peroxide.

An enzyme-catalyzed lipid peroxidative system that requires NAD(P)H and is enhanced by ADP has been reported in trout muscle mitochondria (*91*). This system behaves in a similar manner to that of the sarcoplasmic reticulum and has a comparable specific activity. The lipid peroxidative enzymes are associated with the inner membrane. The lipid oxidative activity of the mitochondria increased with time of storage *in vitro*; this may be the result of disintegration of the outer membrane since sonication produced the same effect. Mitochondria can be isolated from fish muscle tissue only up to about 24 hr post mortem; this may be a reflection of mitochondrial disintegration *in situ*. Lipid oxidation in fish

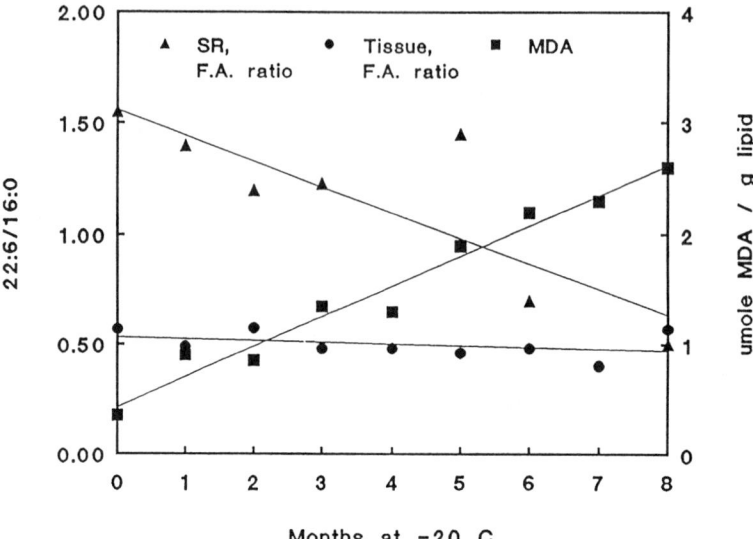

Figure 2. Changes in the ratio of docosahexenoic acid (22:6) to palmitic acid (16:0) in tissue and sarcoplasmic reticulum (SR) lipids and formation thiobarbituric acid reactive substances (expressed as malonaldehyde, MDA) during the storage (−20 °C) of minced herring muscle.

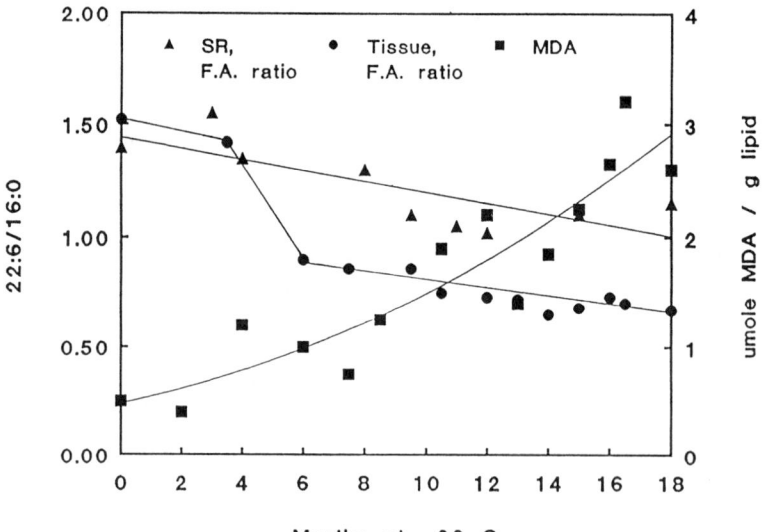

Figure 3. Changes in the ratio of docosahexenoic acid (22:6) to palmitic acid (16:0) in tissue and sarcoplasmic reticulum (SR) lipids and formation thiobarbituric acid reactive substances (expressed as malonaldehyde, MDA) during the storage (−20 °C) of minced winter flounder muscle.

muscle mitochondria was competitive with mitochondrial respiration; respiratory substrates inhibited lipid peroxidation. Ubiquinol played an important role in mitochondrial lipid oxidation. Whether the ubiquinol reacts directly with lipid free radicals or functions to regenerate the tocopheroxyl radical to tocopherol is not known. A similar lipid oxidative system has been described in rat liver mitochondria (92). Mitochondrial oxidations may be of particular importance in muscle tissues which have a high content of mitochondria such as the red muscles of beef and lamb and the dark muscle of fish (93).

Strategies to Prevent Lipid Oxidation by Redox Iron

When developing strategies to design antioxidant systems for foods it is important to know the nature of the lipid oxidation catalyst. For instance, some of the protection systems which work for redox iron will be ineffective in foods in which the primary lipid oxidation catalysts are heme iron and lipoxygenases.

To prevent oxidation by iron its is important to control the active iron species. This can most easily be done by the use of chelators; however not all types and concentrations of chelators will be effective. Effective chelators inhibit iron-catalyzed lipid oxidation by: 1) occupying all reactive coordination sites making the iron catalytically inert. 2) prevention of iron redox cycling (e.g. maintaining either the ferric or ferrous state). 3) prevention of the interaction of iron with substrates by stearic hinderance or formation of insoluble complexes (94). EDTA is commonly added to foods to prevent iron-catalyzed lipid oxidation. However, EDTA can also promote lipid oxidation since EDTA-iron complexes have greater solubility and oxidation-reduction potential than iron alone (36). When EDTA concentration is greater than iron, the iron is completely surrounded and thus it is prevented from interacting with peroxides (95). Phosphates are also used to prevent iron-catalyzed lipid oxidation in foods; however, not all phosphates will provide equal antioxidant activity in all meats. For example, di- and tri-phosphates (0.5%) inhibited the formation of TBARS in frozen salted and unsalted ground beef while trimetaphosphate had little to no antioxidant activity (96). Variability in the activity of phosphates were also found in cooked pork where the antioxidant effectiveness was in the order of sodium pyrophosphate > sodium tripoly-phosphate > sodium hexametaphosphate > disodium phosphate > monosodium phosphate (97).

Another chelator used in foods to inhibit iron-catalyzed lipid oxidation is citric acid. While citric acid does not prevent iron from participating in the Haber-Weiss cycle (36), it has been shown to be effective in inhibiting the development of warmed-over flavor in pork (97) and beef (98). Phytate has been suggested to act as a natural chelating agent in cereals and oil seeds (94) and is an effective antioxidant in chicken (99). Phytate inhibits iron-catalyzed lipid oxidation by blocking the ability of iron to participate in Haber-Weiss-type reactions and by increasing oxidation of ferrous ions to the inactive ferric state (94).

Ceruloplasmin is a copper-containing enzyme found in blood plasma which catalyzes the conversion of ferrous ions to ferric ions in the presence of oxygen (*100*). Ceruloplasmin will inhibit lipid oxidation by maintaining iron in the inactive oxidized state. Addition of ceruloplasmin to turkey muscle decreased lipid oxidation during refrigerated storage (*37*). Whole and spray-dried porcine blood plasma, which contains active ceruloplasmin, will inhibit lipid oxidation *in vitro* and in ground pork during frozen (-15°C) storage (*101,102*). In addition to ceruloplasmin, chelation of iron by transferrin could also be involved in the ability of plasma to inhibit iron-catalyzed lipid oxidation since iron bound to transferrin is not an active catalyst (*29*).

Another way to decrease catalysis of lipid oxidation via redox iron is to decrease the amount of iron in foods. This can be accomplished by avoiding iron contamination during processing and possibly by deceasing the amount of iron in muscle foods by alteration of the diet of the animal (*103*). More research is needed in this area to determine how the diet effects the free and protein-bound iron concentrations of foods. Redox iron could also potentially be inhibited by decreasing the concentration of reducing agents such as NAD(P)H and ascorbate; however, many of these compound are involved in the maintenance of the reduced states of myoglobin and hemoglobin and their elimination would result in product discoloration and could increase the formation of hydrogen peroxide-activated myoglobin which can promote lipid oxidation (*34*).

Other possible mechanism to decrease the amount of iron-catalyzed lipid oxidation in foods would be to decrease the activity of enzymes systems which can reduce iron or generate superoxide and hydrogen peroxide. These methods would include heat treatment or genetic manipulation of the expression of these enzymes.

Acknowledgments

This work was supported in part by the Massachusetts Agricultural Experiment Station, the Univ. of Kentucky Experiment Station, the National Marine Fisheries Institute, Grant No. I-801-84 from BARD-the United States-Israel Binational Agric. Research and Development Fund, a grant from the MIT Sea Grant College Program and by NOAA Contract 88-NER-205 of the Saltonstall-Kennedy Program.

Literature Cited

1. Haber, F.; Weiss, J. *Proc. Roy. Soc. Load. (A).* **1934**, *147*, 332-335.
2. Halliwell, B.; Gutteridge, J.M.C. *Arch. Biochem. Biophys.* **1986**, *246*, 501-514.

3. Graf, E.; Mahoney, J.R. Bryant, R.G.; Eaton, J.W. *J. Biol. Chem.* **1984**, *259*, 3620-3624.
4. Dunford, H. B. *Free Rad. Biol. & Med.* **1987**, *3*, 405-421.
5. Bakkeren, D.L.; de Jeu-Jaspars, C.M.H.; van der Heul, C.; van Eijk, H.O. *Int. J. Biochem.* **1985**, *17*, 925-930.
6. Rush, J.D.; Koppenol, W.H. *Am. Soc. of Biol. Chem.* **1986**, *261*, 6730-6733.
7. Rowley, D.A.; Halliwell, B. *FEBS Lett.* **1982**, *142*, 39-41.
8. Floyd, R.A.; Lewis, C.A. *Biochem.* **1983**, *22*, 2645-2649.
9. Erickson, M.C.; Hultin, H.O. In *"Oxygen Radicals in Biology and Medicine"*, M.G. Simic; K.A. Taylor; J.F. Ward, and C. Sonntag, Eds., Plenum Press, NY. **1988**, 307-312.
10. Erickson, M.C.; Hultin, H.O.; Borhan, M. *J. Food Biochem.* **1990**, *14*, 407-419.
11. Hazell, T. *J. Sci. Food Agric.* **1982**, *33*, 1049-1056.
12. Decker, E.A.; Huang, C.-H.; Osinchek, J.E.; Hultin, H.O. *J. Food Biochem.* **1989**, *13*, 179-186.
13. Decker, E.A.; Hultin, H.O. *J. Food Sci.* **1990**, *55*, 947-950, 953.
14. Kanner, J.; Hazan, B.; Doll, L. *J. Agric. Food Chem.* **1988**, *36*, 412-415.
15. Schricker, B.R; Miller, D.D.; Stouffer, J.R. *J. Food Sci.* **1982**, *47*, 740-743.
16. Igene, J.O.; King, J.A.; Pearson, A.M.; Gray, J.I. *J. Agric. Food Chem.* **1979**, *27*, 838-842.
17. Schricker, B.R.; Miller, D.D. *J. Food Sci.* **1983**, *48*, 1340-1342, 1349.
18. Chen, C.C.; Pearson, A.M.; Gray, J.I.; Fooladi, M.H.; Ku, P.K. *J. Food Sci.* **1984**, *49*, 58-584.
19. Rhee, K.S.; Zibru, Y.A.; Ordonez, G. *J. Agric. Food Chem.* **1987**, *35*, 1013-1017.
20. Puppo, A.; Halliwell, B. *Biochem. J.* **1988**, *249*, 185-190.
21. LaCross, J.M.; Linder, M.C. *Biochem. Biophys. Acta.* **1980**, *633*, 45-55.
22. Boyer, R.F.; McCleary, C.J. *Free Radical Biology & Medicine.* **1987**, *3*, 389-395.
23. Decker, E.A.; Welch, B. *J. Agri. Food Chem.* **1990**, *38*, 674-677.
24. Kanner, J.; Doll, L. *J. Agric. Food Chem.* **1991**, *39*, 247-249.
25. Decker, E.A.; Hultin, H.O. *J. Food Sci.* **1990**, *55*, 951-953.
26. Seman, D.L.; Decker, E.A.; Crum, A. *J. Food Sci.* **1991**, *56*, 356-358.
27. Apte, S.; Morrissey, P.A. *Food Chem.* **1987**, *25*, 127-134.
28. Winterbourn, C.C. *Biochem. J.* **1979**, *182*, 625-628.
29. Baldwin, D.A.; Jenny, E.R.; Aisen, P. *J. Biol. Chem.* **1984**, *259*, 13391-13394.
30. Kontoghiorgles, G.J.; Chambers, S.; Hoffbrand, A.V. *Biochem. J.* **1987**, *241*, 87-92.
31. Ozaki, M.; Kawabata, T.; Awai, M. *Biochem. J.* **1988**, *250*, 589-595.

32. O'Connell, M.; Halliwell, B; Moorhouse, C.P.; Aruoma, O.I.; Baum, H.; Peters, T.J. *Biochem J.* **1986**, *234*, 727-731.
33. O'Connell, M.J.; Ward, R.J.; Baum, H.; Peters, T.J. *Biochem J.* **1985**, *229*, 135-139.
34. Harel, S.; Kanner, J. *J. Agric. Food Chem.* **1985**, *33*, 1188-1192.
35. Morehouse, M.; Flitter, W.D.; Mason, R.P. *FEBS Letters.* **1987**, *222*, 246-250.
36. Mahoney, J.R.; Graf, E. *J. Food Sci.* **1986**, *51*, 1293-1296.
37. Kanner, J.; Sofer, F.; Harel, S.; Doll, L. *J. Agric. Food Chem.* **1988**, *36*, 415-417.
38. Gebicki, J.M.; Bielski, B.H.J. *J. Am. Chem. Soc.* **1981**, *103*, 7020-7023.
39. Kuppusamy, P; Zweier, J.L. *J. Bio. Chem.* 264, **1989**, pp. 9880-9884.
40. Lloyd, R.V.; Mason, R.P. *J. Biol. Chem.* **1990**, *265*, 16733-16736.
41. Britigan, B.E.; Pou, S.; Rosen, G.M.; Lilleg, D.M.; Buettner, G.R. *J. Biol. Chem.* **1990**, *265*, 17533-17538.
42. Donnelly, J.K.; McLellan, K.M.; Walker, J.L.; Robinson, D.S. *Food Chemistry.* **1989**, *33*, 243-270.
43. Scarpa, M.; Stevanato, R.; Viglino, P.; Rigo, A. *J. Biol. Chem.* **1983**, *258*, 6695-6697.
44. Wallace, W.J.; Houtchens, R.A.; Maxwell, J.C.; Caughey, W.S. *J. Biol. Chem.* **1982**, *257*, 4966-4977.
45. Kanner, J.; Mendel, H. *J. Food Sci.* **1977**, *42*, 60-64.
46. Winterbourn, C.C. *Biochem. J.* **1983**, *210*, 15-19.
47. Kanner, J.; Harel, S.; Hazan, B. *J. Agric. Food Chem.* **1986**, *34*, 506-510.
48. Kanner, J.; Salan, M.A.; Harel, S.; Shegalovich, J. *J. Agric. Food Chem.* **1991**, *39*, 242-246.
49. Decker, E.A. *Ph.D. Dissertation.* University of Massachusetts, Amherst. **1988**.
50. Murata, M.; Sakaguchi, M. *J. Food Sci.* **1986**, *51*, 321-326.
51. Murata, M.; Sakaguchi, M. *J. Agric. Food Chem.* **1988**, *36*, 595-599.
52. Powers, J.M.; Mast, M.G. *J. Food Sci.* **1980**, *45*, 760-764.
53. Rhee, K.S.; Smith, G.C.; Rhee, K.C. *J. Food Sci.* **1983a**, *48*, 351-352.
54. Buckley, D.J.; Gray, J.I.; Asghar, A.; Price, J.F.; Krackel, R.L.; Booren, A.M.; Pearson, A.M.; Miller, E.R. *J. Food Sci.* **1989**, *54*, 1193-1197.
55. Cuppett, S.L.; Gray, J.I.; Booren, A.M.; Price, J.F.; Stachiw, M.A. *J. Food Sci.* **1989**, *54*, 52-54.
56. Salih, A.M.; Price, J.F.; Smith, D.M.; Dawson, L.E. *J. Food Qual.* **1989**, *12*, 71-73.
57. Takiguchi, A. *Nippon Suisan Gakkaishi.* **1989**, *55*, 1549-1654.
58. Chang, I.; Watts, B.M. *Food Res.* **1950**, *15*, 313-321.
59. Nambudity, D.D. *J. Food Sci. and Technol.* **1980**, *17*, 176-178.
60. Rhee, K.S.; Smith, G.C.; Terrell, R.N. *J. Food Protect.* **1983b**, *46*, 578-581.

61. Osinchak, J.E. *M.S. Thesis*, Univ. of Mass, Amherst. **1989.**
62. Kanner, J.; German, J.B.; Kinsella, J.E. *CRC Crit. Rev. Food Sci. Nutrition.* **1987,** *25,* 317-364.
63. Kanner, J.; Harel, S.; Jaffe, R. *J. Agric. Food Chem.* **1991,** *39,* 1017-1021.
64. Hochstein, P.; Mordenbrand, K.; Ernsten, L. *Biochem. Biophys. Res. Comm.* **1964,** *14,* 323-328.
65. Bidlack, W.R.; Okita, R.T.; Hochstein, R. *Biochem. Biophys. Res. Commun.* **1973,** *53,* 459-465.
66. Hochstein, P.; Ernster, L. *Biochem. Biophys. Res. Commun.* **1963,** *12,* 388-394.
67. Lin, T.-S.; Hultin, H.O. *J. Food Sci.* **1977,** *42,* 136-140.
68. Govindarajan, S.; Kotula. A.W.; Hultin, H.O. *J. Food Sci.* **1977,** *42,* 571-582.
69. McDonald, R.E.; Hultin, H. O. *J. Food Sci.* **1987,** *52,* 15-21,27.
70. Rhee, K.S.; Dutson, T.-R.; Smith, G.C. *J. Food Sci.* **1984,** *49,* 675-679.
71. Rhee, K.S.; Ziprin, Y.A. *J. Food Biochem.* **1987,** *11,* 1-15.
72. Slabyj, E.M.; Hultin, H.O. *J. Food Sci.* **1982,** *47,* 1395-1398.
73. Shewfelt, R.L.; McDonald, R.E.; Hultin, H.O. *J. Food Sci.* **1981,** *46,* 1297-1301.
74. Decker, E.A.; Erickson, M.C.; Hultin, H.O. *Comp. Biochem. Biophys.* **1988,** *91B,* 7-9.
75. Han, T.J.; Liston, J. *J. Food Sci.* **1989,** *54,* 809-813.
76. Huang, C.-H. *Ph.D. Dissertation,* University of Massachusetts, Amherst. **1991,** 128 pp.
77. McDonald, R.E.; Kelleher, S.D.; Hultin, H.O. *J. Food Biochem.* **1979,** *3,* 125-134.
78. Phillippy, B.Q. *Ph.D. Dissertation.* University of Massachusetts, Amherst. **1984.** 165 pp.
79. Ikeda, S. In *"Advances in Fish Science and Technology,"* J.J. Connell, Ed.; Fishing News Books, Ltd., **1980,** Surrey, England. pp. 111-124.
80. Uchida, K.; Kawakishi, S. *Biochem. Biophys. Res. Comm.* **1986,** *138,* 659-665.
81. Shewfelt, R.L. *J. Food Biochem.* **1981,** *5,* 79-100.
82. Shewfelt, R.L.; Hultin, H.O. *Biochim. Biophys. Acta.* **1983,** *751,* 432-438.
83. Balasubramanian, K.A.; Nalini, S.; Cheeseman, K.H.; Slater, T.S. *Biochem. Phys. Acta* **1989,** *1003,* 232-237.
84. Apgar, M.E.; Hultin, H.O. *Cryobiology.* **1982,** *19,* 154-162.
85. Slabyj, B.M.; Hultin, H.O. *J. Food Biochem.* **1983,** *7,* 105-112.
86. Borhan, M.; Hultin, H.O.; Rasco, B.A. *J. Food Biochem.* **1990,** *14,* 307-317.
87. McDonald, R.E.; Apgar, M.E.; Hultin, H.O. *Rec. Adv. in Food Sci. & Technol.* **1981,** *1,* 84-91.

88. Halpin, B.E. *M.S. Thesis.* University of Massachusetts, Amherst. **1984**, 119 pp.
89. Slabyj, B.M.; Hultin, H.O. *J. Food Sci.* **1984**, *49*, 1392-1393.
90. Lin, T.-S; Hultin, H.O. *J. Food Sci.* **1976**, *41*, 1461-1465.
91. Luo, S.-W,H, *Ph.D. Dissertation.* University of Massachusetts, Amherst. **1987**.
92. Tretter, L.; Szabados, G.; Ando, A.; Horvath, I. *J. Bioenerg. Biomembranes.* **1987**, *19*, 41-44.
93. Love, R.M. *The Chemical Biology of Fish*; Academic Press: London. **1980**, Vol. 2, p. 85.
94. Graf, E.; Eaton, J.W. *Free Radical Biology Medicine.* **1990**, *8*, 61-69.
95. Waters, W.A. *J. Am. Oil Chemists' Soc.* **1971**, *48*, 427-433.
96. Mikkelsen, A.; Bertelsen, G. Skibsted, L.H. *Z Lebensm Unters Forsch.* **1991**, *192*, 309-318.
97. Shahidi, F.; Rubin, L.J.; Diosady, L.L.; Kassam, N.; Fong, J.C.-L-S. *Food Chemistry.* **1986**, *21*, 145-152.
98. Roozen, J.P. *Food Chemistry.* **1987**, *24*, 167-185.
99. Empson, K.L.; Labuza, T.P.; Graf, E.; *J. Food Sci.* **1991**, *56*, 560-563.
100. Osaki, S.; Johnson, D.A.; Frieden, E. *J. Biol. Chem.* **1966**, *241*, 2746-2751.
101. Faraji, H.; Decker E.A. *J. Food Sci.* **1991**, *56*, 1038-1041.
102. Faraji, H; Decker, E.A.; Aaron, D.K. *J. Agri. Food Chem.* **1991**, *39*, 1288-1290.
103. Kanner, J.; Bartov, I.; Salan, M.; Doll, L. *J. Agric. Food Chem.* **1990**, *38*, 601-604.

RECEIVED February 19, 1992

Chapter 4

Mechanism of Nonenzymic Lipid Peroxidation in Muscle Foods

Joseph Kanner[1]

Department of Food Science, Agricultural Research Organization, The Volcani Center, Bet Dagan 50250, Israel

When cells are injured, such as in muscle foods after slaughtering, lipid peroxidation is favored, and traces of O_2^-, H_2O_2, as well as traces of lipid peroxides, are formed. The stability of a muscle food product will depend on the "tone" of hese peroxides and especially the catalytic involvement of metal ions in the process. Ferrylmyoglobin is generated in muscle tissues, it oxidizes cytosolic-reducing compounds but not membrane lipids. Free iron ions are the main non-enzymic catalyzers of muscle lipid peroxidation. The main source of free iron seems to be ferritin, but myoglobin is also a source. Ascorbic acid is the main electron donor for iron-redox cycle in turkey muscle. The cytosol contains prooxidants and antioxidants and the "tone" of both affect lipid peroxidation.

Lipid peroxidation is one of the primary mechanisms of quality deterioration in stored foods, especially in muscle tissues. The changes in quality can be manifested by deterioration in flavor, color, texture, nutritive value and the production of toxic compounds (*1-5*). Evidence also exists, that dietary lipid peroxidation products, and especially cholesterol oxides, are involved in arterial injury and the atherosclerotic process (*5*). The oxidation of muscle lipids involves peroxidation of the unsaturated fatty acids, in particular the polyunsaturated fatty acids (PUFA) (*2-4*). The PUFA are associated with phospholipids, which are critical to the development of the off-flavor in muscle products (*1-4,6,7*). Most of these highly-unsaturated fatty acids are located in the membranes of muscle foods. The mechanism of lipid peroxidation in muscle foods was studied by several researchers, utilizing model systems of linoleate emulsion (*8-11*). These model systems could be conducted for studying general problems of lipid peroxidation but not for simulating lipid peroxidation in muscle foods (*3,12*). We adopted the utilization of muscle microsome membrane (sarcosomes) (*13*) or washed membrane residue (*14*) to study lipid peroxidation

[1]Current address: Department of Food Science and Technology, University of California, Davis, CA 95616

in muscle foods. During these studies we evaluated membrane lipid peroxidation by three possible pathways (i) microsome enzymic lipid peroxidation dependent on NADPH or NADH and iron ions; (ii) non-enzymic membrane lipid peroxidation stimulated by H_2O_2-activated myoglobin, or ferryl ion; (iii) non-enzymic lipid peroxidation catalyzed by the iron-redox cycle system.

Lipid peroxidation in raw muscle foods may be stimulated also by the enzymes lipoxygenase (15-17) or cyclooxygenase, only if the enzymes are activated by preformed peroxides and the fatty acids are in free form (4). However, in cooked muscle foods, during refrigerated storage which produces the warmed-over flavor, lipid peroxidation is totally dependent on non-enzymic reactions. Figure 1 schematically presents the possible pathways for initiating and propagating lipid peroxidation in muscle foods. This chapter will deals only with the mechanism of non-enzymic lipid peroxidation in muscle foods.

Activation of Oxygen Species and Metal Compounds

Singlet Oxygen. Singlet oxygen can be generated by microwave discharges (18), chemical (19) and photochemical reactions (20). During propagation of lipid oxidation, hemeproteins can accelerate the generation of peroxyl radicals, by disproportionation form singlet oxygen and electronically excited states of carbonyl (21), by the following reaction:

$$LOO\cdot + LOO\cdot \longrightarrow LOH + LO + {}^1O_2 \qquad [1]$$

$$LOO\cdot + LOO\cdot \longrightarrow LOH + LO^* + O_2 \qquad [2]$$

Singlet oxygen can initiate lipid peroxidation (4), however no strong evidence of this pathway was found in muscle foods. Most recently, it was shown that light may accelerate lipid peroxidation in pork and turkey, which was inhibited by ${}^1O_2$ quenchers (22).

Superoxide and Perhydroxyl Radical. Under biological conditions, significant amount of O_2^- can be generated. In food muscle tissue, there is no direct evidence that O_2^- is generated, however many studies done with other biological tissues support the presence of O_2^- in muscle foods. The sources for O_2^- in muscle food could arise from membrane electron transfer systems, autoxidation of oxymyoglobin to metmyoglobin, activation of several leukocytes presented in the vasculature of the muscle tissue, and oxidation of ascorbic acid and other reducing components by "free" iron (4).

Perhydroxyl radical (HO_2), whose pKa is 4.8 in water, is a much stronger oxidant than O_2^-. HO_2, but not O_2^-, could initiate lipid peroxidation (23). The loss of charge during formation of HO_2 from O_2^- allows the radical to penetrate into the membrane lipid region more easily, where it could initiate lipid peroxidation (24). Under physiological conditions, nearly 0.3% of the O_2^- formed exists in the protonated form. However, near the membrane the pH drop 3 pH units (25), in muscle tissue the pH decreased from 6.5-7.0 to 5.5-6.0 and the amount of HO_2 could reach 10-20% of O_2^-. However, HO_2 has not yet been proved to initiate lipid peroxidation of cell membranes.

Hydrogen Peroxide. Hydrogen peroxide is normally present as a metabolite at low concentration in aerobic cells. A system generating O_2^- would be expected to produce H_2O_2 by non-enzymatic dismutation or by superoxide dismutase (SOD) catalyzed dismutation.

Mitochondria, microsomes, peroxisomes and cytosolic enzymes have all been recognized as effective H_2O_2 generators when fully provided with their substrates. H_2O_2 can be generated directly by several enzymes, such as aldehyde oxidase or glucose oxidase (4). H_2O_2 is produced from autoxidation of flavins, thiols, phenoleates, or ascorbic acid, by O_2^- generation and spontaneous dismutation to H_2O_2, or by the interaction of the O_2^- with the semiquinone radical, such as that of ascorbic acid (12). The rate of these reactions is 2.6×10^8 M^{-1} sec^{-1} (26), the overall rate of O_2^- dismutation is only 5×10^5 M^{-1} sec^{-1} (27). H_2O_2 has limited reactivity and has not been shown to react directly with polyunsaturated fatty acids, however it can cross biological membranes (24). The generation of H_2O_2 in turkey muscle tissues was determined. Incubation of muscle tissues at 37°C shows H_2O_2 generation of 1 nmole/g min of fresh weight. Muscle tissues aging at 4°C increased the generation of H_2O_2 (28).

Hydroxyl radical. Hydroxyl radical, HO·, is produced when water is exposed to high-energy ionizing radiation, and its properties have been documented (1, 29). One electron reduction of H_2O_2 decompose it to HO$^-$ and HO·, a highly reactive ($E°_{pH\ 7}$ = +2.18) capable of oxidizing lipids and any other biological molecule. Hydroxyl radical produced *in vivo* or *in situ* would react at or close to its site of formation.

Most of the HO· generated *in vivo* or *in situ* comes from the metal-dependent breakdown of H_2O_2, according to the following reaction:

$$M^{n+1} + H_2O_2 \longrightarrow M^{(n+1)+} + HO\cdot + HO^- \qquad [3]$$

in which M^{n+1} is a transition metal. Fe^{+2} is known to form in the same reaction, which is called the Fenton reaction (30). *In vivo* or *in situ* muscle tissues, only the iron(II)-dependent formation of HO· actually happens under normal conditions (4). It may be possible that in several muscle foods and especially in fish, Cu(I) also play an important role in HO· generation (31). An interesting calculation of HO· formed in one cell if free iron and H_2O_2 is in the range of 1 μM, was found to be 46 sec^{-1} (32). Such an amount could produce an enormous effect on food, and especially to biological systems. Hydroxyl radicals are detected in beer (33). We have adopted a method to determine HO· in muscle homogenate and cytosol using benzoate as a scavenger of the radical. During this reaction, benzoate is hydroxylated to monohydroxy compounds, mostly to salicylic acid, the compounds are separated by HPLC and the hydroxylated compounds are detected fluorometrically (34) (Table I).

The study demonstrated that HO· radicals are formed especially during heating. The addition of EDTA, which increases the yield of HO· produced by iron-ascorbate (12) was found to also enhance the generation of HO· in muscle food. Hydroxyl radicals attack every biological molecule. The homogenate, which contains proteins and other compounds, that compete with benzoate to for HO·, shows lower results the to cytosol. Similar differences were obtained when the similar homogenate and the cytosol were gamma-irradiated. The results demonstrated that the potential of the cytosol and homogenate to generate HO· is 3-fold less than gamma-irradiation of 100 krad. Hydroxyl radicals could be determined also in biological tissues and foods by the accumulation of o-tyrosine, a non-metabolite product of phenylalanine (35).

Table I. Generation[a] of HO· in Muscle Homogenate and Cytosol During their Incubation for 15 min at 37°C, 98°C or by Gamma-Irradiation (Results are expressed in nmole salicylic acid eq/gr F.W.)

Treatment	37°C	98°C	Gamma-Irradiated (Krad)	
			100	700
H_2O	0	-	60.0	160.0
H_2O-EDTA	0	-	62.0	155.0
Homogenate	0	1.0	15.5	51.2
Homogenat-EDTA	0	3.5	15.0	52.0
Cytosol	0	3.0	17.5	56.0
Cytosol-EDTA	0	11.5	18.0	57.0

[a]Hydroxyl radical generation was determined by a method which used the scavenging properties of benzoate (2 mM). The monohydroxy benzoate compounds generated, and especially salicylic acid, are separated by HPLC and detected fluorometrically (34).

Ferryl Ion. Methemoglobin and metmyoglobin, the ferric states of these proteins, are activated by H_2O_2 producing a short-lived intermediate with one oxidizing equivalent on the heme, and one on the globin, giving an oxene-ferryl hemoglobin radical (13). Studies of H_2O_2-activated metmyoglobin showed that although it is not identical with Compound I or II of horseradish peroxidase, it has some structural features in common with both (36). Recently, it was demonstrated that the oxidizing equivalent of the globin is associated with a tyrosine radical (37).

Hydrogen peroxide-activated metmyoglobin and methemoglobin were found to oxidize a series of phenols (38), reducing agents (4) uric acid (39) and to cause protein cross-linking (40). We have reported that H_2O_2-activated myoglobin and H_2O_2-activated hemoglobin could initiate membrane lipid peroxidation (13). It was also found that ferryl myoglobin and hemoglobin could oxidize ß-carotene, methional (41), KTBA (42), desferrioxamine (43), salicylic acid (13,44) but not formate, methionine or benzoate (34,41,42). There have been several claims that myoglobin and hemoglobin are catalysts of the Fenton reaction (45). However, such claims have been based on non-specific scavengers of HO. Recently, we demonstrated that oxy and met forms of myoglobin and hemoglobin in the presence of a low concentration of H_2O_2 form the ferryl species and not free hydroxyl radical (*42*). The availability of a free coordination site is a stringent requirement for H_2O_2 activation of ferric-hemeproteins to ferryl species (*4*). The iron in cytochrom-C (cyt-C) is bound covalently to the polypeptide chain by all six coordinations. Iron cyt-C interacts with H_2O_2 without producing an intermediate Compound I. Incubation of heme proteins with a molar excess of H_2O_2 can cause heme degradation and release of iron ion (*46*). Recent studies have measured this iron release and its mechanism in myoglobin and hemoglobin vis cyt-C was presented (*47*).

Ferryl myoglobin has been found in *in vivo* systems under certain experimental conditions. Most recently, it was shown that ferryl myoglobin is generated in an isolated ischemic rat heart (*48*). Generation of a ferryl hemoglobin was detected in intact human red blood cells and myocytes, exposed to a continuous flux of peroxides (*49,50*). In order to determine if ferryl myoglobin is formed in muscle tissue, we performed spectrophotometric studies in the presence of sodium sulfide (Na_2S). Sodium sulfide adds to a double bond of heme pyrrol and disrupts the porphyrin conjugation forming a chlorin-type structure (*51*). Addition of 1 mM Na_2S to a reaction mixture, containing muscle homogeneate and a source of H_2O_2, produced an absorption maximum at 617 nm (Figure 2) as expected for the formation of oxysulfomyoglobin (*52*). However, if Na_2S was added to the homogenate 3 min after the addition of H_2O_2, S-Mb was not detected. The results indicate that ferryl ion generated is rapidly reduced by electron donors compounds in muscle cytosol.

Free Metal Ions. Transition metals, e.g. iron and copper, with their labile d-electron system, are well-suited to calalyze redox reactions. Iron is an important catalyst in biological systems (*4*). About two-thirds of body iron is found in hemoglobin, and smaller amounts in myoglobin. A very small amount of prosthetic components is found in various iron-containing enzymes, and in the transport protein transferrin. The remainder is present in intracellular storage proteins ferritin and hemosiderin (*52*). A small pool of non-protein non-heme iron provides "free" iron at micromolar concentration in tissues.

The small "transit pool" of iron seems to be chelated to small molecules. The exact chemical nature of this pool is not clear, but it may represent iron ions attached to phosphate esters (ATP, ADP), organic acids (citrate), and perhaps to membrane lipids or DNA (*53,54*). All these iron compounds are capable of decomposing H_2O_2 or ROOH to form free radicals (*53*). Gutteridge et al. (*55*) developed the bleomycin assay to measure the availability of iron in biological tissues for radical reaction. Using this method and others (*56-58*), we found that turkey muscle tissue contains between 0.5-2.0 μg/g F.W. of "free" chelatable iron ions, Table II.

The main source of free iron in cells seems to be ferritin. Ferritins are the major iron storage protein in cells (*52*). Iron can be released from ferritin and utilized by mitochondria for the synthesis of hemeproteins. In muscle cells, mitochondria synthesize myoglobin (*59*).

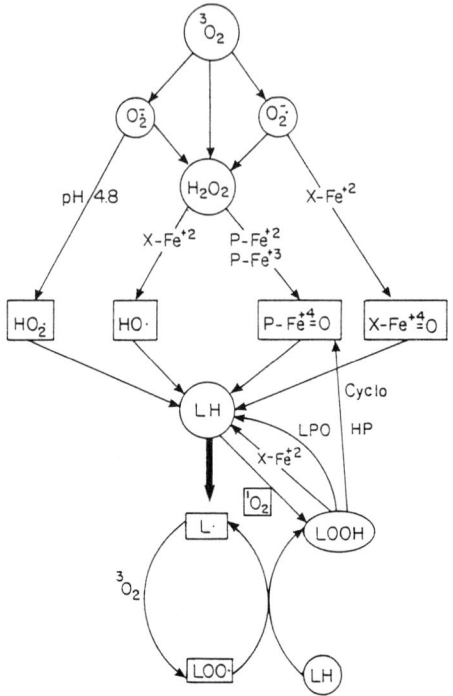

Figure 1. Schematic outline of several pathways for intiation of lipid peroxidation in biological systems.

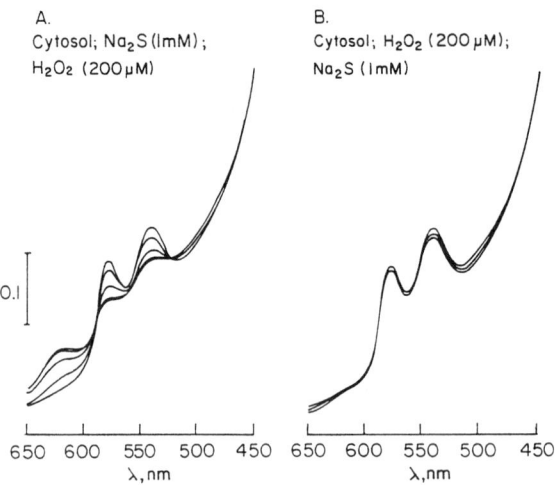

Figure 2. Generation of ferryl ion in turkey muscle cytosol by addition of H_2O_2 and Na_2S. A) Addition of Na_2S was before H_2O_2. B) Addition of Na_2S was 3 min after H_2O_2. Each line denotes scanning intervals of 2 min.

Table II. "Free" Iron Ions in Turkey Red Muscles

Method	μg/g F.W.	μM
[a]Bleomycin/DNA	1.5-2.5	30-50
[b]Ferrozine-Fe^{+2}	0.8-1.1	16-22
[c]Desferal-Fe^{+3}	0.5-1.0	10-20

[a](ref.56); [b](ref. 14); [c](ref. 102)

Ferrous ions can be released from ferritin by reductants small enough to pass through channels in the protein shell (60). Recently, it was found that O_2^- releases iron from ferritin (61) and O_2^- is the primary reductant in ascorbate-mediated ferritin iron release (61). Most recently, ferritins were separated from turkey muscle tissue (62). New data on beef muscle ferritin was also reported (63). During storage of muscle, ferritin lost iron at a significant rate (Table III).

Table III. Loss of Iron from Ferritin[a] Isolated from Dark and Light Turkey after Storage at 4°C

Treatment	Fe-Ferritin μg/g/Wet Tissue Storage Time (in days)		
	0	2	9
Dark muscle	1.50	1.14	0.60
Light muscle	0.55	0.42	0.08

[a]Ferritin was isolated by the standard procedure and ammonium sulfate, Kanner et al. (62) (Adapted from ref. 62).

Ceruloplasmin and albumin are the main protein molecules which chelate copper ions (64). Copper in mammalian and avian is also chelated in skeletal muscles and brain by carnosine and other histidine-di-peptides (65), or in fish by histidine (66). The biological role of ceruloplasmin has been suggested to be that of a "ferroxidase" which catalyzes the oxidation of ferrous ions to ferric ions, and facilitates the binding of iron to transferrin (67). Because of its ability to inhibit Fe^{+2} dependent radical reactions, ceruloplasmin is an important extracellular antioxidant (68,69). Human skeletal muscle contains one-third of the total copper in the body 20-47 μmole/kg. Most of it seems to be chelated by carnosine and other histidine-di-peptides, such as anserine and hemo-carnosine. All these compounds were found to work as antioxidants (65).

The amount of free copper seems to be very low and most of it chelated to proteins, such as ceruloplasmin, albumin or histidine-di-peptides, which stop them from catalysis of lipid peroxidation (65). However, in fish, it was published that the amount of low molecular weight complexes of copper is in the range of 100-300 ppb, most of it probably bonded to histidine (*31*).

Non-Enzymic Lipid Peroxidation in Model System and in *In Situ* Muscle Foods
Many studies have been directed toward the identification of the catalysts that promote muscle lipid peroxidation, In the beginning, lipid peroxidation was attributed to heme catalysts (*70,71*). The involvement of heme proteins as catalyzers of lipid peroxidation was first described by Robinson (*72*), who found that hemoglobin accelerates the peroxidation of unsaturated fatty acids. Lipid peroxidation by heme compounds exhibits an induction period which was postulated to be dependent on preformed hydroperoxides (*73*). Heme proteins, such as hemoglobin or myoglobin, accelerate the decomposition of hydroperoxides to free radicals which, in the presence of oxygen, propagate lipid peroxidation (*74-76*). Studies employing model systems of linoleate emulsions (*77*) showed that both heme and non-heme iron are important catalysts of lipid peroxidation in muscle products. Sato and Hegarty (*78*) directed a study of muscle lipid peroxidation using a water-washed muscle residue. They indicated that the substances responsible for initiating the reaction were water-soluble. Sato and Hegarty (*78*) and others (*79-81*) proposed that non-heme iron plays a major role in the catalysis of muscle lipid peroxidation and that myoglobin is not directly responsible for off-flavors developed during storage of cooked muscle foods. However, all these model systems suffer from many errors. Model systems using linoleate emulsion contain preformed hydroperoxides at different levels (*76*) they could not simulate lipid peroxidation in muscle cells, which are mostly of a membranal nature, containing unoxidized phospholipids. Model systems containing washed muscle tissues better simulate the situation in muscle tissues. However, most of the researchers using this model system had omitted from it several important compounds which, during the exhaustive washing or dialysis of the muscle samples, were removed (*3,78-82*). During incubation of minced muscle tissue, H2O2 is generated (*28*) and this could activate oxy- or metmyoglobin to a ferryl compound. We have demonstrated, for the first time, that an H_2O_2-activated metmyoglobin or oxymyoglobin, can initiate membrane lipid peroxidation (*13*) (Figure 3). These results were confirmed by many other researchers (*3,82-85*). Rhee et al. (*82*) and Asghar et al. (*3*) conducted experiments to determine whether H_2O_2-activated metmyoglobin could initiate lipid peroxidation in water-extracted muscle residue from raw and cooked beef and chicken stored at 4°C. While MetMb alone showed little or extremely low catalytic activity, H_2O_2-activated myoglobin accelerated lipid peroxidation, and its prooxidative effects were greater than ferrous ion (*3*).

H_2O_2 is only one of the cytosolic compounds that could affect lipid peroxidation in muscle foods. One should know that the cytosol contains pro- and antioxidants, all these compounds will affect *in situ* lipid peroxidation. The cytosolic extract from turkey muscles totally inhibited membranal lipid peroxidation catalyzed by ferryl ion. Both low (< 10 KD), and high (> 10 KD), molecular weight fractions separated from the cytosol exhibited inhibitory effects on ferryl-dependent membranal lipid peroxidation (*14*) (Figure 4).

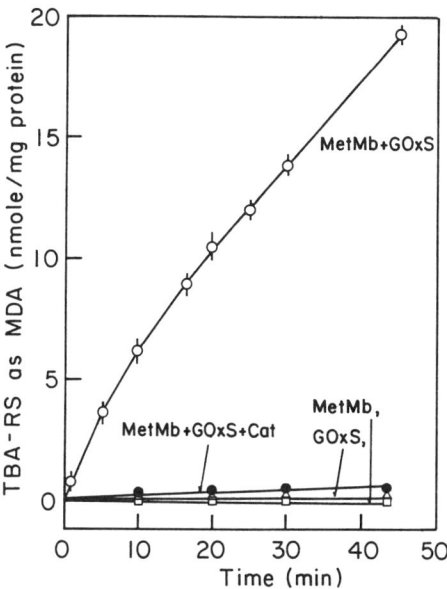

Figure 3. Effect of glucose oxidase system (GOxS) on membranal lipid peroxidation by activated MetMb. The reaction mixture contained glucose oxidase, 0.04 units, glucose, 400 uM, microsomes, 1 mg protein, KCl (0.12 M), EDTA (0.1 mM), and MetMb (30 uM) in 1 mL of histidine buffer (5 mM), pH 7.3, at 37°C. Error bars denote standard deviation (n=3); 0, MetMb + GOxS; 0, MetMb + catalase (40 units) and GOxS; MetMb or GOxS alone. (Adapted from ref. *103*).

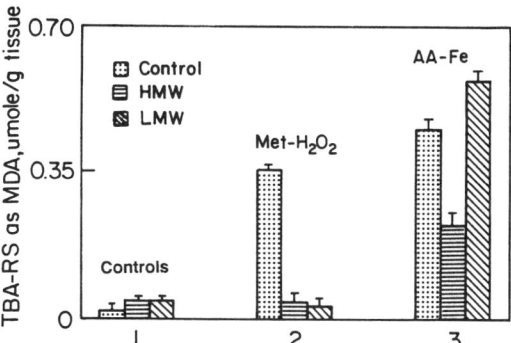

Figure 4. Effect of cytosolic fractions on lipid peroxidation of washed muscle residue by several systems. (1) Washed muscle residue in the presence of cytosolic fractions without catalyzers. (2) Washing muscle residue in the presence of LMW and HMW reconstituted to the cytosolic volume (1:3 w/v) plus activated H_2O_2-metmyoglobin (30 μM each). (3) Washed muscle residue in the presence of LMW and HMW reconstituted to the cytosolic volume (1:3 w/v) in the presence of ascorbic acid (200 μM) and $FeCl_3$ (10 μM).(Adapted from ref. *14*).

Turkey muscle cytosol contains reducing compounds at a level equivalent to ca 3 mg of ascorbic acid equivalent/100 g of fresh weight, or at a concentration of 150 μM. The amount of ascorbic acid is 80% of the reducing compounds (14). Recently we have assumed that ascorbic acid is the main inhibitor of ferryl-dependent membranal lipid peroxidation. However, most recent results indicate that destruction of ascorbic acid by ascorbate oxidase does not prevent the inhibitory effect of the cytosol (results not shown). Carnosine present in the muscle tissue at high concentration was found by us to reduce ferryl to oxymyoglobin (Figure 5).

In spite of recent studies, we do not know which are the main inhibitors in the cytosol that prevent ferryl oxidation of membrane lipids. Our data indicate that *in situ* turkey muscle, the iron-redox cycle-dependent lipid peroxidation, and not ferryl, may play a major role in the catalysis of lipid peroxidation in muscle foods. It is interesting to note that ferryl-dependent membrane lipid peroxidation and oxidation of other compounds are implicated in the "ischemic heart reperfusion injury" (84,85), and in the pathology of oxidized hemoglobin in erythrocytes of thalassemia, sickle cell anemia and GGPD deficiency (85).

Iron Redox Cycle-Dependent Lipid Peroxidation

Biological oxidation is due almost exclusively to metal ion-promoted reactions, of which iron is the most abundant (52). It has long been known that iron can catalyze peroxidation and that this can be stimulated by the presence or addition of thiols and ascorbic acid (76,86,87). However, only recently has the involvement of hydroxyl radicals in these reactions been postulated along with the contribution to the initiation of lipid peroxidation (88,89).

Ferrous ions in aerobic aqueous solution produce superoxide, hydrogen peroxide and hydroxyl radical (4,90) by the following reactions:

$$Fe^{+2} + O_2^- \longrightarrow Fe^{+3} + O_2 \qquad [4]$$

$$2O_2^- + 2H^+ \longrightarrow H_2O_2 + O_2 \qquad [5]$$

$$Fe^{+2} + H_2O_2 \longrightarrow HO^\cdot + HO^- + Fe^{+3} \qquad [6]$$

These reactions are cycled by O_2^- (so called Haber-Weiss), or by reducing agents, e.g. ascorbic acid or thiols (so called Redox-Cycle). Ascorbic acid, NAD(P)H, glutathione and cysteine are known reducing compounds present in biological tissues. They very rapidly reduce ferric to ferrous ions, producing a redox cycle (12). In fact, reaction [6] is Fenton's reagent (30). The first redox cycling Fenton's reagent was developed by Udenfriend et al. (91) using ascorbic acid, Fe-EDTA, oxygen and H_2O_2 for hydroxylation of aromatic substances.

Ferrous ions can stimulate lipid peroxidation by generating the hydroxyl radical from H_2O_2 but also by the breakdown of performed lipid peroxides (LOOH) to form the alkoxyl (LO$^\cdot$) radical [Equation 9].

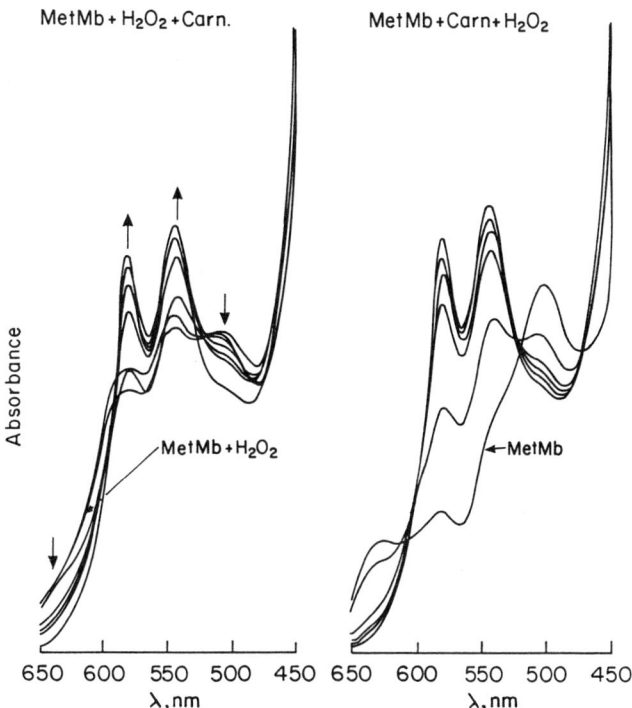

Figure 5. Reduction of ferryl ion by carnosine. Metmyoglobin (30 μM), H_2O_2 (50 μM) and carnosine (10 mM). Each line denotes a scanning interval of 2 min.

$$LH + HO· \longrightarrow L· + H_2O \qquad [7]$$

$$L· + O_2 \xrightarrow{LH} LOOH + L· \qquad [8]$$

$$LOOH + Fe^{+2} \longrightarrow LO· + HO^- + Fe^{+3} \qquad [9]$$

$$LH + LO· \longrightarrow L· + LOH \qquad [10]$$

The formation of HO· radicals observed in O_2^--generating model systems is usually inhibited by the addition of SOD and catalase (*4,2,24*). The inhibitory action of SOD clearly shows that $O2^-$ is also involved in the observed HO· production. The simplest explanation of its role would be that $O2^-$ reduced oxidized metal ions and so promotes the reduction of ferric to ferrous ion. When ascorbate is the reducing agent, SOD prevents the HO· production at a level of 50% (*12*) (Tables IV and V).

Table IV. Effect of Several Enzymes, Proteins, and Hydroxyl Radical Scavengers on the Generation of Ethylene from KTBA by Ferric Chloride, Ascorbate, and Cysteine

Treatment	Ethylene µmole/30 min	% Inhibition
control (Fe^{+3} 10 µM; ascorbate 200 µM)	0.43	
control + microsomes (1 mg protein/mL)	0.27	37.2
control + superoxide dismutase (200 units)	0.11	74.4
control + catalase (200 units)	0.01	97.7
control + mannitol (100 mM)	0.13	69.8
control + ethanol (400 mM)	0.08	81.4
control (Fe^{+3} 10 µM; cysteine 200 µM)	0.19	
control + catalase (200 units)	0.04	90.7
control + mannitol (100 mM)	0.03	93.0

(Adapted from ref. *12*).

As superoxide is generated during ascorbic acid oxidation, SOD inhibits HO·-dependent oxidation by decreasing the amount of H_2O_2 formed in the reaction by almost 50%. Ascorbic acid semiquinone reacts with O_2^- producing H_2O_2 by the following reaction (*26*):

$$A· + O_2^- \xrightarrow[2H^+]{2.6 \times 10^8 M^{-1}s^{-1}} A + H_2O_2 \qquad [11]$$

$$O_2^- + O_2^- \xrightarrow[2H^+;\ SOD]{10^9 M^{-1} s^{-1}} H_2O_2 + O_2 \qquad [12]$$

Table V. Effect of Several Enzymes, Proteins, and Hydroxyl Radical Scavengers on the Generation of Ethylene from KTBA by Ferric Chloride, Ascorbate and EDTA

Treatment	Ethylene µmole/30 mi	% Inhibition
control (Fe^{+3} 10 µM; ascorbate 200 µM; EDTA 100 µM)	10.0	
control + microsome (1 mg protein/mL)	7.89	21.1
control + bovine serum albumin (0.5 mg/mL)	8.57	14.3
control + superoxide dismutase[a] (200 units)	5.68	43.1
control + catalase[a] (200 units)	0.19	98.1
control + L-histidine (25 mM)	1.35	86.5
control + Na formate (50 mM)	1.43	85.7
control + Me_2SO (200 mM)	0.14	98.5
control + ethanol (200 mM)	0.35	96.4

[a]Autoclaved SOD and catalase gave results similar to control. (Adapted from ref. 12)

During reaction [11] two moles of O_2^- will generate 2 moles of H_2O_2, while SOD, from the same amount of O_2^-, generates only 1 mole H_2O_2. EDTA enhances iron-ascorbate generation of HO·, and both reactions are inhibited partially by SOD, and totally by catalase (12). Hydroxyl radical scavengers, such as mannitol, formate, benzoate, DMSO and ethanol, inhibit HO·-dependent oxidation of target molecules (12,24). Iron-dependent, hydroxyl radical-initiated peroxidation has sometimes been demonstrated in fatty acid systems solubilized by detergents (92), and these reactions are inhibited by SOD, catalase and HO· scavengers.

However, most scientists (12,32), find that when catalase, SOD or scavengers are added to isolated cellular membrane fraction, in the presence of Fe^{+2} and ascorbate, prooxidation is not inhibited. EDTA, which enhances iron-ascorbate-dependent HO· oxidation, prevents membrane lipid peroxidation by the same catalysts (12), Table VI.

How can we explain all of these results? Membrane bound iron certainly does participate in lipid peroxidation. Ascorbate, which enhances iron-dependent membranal peroxidation, forms a system with the metal which generates O_2^- and H_2O_2 close to the target molecule, the lipids. The oxidation by HO· is site-specific. The constant rate reaction of HO· with most of the biomolecule is about $10^9 M^{-1} s^{-1}$. For this reason, the radical will attack on its site of formation. A site-specific attack is poorly inhibited by HO· scavengers, SOD or catalase.

Table VI. Membrane Lipid Peroxidation Stimulated by Iron-Ascorbic Acid as Affected by Several Hydroxyl Radical Scavengers and Antioxidants

Treatment	TBA-RS as MDA nmole/mg Protein (20 min)	% Inhibition
Fe^{+3} (10 μM)	0.6	
control (Fe^{+3} 10 μM; ascorbate 200 μM)	31.5	
control + D-mannitol (100 mM)	32.5	0
control + L-histidine (25 mM)	28.0	11.0
control + ethanol (400 mM)	30.0	4.8
control + Me_2SO (200 mM)	29.7	5.7
control + α-tocopherol (25 μM)	27.1	14.1
control + BHT (25 μM)	1.1	96.4
control + ascorbyl palmitate (25 μM)	0.6	98.0
control + EDTA (100 μM)	1.3	95.9

(Adapted from ref. *12*).

In the presence of EDTA, however, iron is removed from these binding sites (*93*) to produce HO· in the free solution or cytosol. Hydroxyl radicals formed in the exogenous microsomal environment do not have the capability to initiate membranal lipid peroxidation (*12*).

It is well known that ferrous ion could stimulate peroxidation by catalyzing the decomposition of preformed hydroperoxides to alkoxyl radicals that can "initiate" lipid peroxidation (*76*). The reaction of ferrous ions with LOOH are an order of magnitude faster than their reaction with H_2O_2, 10^3M and 76 $M^{-1}s^{-1}$, respectively (*94*).

Iron redox-cycle-dependent membrane lipid peroxidation needs to attack unsaturated fatty acids only by a few HO· radicals. Once the first hydroperoxides are formed, propagation of lipid peroxidation will be catalyzed by the breakdown of LOOH to free radicals, and these reactions are not affected by SOD, catalase or HO· scavengers. The importance of "free" iron ions in the catalysis of turkey muscle lipid peroxidation was demonstrated by us using ceruloplasmin as a specific inhibitor of ferrous-dependent lipid peroxidation (*95*) (Table VII). These results were also confirmed by others (*96*).

Reducing compounds are the driving forces for the catalysis of lipid peroxidation by metal ions. In turkey muscle, ascorbic acid is the main reducing compound which affects iron-redox cycle. Destruction of the cytosolic ascorbic acid by ascorbic acid oxidase, prior to the addition of the cytosol to the membrane washed model system, totally inhibited lipid peroxidation (Figure 6). However in fish, other compounds, and not ascorbic acid, seem to affect the iron-redox cycle (*96*).

The cytosol contains compounds acting to stimulate, but also to inhibit, the process of lipid peroxidation. Turkey muscle cytosol contains compounds which totally inhibit membrane lipid peroxidation by ferryl ion. However, it contains compound which only partially-inhibited lipid peroxidation by an iron-redox cycle (14). The inhibitory effect of the cytosolic extract was also shown by other researchers (97,98). Many naturally-occurring compounds in the cytosol may affect non-enzymic lipid peroxidation. These include: a) superoxide dismutase; b) catalase; c) glutathione peroxidase; d) glutathione S-transferase; e) phospholipid hydroperoxidase glutathione peroxidase; f) myoglobin; g) chelating compounds, such as proteins, carnosine, anserine; and h) reducing compounds, such as ascorbic acid, NAD(P)H, SH-proteins, glutathione. In spite of numerous publications on this topic, we need to more fully elucidate the antioxidative contribution of cytosolic compounds to the process of lipid peroxidation.

Dietary supplements to feedstuffs of animals could affect lipid composition, such as the amount and degree of the unsaturated fatty acid, and α-tocopherol; both affect lipid peroxidation (99). More recently, we have shown that turkey muscle lipid peroxidation in situ could be affected by dietary iron. Our results demonstrated that the normal turkey feedstuff contains iron at a high concentration and this seems to affect meat quality and its stability during storage. The removal of Fe supplementation several weeks prior to slaughtering had no significant effect on the body weight, but it did significantly decrease muscle lipid peroxidation (100) (Figure 7).

The control and prevention of muscle lipid peroxidation in situ may be divided into three categories, i.e. those acting to control the level of the pro- and antioxidants by dietary supplements, those acting to control the level of oxygen and active oxygen species, and those acting to control activated catalysts and free radicals. However, discussion on this subject would go beyond the goal of this chapter (for reviews, see 4,101).

Table VII. Effect of Ceruloplasmin (150 U/g) on in situ Lipid Peroxidation in Minced Turkey Muscle Stored at 4°C for 7 days

Treatment	TBA-RS as MDA nmole/mg Tissue	% Inhibition
Control (run 1)	120	-
Ceruloplasmin	20	84
Control (run 2)	25	-
Ceruloplasmin	10	60

(Adapted from ref. 95).

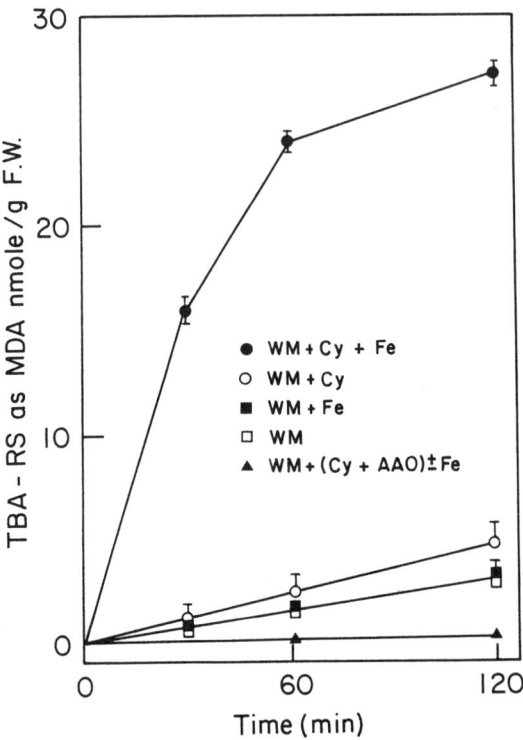

Figure 6. Effect of cytosol reducing factor on Fe-mediated lipid peroxidation of washed muscle membranes. WM = washed muscle membrane. Cy = cytosol. AAO = ascorbate oxidase (preincubation, 30 min at 25 °C). Fe = ferric chloride (20 μM).

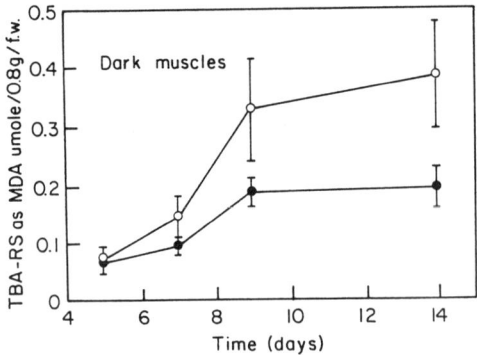

Figure 7. Effect of removal of iron supplementation, 5 weeks prior to slaughtering, on dark turkey muscle lipid peroxidation: control; O, without Fe (5 weeks) supplementation. (Adapted from ref. *100*).

Literature Cited

1. Simic, M.G.; Karel, M. *Autoxidation in Food and Biological Systems*; Plenum: New York, 1980.
2. St. Angelo, J.A.; Bailey, M.E. *Warmed-Over Flavor of Meat*; Academic Press: New York, 1987.
3. Asghar, A.; Gray, J.I.; Buckley, D.J.; Pearson, A.M.; Boosen, A.M. *Food Technol.*, **1988**, *42(6)*, 102.
4. Kanner, J.; German, J.B.; Kinsella, J.E. *CRC Critical Review in Food Science and Nutrition*, **1987**, *25*, 317.
5. Addis, P.B.; Park, S.W. In *Food Toxicology: A Perspective on the Relative Risk*; Scanian, R.A., Taylor, S.L. Eds.; Marcel Dekker Inc.: New York, 1989, pp. 297-330.
6. Keller, J.P.; Kinsella, J.E. *J. Food Sci.* **1973**, *38*, 1200.
7. Igene, J.O.; King, J.A.; Pearson, A.M.; Gray, J.I. *J. Agric. Food Chem.* **1979**, *27*, 838.
8. Liu, H.P. *J. Food Sci.* **1970**, *35*, 964.
9. Lee, Y.B.; Hargus, G.L.; Kirkpatrick, J.A.; Berner, D.L.; Forsythe, R.H. *J. Food Sci.* **1975**, *40*, 596.
10. Fisher, J.; Deng, J.C. *J. Food Sci.* **1977**, *42*, 610.
11. Sklan, D.; Tenne, Z.; Budowski, P. *J. Sci. Food Agric.* **1983**, *34*, 93.
12. Kanner, J.; Harel, S.; Hazan, B. *J. Agric. Food. Chem.* **1986**, *34*, 506.
13. Kanner, J.; Harel, S. *Arch. Biochem. Biophys.* **1985**, *237*, 314.
14. Kanner, J.; Salan, M.A.; Harel, S.; Shigalovich, I. *J. Agric. Food Chem.* **1991**, *39*, 242.
15. German, J.B.; Kinsella, J.E. *J. Agric. Food Chem.* **1985**, *33*, 680.
16. Hsieh, R.J.; Kinsella, J.E. *J. Agric. Food Chem.* **1989**, *37*, 279.
17. Grossman, S.; Bergman, M.; Sklan, D. *J. Agric. Food Chem.* **1988**, *36*, 1268.
18. Arnol, S.J.; Ogryzto, E.A.; Witzke, A. *J. Chem. Phys.* **1964**, *40*, 1769.
19. Khan, A.V.; Kasha, M. *J. Chem. Phys.* **1964**, *40*, 650.
20. Foote, C.S.; Wexler, S. *J. Am. Chem. Soc.* **1964**, *86*, 3879.
21. Hawco, F.J.; O'Brien, C.R.; O'Brien, P. *J. Biochem. Biophys. Res. Commun.* **77**, *76*, 304.
22. Whang, K.; Peng, I.C. *J. Food Sci.* **1988**, *53*, 1596.
23. Gebicki, J.M.; Bielski, B.H.J. *J. Am. Chem. Soc.* **1981**, *103*, 7020.
24. Halliwell, B.; Gutteridge, M.C. *Arch. Biochem. Biophys.* **1986**, *246*, 501.
25. Etherington, D.J.; Pugh, G.; Silver, I.A. *Acta Biol. Med. Ger.* **1981**, *40*, 1625.
26. Bielski, B.H.J.; Cohen, G.; Greenwald, R.A. Eds. In *Oxyradicals and their Scavenger Systems*; Elsevier: New York, 1983, Vol. 1, p.1.
27. Bielski, B.H.J.; Allen, A.O. *J. Phys. Chem.* **1977**, *81*, 1048.
28. Harel, S.; Kanner, J. *J. Agric. Food Chem.* **1985**, *33*, 1186.
29. Anbar, M.; Neta, P. *Int. J. Appl. Radiat. Inst.* **1967**, *18*, 495.
30. Fenton, H.J.H.; Jackson, H. *J. Chem. Soc. (London)* **1899**, *75*, 1.
31. Decker, E.A.; Hultin, O.H. *J. Food Sci.* **1990**, *55*, 947.
32. Halliwell, B.; Gutteridge, M.C. *Methods in Enzymology*, **1990**, *186*, 15.

33. Kaneda, H.; Kano, Y.; Osawa, T.; Ramarathnam, N.; Kawakishi, S.; Kamada, K. *J. Food Sci.* **1988**, *53*, 885.
34. Kanner, J.; Harel, S.; Granit, R. *Arch. Biochem. Biophys.* **1991**, *289*, 130.
35. Karam, L.R.; Simic, M.G. *Anal. Chem.* **1988**, *80*, 117A.
36. Chance, M.; Power, L., Kirmar, C.; Chance, B. *Biochemistry* **1986**, *215*, 1259.
37. Tew, D.; Ortizde Montellano, P.R. *J. Biol. Chem.* **1988**, *263*, 17880.
38. Shiga, T.; Imiazumi, K. *Arch. Biochem. Biophys.* **1975**, *167*, 469.
39. Ames, B.N.; Catheart, R.; Schwiers, E.; Hochstein, P. *Proc. Natl. Acad. Sci. U.S.A.* **1981**, *78*, 6858.
40. Rice, R.H.; Lee, Y.M.; Brown, W.D. *Arch. Biochem. Biophys.* **1983**, *221*, 471.
41. Kanner, J.; Harel, S. *Lipids*, **1985**, *237*, 314.
42. Harel, S.; Kanner, J. *Free Rad. Res. Comms.* **1988**, *5*, 21.
43. Kanner, J.; Harel, S. *Free Rad. Res. Comms.* **1987**, *3*, 309.
44. Galaris, D.; Mira, D.; Sevanian, A.; Cadenas, E.; Hochstein, P. *Arch. Biochem. Biophys.* **1988**, *362*, 221.
45. Sadrzadeh, S.M.; Graf, E.; Panter, S.S.; Halloway, P.E.; Eaton, J.W. *J. Biol. Chem.* **1985**, *259*, 14354.
46. Cantoni, L.; Gibbs, A.H.; De Matteis, F. *Int. J. Biochem.* **1981**, *13*, 823.
47. Harel, S.; Salan, M.A.; Kanner, *J. Free Rad. Res. Comms.* **1988**, *5* 11.
48. Arduino, A.,; Eddy, L.; Hochstein, P. *Free Rad. Biol. Med.* **1990**, *9*, 511.
49. Giulivi, C.; Davis, K.J.A. *J. Biol. Chem.* **1990**, *265*, 19453.
50. Walter, F.P.; Kennedy, F.G.; Jones, D.P. *FEBS Lett.* **1983**, *163*, 292.
51. Nicholls, P. *Biochem. J.* **1961**, *81*, 374.
52. Aisen, P.; Liskowsky, I. *Ann. Rev. Biochem.* **1980**, *49*, 357.
53. Spiro, T.G.; Saltman, P. *Struct. Bonding*, **1969**, *6*, 116.
54. Crichton, R.R.; Charloteaux-Water, B. *Eur. J. Biochem.* **1987**, *164*, 485.
55. Gutteridge, J.M.C.; Rowley, D.A.; Halliwell, B.*Biochem. J.* **1981**, *199*, 263.
56. Kanner, J.; Hazan, B.; Doll, L. *J. Agric. Food Chem.* **1988**, *36*, 412.
57. Kanner, J.; Harel, S.; Jaffe, R. *J. Agric. Food Chem.* **1991**, *39*, 1017.
58. Kanner, J.; Harel, S. *Free Rad. Res. Comms.* **1987**, *3*, 309.
59. Flatmark, T.; Romslo, *J. Biol. Chem.* **1975**, *250*, 6433.
60. Ulvik, R.J. *Biochim. Biophys. Acta*, **1982**, *715*, 42.
61. Boyer, R.F.; McCleary, C.J. *Free Rad. Biol. Med.* **1987**, *3*, 389.
62. Kanner, J.; Doll, L. *J. Agric. Food Chem.* **1991**, *39*, 247.
63. Decker, E.A.; Welch, B. *J. Agric. Food Chem.* **1990**, *38*, 674.
64. Marceau, N.; Aspin, N. *Biochim. Biophys. Acta*, **1973**, *293*, 338.
65. Kohen, R.; Yamamoto, Y.; Cundy, K.C.; Ames, B.N. *Proc. Natl. Acad. Sci. U.S.A.* **1988**, *85*, 3175.
66. Decker, E.A.; Huang, C.H.; Osinchak, J.E.; Hultin, O.H. *J. Food Biochem.* **1989**, *13*, 179.
67. Osaki, S., Johnson, D.A.; Friden, E. *J. Biol. Chem.* **1966**, *241*, 2745.
68. Barber, A.A. *Arch. Biochem. Biophys.* **1961**, *96*, 38.
69. Vidlakova, M.; Erazimova, J.; Hoski, J.; Placer, Z. *Clin. Chim. Acta.*, **1972**, *36*, 61.
70. Watts, B.M.; Peng, D. *J. Biol. Chem.*, 1947, 170, 441.

71. Tappel, A.L. Arch. Biochem. Biophys., 1953, 44, 378.
72. Robinson, M.E. *Biochem. J.* **1924**, *18*, 255.
73. Tappel, A.L. In *Symposium on Foods: Lipids and their Oxidation*; Schultz, H.W., Day, E.A., Sinnhuber, R.O., Eds.; Avi Publishing: Westport, CT, 1962, p. 123-138.
74. O'Brien, P.J. *Can. J. Biochem.* **1969**, *47*, 486.
75. Kendrick, J.; Watts, B. *Lipids*, **1969**, *4*, 404.
76. Kanner, J. *Ph.D. Thesis*, Hebrew Univ. of Jerusalem, Israel, 1974.
77. Lin, H.P.; Watts, B.M. *J. Food Sci.* **1970**, *35*, 596.
78. Sato, K.; Hegarty, G.R. *J. Food Sci.* **1971**, *36*, 1098.
79. Love, J.D.; Pearson, A.M. *J. Agric. Food Chem.* **1974**, *22*, 1032.
80. Igene, J.O.; King, J.A.; Pearson, A.M.; Gray, J.I. *Agric. Food Chem.* **1979**, *27*, 838.
81. Tichivangana, J.Z.; Morrisey, P.A. *Meat Sci.* **1985**, *15*, 107.
82. Rhee, K.S.; Ziprin, Y.A.; Ordonez, G. *J. Agric. Food Chem.* **1987**, *35*, 1013.
83. Galaris, D.; Cadenas, E.; Hochstein, P. *Arch. Biochem. Biophys.* **1989**, *273*, 97.
84. Galaris, D.; Eddy, L.; Arduino, A.; Cadenas, E.; Hochstein, P. *Biochem. Biophys. Res. Commun.* **1989**, *160*, 1162.
85. Solar, I.; Dulitzky, J.; Shaklai, N. *Arch. Biochem. Biophys.* **1990**, *283*, 81.
86. Ottolengh, A. *Arch. Biochem. Biophys.* **1959**, *77*, 355.
87. Wills, E.D. *Biochem. Biophys. Acta*, **1965**, *98*, 238.
88. Borg, D.C.; Schaich, K.M.; Elmore, J.J. Jr.; Beld, J.A. *Photochem. Photobiol.* **1978**, *28*, 887.
89. Tien, M; Svingen, B.A.; Aust, J.D. *Arch. Biochem. Biophys.* **1982**, *216*, 142.
90. Cohen, G.; Sintel, P.M. In *Chemical and Biochemical Aspects of Superoxide and Superoxide Dismutase*; Bannister, J.V., Hill, H.A.O., Eds.; Elsevier: Amsterdam, 1980, p. 27.
91. Udenfriend, S.; Clorte, C.T.; Axelrod, J.; Broadie, B.B. *J. Biol. Chem.* **1954**, *208*, 731.
92. Girotti, A.W.; Thomas, J.P. *Biochem. Biophys. Res. Commun.* **1984**, *118*, 474.
93. Vile, G.F.; Winterbourn, C.C. *FEBS Lett.* **1987**, *215*, 151.
94. Garnier-Suillero, A.; Tosi, L.; Paniago, E. *Biochim. Biophys. Acta*, **1984**, *794*, 307.
95. Kanner, J.; Sofer, F.; Harel, S.; Doll, L. *J. Agric. Food Chem.* **1988**, *36*, 415.
96. Decker, E.A.; Hultin, O.H. *J. Food Sci.* **1990**, *55*, 851.
97. Slabyj, B.M.; Hultin, O.H. *J. Food Biochem.* **1983**, *7*, 105.
98. Han, T.J.; Liston, J. *J. Food Sci.* **1989**, *54*, 809.
99. Asghar, A.; Lin, C.F.; Gray, J.I.; Buckley, D.J.; Booren, A.M.; Flegal, C.J. *J. Food Sci.* **1990**, *55*, 46.
100. Kanner, J.; Bartov, I.; Salan, M.O.; Doll, L. *J. Agric. Food Chem.* **1990**, *38*, 601.
101. Hsieh, R.J.; Kinsella, J.E. *Adv. Fd. and Nutr. Res.* **1989**, *33*, 233.
102. Grower, J.D.; Healing, G.; Green, C.J. *Anal. Biochem.* **1989**, *180*, 126.
103. Harel, S.; Kanner, J. *J. Agric. Food Chem.* **1985**, *33*, 1188.

RECEIVED February 19, 1992

Chapter 5

Role of Lipoxygenases in Lipid Oxidation in Foods

J. Bruce German[1], Hongjian Zhang[1], and Ralf Berger[2]

[1]Department of Food Science and Technology, University of California, Davis, CA 95616
[2]Institut für Lebensmitteltechnologie und Analytische Chemie der Technischen Universität München, Munich, Germany

Although lipid oxidation is considered a deteriorative process responsible for generating off-flavors, however, specific oxidation products are desirable flavor compounds particularly when formed in more precise i.e. less random reactions. Lipoxygenases confer positional and stereospecificity to the initiation reactions of lipid peroxidation. As a result it has been proposed that some of the volatiles produced in certain food systems reflect the activity of these enzymes. We have investigated several tissues to determine the activity, specificity and stability of endogenous lipoxygenases and the oxidation products and volatiles generated. Gills of marine and freshwater fish were shown to contain two distinct lipoxygenases differing in specificity and stability. The products and volatile patterns were found to respond to changing the activities of these enzymes. This behavior is consistent with these enzymes being important to the biogeneration of fresh fish flavors. Understanding these enzyme systems will facilitate the development of potential technologies for the biogeneration of fresh fish flavors for seafood products.

The oxidation of polyunsaturated fatty acids (PUFA) is well recognized as a source of volatile products which are perceived as off-flavors (*1*). The potency of these lipid oxidation derived volatile off-flavors is underscored by the significant quality losses of PUFA-containing foods attributed in large part to these processes (*2-4*) and the substantial measures taken to minimize them (*5*). Somewhat paradoxically, then, is the developing appreciation that in many foods some of these same volatiles are critical components of the overall impact of 'fresh' flavor. That is, certain molecules which are part of the collection of volatile breakdown products of autoxidation when detected separately are not off-flavors per se but in fact contribute the impact flavors associated with freshness itself. Current research in a variety of laboratories is attempting to resolve this paradox. This is best considered from the perspective of what fundamentally distinguishes freshness (6). Freshness is not simply a lack of off-flavor. There are distinct compounds which lead to this perception. Similarly, this is not a simple response to the sensory detection and transduction of olfactory signals but rather a cognitive decision based on the overall spectrum of sensory inputs. The mechanisms underlying the development of this cognitive decision relate to the ephemeral nature of certain volatiles and especially their production. Thus, starting from the same precursors, the products of a non-specific or random breakdown process are generally perceived collectively as deteriorative (off) flavors while controlled or specifically catalyzed reactions yield a different mixture which is ultimately perceived as flavor. In the case of 'fresh' this is arguably related to the inherently biological origin of these compounds which relates to the temporary or transitory nature of certain volatiles and notably their biosynthesis (6). The loss of freshness is both in increase in deteriorative compounds as well as a loss of the volatiles themselves or the biosynthetic capacity which gives rise to them. Therefore, in searching for the candidates for 'fresh' flavors one must be drawn to those mechanisms which are particularly transient. In this regard the lipoxygenases are increasingly recognized.

Autoxidation

The specificity and kinetics of lipoxygenases are conspicuous when contrasted with the known processes of autoxidation. In autoxidation of, for example, linoleic acid 18:2, the methylene carbon interrupting two non-conjugated double bonds is relatively easily oxidized by a variety of single electron oxidants yielding an allyl radical which is very rapidly quenched by ground state triplet oxygen (7). The resulting peroxy radical is itself an excellent single electron oxidant for methylene interrupted 1,4 diene carbon systems (7). This is the fundamental driving factor for propogation of autoxidation. The acyl hydroperoxides formed by this reduction are substrates for subsequent scission reactions leading to hydrocarbons, carbonyls and alcohols, many of which are volatile some with very low flavor thresholds (8,9).

This chemistry is well described and one feature of importance is the relative non-specificity of oxygen addition sites. In linoleic acid both the 9 and 13 positions are attacked with similar frequency and also while a chiral center is formed a racemic mixture of products results (1), Figure 1.

As the number of double bonds increases, the number of oxidation and oxygen addition sites increases proportionately. Arachidonic acid with four methylene interrupted double bonds is a source of 6 major oxygen addition sites each leading to a spectrum of scission products and commensurately abundant volatiles, Figure 2.

It is also important to flavor considerations that once a single fatty acid contains greater than two double bonds, internal peroxy radicals can attack adjacent double bonds intramolecularly leading to a variety of hydroxy epidioxides analogous to those shown for linolenic acid (1), Figure 3.

Since many foods contain a mixture of polyunsaturated fatty acids the

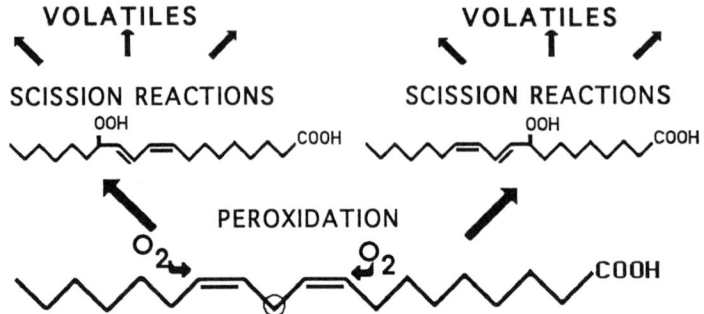

Figure 1. Peroxidation of Linoleic Acid.

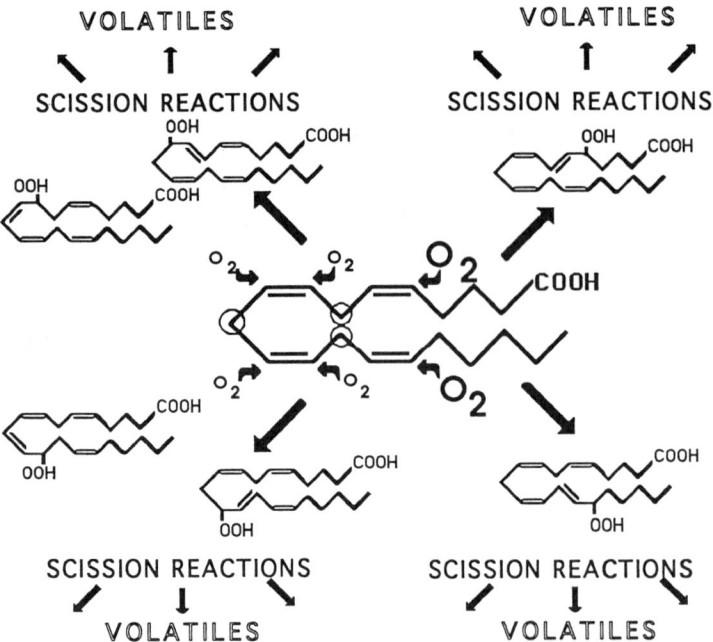

Figure 2. Peroxidation of Arachidonic Acid.

possible number of products formed by purely autocatalytic events is thus very large indeed. Fish containing abundant fatty acids with 4, 5 and 6 double bonds per molecule, represent a food system in which the polyunsaturated fatty acids, the primary substrates for lipid oxidation, are the most unsaturated typically encountered in foods. The volatile compounds generated in fish oils have been reported and several mechanisms for their biogenesis have been proposed. Importantly, the volatiles produced from pure fish oils and those from fresh fish tissues are dramatically different as are the odor impressions (*10*). This would imply that while the substrates are common, the mechanisms of volatile generation must be different. Our thesis in this work has been that enzymes as biological and hence ephemeral catalysts profoundly constrain the oxygen attack and subsequent breakdown reactions. This would mean that of all possible volatiles which wouldbe generated by autoxidation, enzymes both increase specific subsets and decrease others and when perceived by the olfactory centre a cognitive decision deems the overall impression to be 'fresh'.

Enzymatic Oxidation via Lipoxygenases

Lipoxygenases can participate in the generation and proliferation of lipid products in several ways. Primarily is the catalytic reaction for which they are designated, the addition of molecular oxygen to a *cis, cis*-4-pentadiene containing unsaturated fatty acid releasing a fatty acyl hydroperoxide. These hydroperoxides can then be broken down by enzyme or non-enzyme catalyzed scission reactions to yield specific chain length volatile compounds (*11*). The lipoxygenases can act at more than one methylene carbon on the substrate molecule to yield double oxygenation sites enzymatically (*12-14*). The fatty acid hydroperoxides can also be broken down via homolytic cleavage pathways and catalyzing further oxidations of the parent molecule intramolecularly (*15-18*). We have been studying the endogenous lipoxygenases in fresh fish tissue to determine the mechanisms by which these enzymes influence the specificity and time course of peroxidation reactions. Our first questions related to what precise reactions of polyunsaturated fatty acids did the lipoxygenases catalyze in fish tissue? We have thus described:

1) the specificity of primary abstraction and oxygen addition sites by the enzymes present leading to monohydroperoxide derivatives of substrate PUFA, 2) the catalysis of secondary sites on the same molecule leading to dihydroxy derivatives and 3) the depletion of antioxidants/peroxidases leading to an increase in trihydroxy derivatives.

Specificity of Fish Lipoxygenases

Lipoxygenases catalyze the initial peroxidation event in the oxygenation of PUFA. Employing a non-heme iron in the high spin Fe III state as the single electron oxidant, the enzyme catalyzes the stereospecific hydrogen removal, oxygen addition and peroxy radical reduction reactions forming a stereospecific conjugated diene hydroperoxy fatty acid product (*19*). The specificity of formation of these hydroperoxides is predicted to be an important determinant of the final flavor profile (20,21) A graphic indication of such specificity is shown by the HPLC separation and UV detection of the products of oxidation of arachidonic acid by fish gill homogenates, Figure 4.

Separation on the same chromatographic system of the autoxidation products yields multiple peaks representing the mixture of the various oxygen addition positions described previously. The single dominant product from the fish gill tissue was the 12 hydroxy derivative of arachidonic acid.

Separation of the stereoisomers of these hydroxy products of arachidonic acid using chiral analyses by normal phase chiral HPLC (22) and circular dichroism spectroscopy (data not shown) and NMR spectroscopy (Figure 5) further identify this to be the 12 (S) 5Z 8Z 10E 14Z HETE formed by the arachidonate 12-lipoxygenase. Further clarification of the substrate dependency and product specificity towards the other abundant PUFA in fish 20:5 n3, 22:4 n6, 22:5 n3 and 22:6 n3 indicated that the enzyme exhibits specificity towards the methyl rather than the carboxyl terminus and is more precisely described as an n-9 lipoxygenase (23).

While these experiments illustrate the potential enzyme activity, *in situ* lipoxygenases oxygenate solely unesterified fatty acids released from membranes by the action of phospholipases. Hence experiments were conducted to determine if the lipoxygenases were a major contributor to the peroxidation of the actual fatty acids available for oxygenation in the tissue. Perfused, homogenized trout gill tissue was used to investigate the release and metabolism of endogenous fatty acids. The n-3 PUFA were most abundant esterified to phospholipids in membranes. After incubation however, the free fatty acids were separated and analyzed and arachidonic, eicosapentaenoic and docosahexaenoic acids represented quantitatively, the potential substrates for lipoxygenase catalysis. The monohydroxy derivatives formed endogenously from these liberated fatty acids were purified over SPE cartridges as methyl esters and derivatized to trimethyl silyl ethers, separated by GC and structures confirmed by GC/MS. Greater than 95 % of metabolites were the respective monohydroxy derivatives expected from the activity of 12-lipoxygenase on these fatty acids, Figure 6.

The obvious question was the following: does this lipoxygenase influence volatile production from the fresh tissue in which the enzyme is active? Rational decomposition mechanisms have been proposed for this position-specific n-9 monohydroperoxide and conclude that the inordinate abundance of three 9 carbon products 3,6-nonadienal, 2,6-nonadienal and 3,6-nonadien-1-ol isolated from fresh fish volatiles are due to this lipoxygenase's activity (21). Similarly 8 carbon decomposition and rearrangement products could be expected from an abundance of lipoxygenase generated hydroperoxides (20). This was tested in the fish gill model using two separate approaches: adding a selective inhibitor of the 12-lipoxygenase, esculetin at 1 uM, and adding exogenous polyunsaturated substrates to the tissue preparation. Results shown in Table I are consistent with formation of 1 octen-3-ol and 2 octen-1-ol from n-6 PUFA such as arachidonic acid and 1,5-octadien-3-ol and 2,5-octadien-1-ol from n-3 PUFA 20:5 and 22:6 all by the n-9 lipoxygenase.

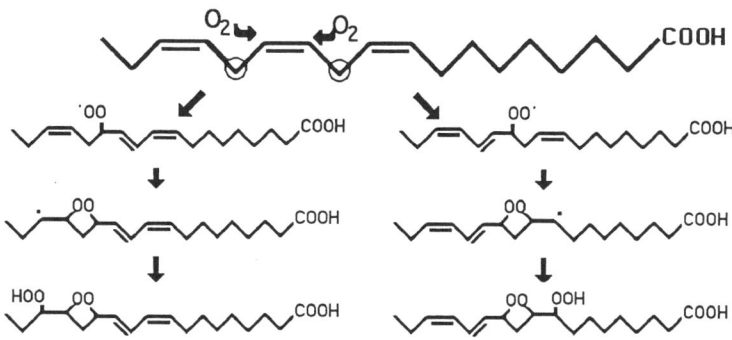

Figure 3. Peroxidation of Linolenic Acid.

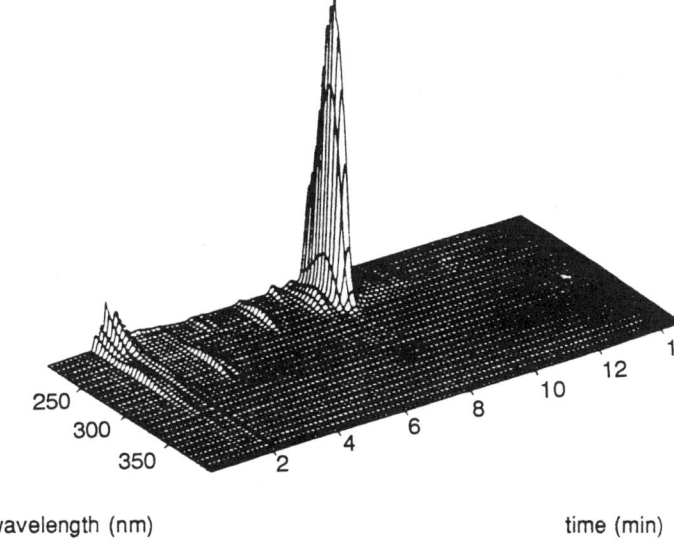

Figure 4. HPLC separation and UV-VIS spectral detection of the products formed from exogenous arachidonic acid by trout gill homogenates. Fresh gill tissue was homogenized in 5 volumes (0.05 M Phosphate buffer pH 7.5) and then incubated with 20 μM arachidonic acid for 20 minutes. Products were extracted with 2 volumes ethyl acetate, concentrated, injected onto a reverse phase C-18 HPLC column and eluted isocratically with a 70% methanol in buffered water solvent system. Spectra recorded at 1 second intervals were converted into the two dimensional chromatogram of elution time and UV-VIS absorption.

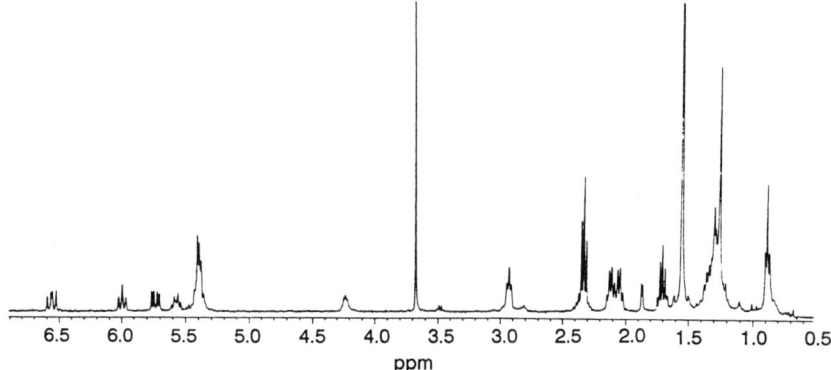

Figure 5. Proton NMR spectrum of the primary metabolite of arachidonic acid formed by trout gill. Products formed as described in figure 1 were methylated using diazomethane and the methyl HETE fraction purified over silica solid phase extraction columns using hexane ether 75:25 as eluting solvent. Samples were dried under nitrogen, redissolved in CCl_3D and proton spectra recorded using a Nicholet NT-360 NMR Spectrometer. Proton resonances were identified from standards and proton decoupling experiments.

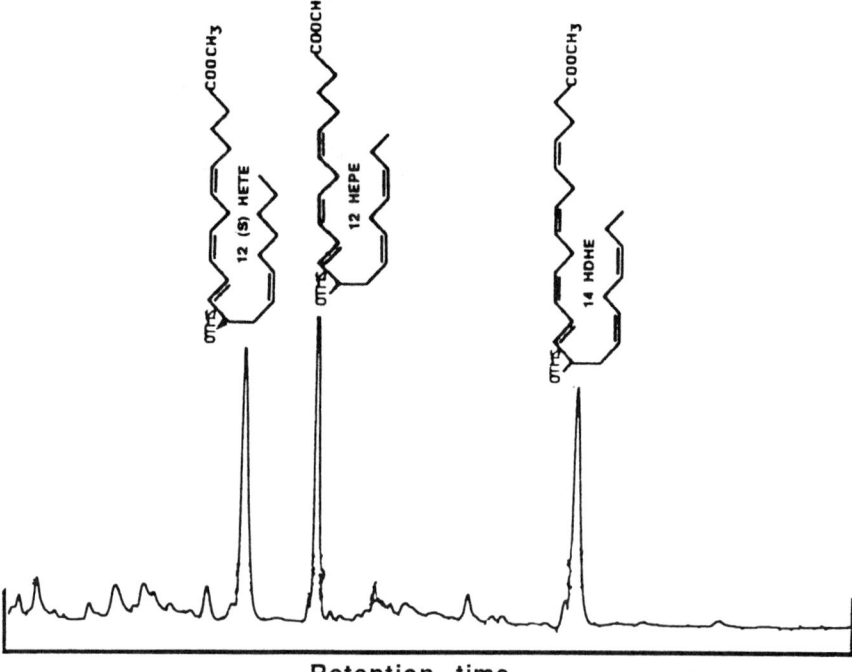

Figure 6. Gas chromatographic separation and flame ionization detection of the major hydroxylated polyunsaturated fatty acids formed from endogenous fatty acids in trout gill. Fatty acid structures shown were confirmed by GC/MS.

Table I. Concentrations of discriminating volatiles in trout gill homogenates

Compound(ug/100g fr.wt)	Control	Esculetin	+5uM 20:4	+5uM 22:6
1-Octen-3-ol	100	58	130	133
1,5-Octadien-3-ol	152	93	155	207
2-Octen-1-ol	45	35	58	58
2,5-Octadien-1-ol	200	113	157	225
1-Penten-3-ol	[a]n.d.	7	10	23

[a]nd, not detected. (From ref. 24).

These are not the only volatile flavor compounds formed in these and other fish however. Similar analyses of volatiles released from fresh teleost fishes have found 5 carbon volatiles in some abundance. These volatile aldehydes and alcohols are the same as those identified in a variety of plants. In plant tissues these 5 carbon products have been shown to be the direct result of an n-6 lipoxygenase activity such as the type 1 soybean lipoxygenase. The observation of large quantities of these analogous compounds from fresh fish suggested that there might be a similar n-6 lipoxygenase present in fish as well (21). Initial attempts to identify this activity were unsuccessful. Previous experiments searching for this enzyme in fresh tissue homogenates had not revealed significant quantities of this activity, recent experiments attempting to purify the 12-lipoxygenase have identified both the presence of the enzyme and at least a partial explanation for its obscurity.

During purification of the n-9 lipoxygenase, an increasing quantity of an additional conjugated diene product with hydroxyl function not at the n-9 but rather at the n-6 position suggesting an n-6 lipoxygenase metabolic activity. These could have arisen via an alteration in the absolute specificity of the single n-9 lipoxygenase enzyme or the appearance of a previously undetected n-6 lipoxygenase enzyme. Tissue samples containing the 12-lipoxygenase enzyme activity applied to hydroxlapatite columns eluted the two lipoxygenase peaks separately (Figure 7). The first peak, eluting at low ionic strength was the enzyme similar in the position and chirality of its catalytic reaction to the 15-lipoxygenase characterized in leukocytes, reticulocytes and soybeans (25). Physical descriptions of the enzyme found an apparent molecular weight of 70 ± 5 kdaltons consistant with that found for mammalian lipoxygenases but significantly smaller than the soybean enzyme (25).

The substrate specificity of this new n-6 specific enzyme was determined by incubating a variety of fatty acids with the partially purified enzyme preparation. Fatty acids with double bonds at both the n-6 and n-9 positions were substrates for activity. Similarly, fatty acids with chain lengths from 18 to 22 carbons with the appropriate orientation of double bonds (n-6 n-9) were effective (though not equally reactive) substrates. The respective substrates tested and products formed as confirmed by HPLC and GC/MS are summarized in Table II (from ref. 26).

Table II Substrate specificity and products generated by the n-6 lipoxygenase of trout

Substrate fatty acid	Product formed
18:2 n6	13(S) Hydroxyoctadecadienoic acid
18:3 n3	13 Hydroxyoctatrienoic acid
20:3 n6	15 Hydroxyeicosatrienoic acid
20:3 n3	15 Hydroxyeicosatrienoic acid
20:4 n6	15(S) Hydroxyeicosatetraenoic acid
20:5 n3	15 Hydroxyeicosapentaenoic acid
22:6 n3	17 Hydroxydocosahexaenoic acid

Thus the fish gill tissue contains two separate lipoxygenases each active towards polyunsaturated fatty acids but exhibiting different hydroperoxide addition sites. If we extrapolate then from the enzymes as catalysts to their activity as producers of volatile precursors, different lipoxygenases present would be predicted to result in distinct volatiles from the same fatty acids. Evidence in favor of this hypothesis came from parallel experiments on the inhibition of the two enzymes and the volatiles produced by trout gills. The phenolic inhibitor esculetin has previously been shown to strongly inhibit the 12-lipoxygenase. In assaying its inhibition of the trout gill system it was found that the 15-lipoxygenase was not as sensitive to esculetin and in fact net product from this enzyme actually increased (Figure 8).

The volatiles from trout gill homogenates were also examined using similar protocols in which esculetin was added to the homogenates prior to incubation and subsequent trapping of the volatiles generated. When the volatiles released from these incubations were analyzed, whereas the 8 carbon volatiles (the putative 12-lipoxygenase-derived compounds) were reduced by esculetin as predicted from the reactions initiated by this enzyme, the 5 carbon product 1 penten-3-ol (a breakdown product of the 15-lipoxygenase reaction) increased (see Table I).

Sequential Reactions of Fish Lipoxygenases

Fatty acids from fish are conspicuous by the number of double bonds in a single fatty acid molecule. Increasing the number of double bonds in a fatty acid increases the number of carbons susceptible to hydrogen abstraction by both autoxidative and enzymatic reactions. This also increases the possibility of multiple oxygenations on a single molecule. Again, the possibility arises that such reactions could be favored by the presence of specific (enzymatic) catalysts. We have found oxidation products in the fish tissues which arise by sequential attack by the two separate enzymes. The critical moiety on a PUFA substrate necessary for enzyme catalysis is the presence of the appropriate 1,4-pentadiene double bond system. In the case of the

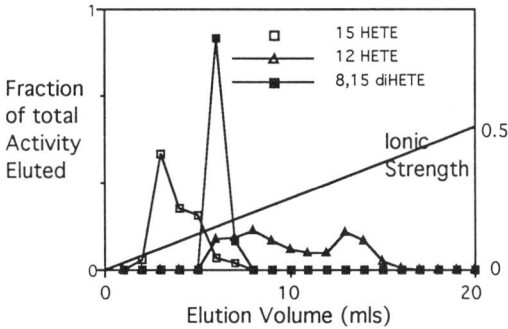

Figure 7. Elution of trout gill lipoxygenase activities from a hydroxylapatite column. The redissolved 45% ammonium sulphate precipitate was placed on a hydroxyapatite column and eluted with a linear gradient of sodium chloride. 1-mL fractions were collected and assayed for ability to convert arachidonic acid to chromophore containing mono- and dihydroxylated products. (Reproduced from ref. 26. Copyright 1990 American Chemical Society.)

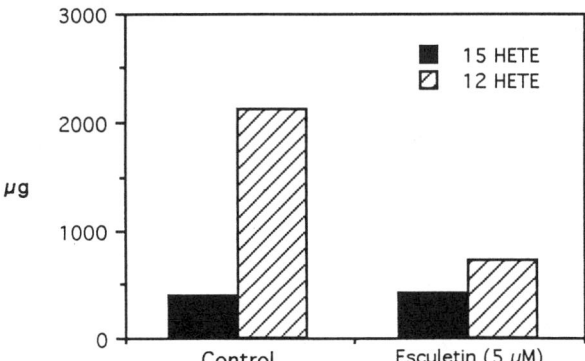

Figure 8. Effect of esculetin on the release of particular hydroxy fatty acids (HETEs) formed from exogenous arachidonic acid by trout gill homogenates. Esculetin or its negative control was incubated for 10 minutes prior to addition of arachidonic acid. Following 20 minutes incubation, the products formed were analyzed by HPLC.

15-lipoxygenase, starting from arachidonic acid 20:4 n6, it is the n-6 n-9 double bond system which is necessary and for the 12-lipoxygenase it is the n-9 n-12 arrangement. Importantly, the 15 HETE product of the 15-lipoxygenase still contains the n-9 n-12 double bonds necessary for 12-lipoxygenase attack as described schematically in Figure 9.

The most convincing evidence of the role of the respective enzymes is that these are both highly selective stereospecific reactions consistent with their enzymatic nature. Both the 15-and 12-lipoxygenases show very strong chiral selectivity, each producing the specific (L_S) monoHETE isomer. Therefore, the initial product of the 15-lipoxygenase is the $15(L_S)$. The second abstraction at carbon 10 and oxygen addition at carbon 8 (-2) retains the specificity of the $12(L_S)$ lipoxygenase for the L hydrogen to yield the $8(L_R)$ product. To resolve the chirality of the second oxygenation, the diHETE from trout was analyzed on RP HPLC relative to standard diHETE's and results indicated an $8(L_R),15(L_S)$ structure. The chiral specificity of the 12-lipoxygenase for abstraction and oxygen addition was retained but interestingly rather than oxygen adding + 2 from the c-10 H abstraction carbon it was reversed to -2 in forming the second hydroperoxide product on carbon 8. The stereospecificity of the arachidonic acid-derived compound was thus determined to be the $8(L_R),15(L_S)$ diHETE. The formation of this dihydroxy class of compounds has been verified for the three main PUFA substrates in fish: arachidonic eicosapentaenoic and docosahexaenoic acids (data not shown).

Depletion of Antioxidants/Peroxidases

A third mechanism by which PUFA products are modified by lipoxygenase actions is somewhat less direct. Biological tissues contain ample catalysts to eliminate hydroperoxides. These enzymatic mechanisms of reductive decomposition of hydroperoxides include catalase, glutathione peroxidase and glutathione transferase. The latter two enzymes have been shown to effectively reduce acyl hydroperoxides using glutathione as the reductant. If these scavenging systems are depleted however, hydroperoxide decomposition can proceed in uncontrolled metal catalyzed reactions yielding in addition to those scission products described previously (1,3,11,20,21,27,28) also intramolecular oxidations resulting in products such as hydroxyepoxides and trihydroxy derivatives. These compounds have been isolated from fish tissues and their origin traced to the lipoxygenase(29). The importance of the reducing substrate glutathione to minimizing decomposition of these hydroperoxides can be illustrated by the lipoxygenase reaction itself. The dependence of random decomposition reactions on the activity of hydroperoxide scavenging activities as shown by the relative generation of the aggregate of trihydroxy derivatives of radiolabeled 22:6 n3 as a function of glutathione depletion (Figure 10).

Once again, the products of these reactions were quite different relative to the products of purely autoxidative reactions due to a) the specificity of the initial hydroperoxides formed by the lipoxygenase and b) by the depletion of scavenging systems due to the rapid & quantitative hydroperoxide release.

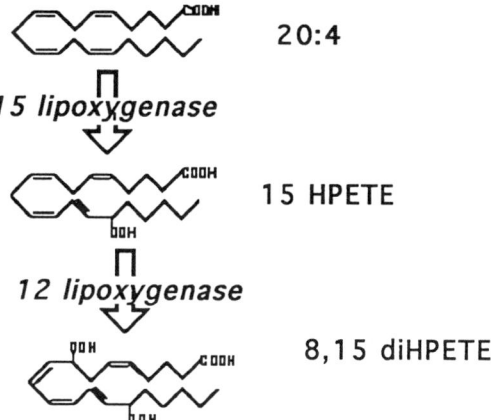

Figure 9. Enzymatic catalysis of polyunsaturated fatty acids by 15- and 12-lipoxygenases.

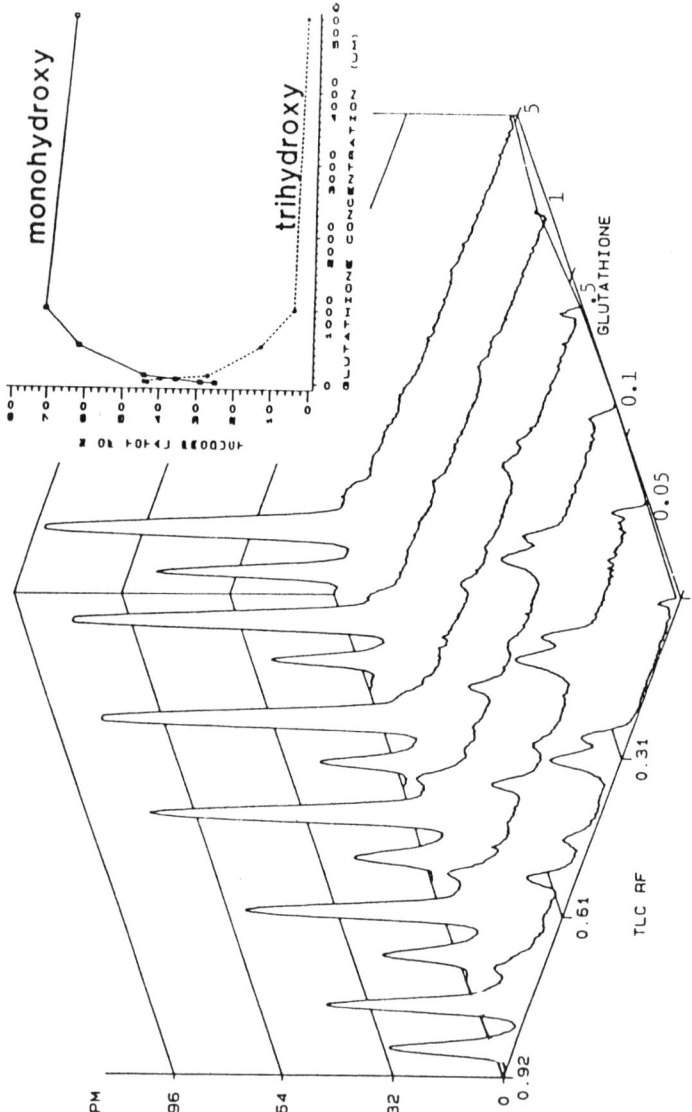

Figure 10. Sequential autoradiograms of the separation by thin layer chromatography of the products formed from radiolabeled 22:6 by trout gill. Glutathione was included in the reaction mixture at levels of 0, 0.05, 0.1, 0.5, 1, and 5 mM as indicated. The relative quantity of monohydroxy derivatives and trihydroxy derivatives of the substrate fatty acid 22:6 n3 are illustrated as a function of glutathione concentration in the inset.

The lipoxygenase enzymes also participate in reactions, historically termed secondary, which are now known to be a variety of distinct catalytic activities. These are: cooxidation reactions producing further autoxidation initiators, cooxidation reactions consuming both enzymatic and nonenzymatic antioxidants, and peroxidative reactions of the product hydroperoxide itself which have been found to yield volatile products (*30-39*).

Properties of the Lipoxygenases relevant to their Role in Flavor Production

With the understanding that lipoxygenase enzymes in animal tissues generate different volatile patterns depending on the substrates, activity of the enzymes and condition of the tissues we have investigated various properties of the catalytic proteins. One aspect particularly relevant to the lipoxygenases and their role in the production of 'fresh' flavors is their stability. These enzymes as a class are remarkable for their lack of stability as catalysts. The lipoxygenases and the dioxygenases in general exhibit a coincidental self inactivation frequently termed suicidal. The cause is the product, in this case a hydroperoxide, which is the substrate for a secondary peroxidase reaction which is (presumably artifactually) a self destructive catalytic activity of the enzyme. The sensitivity of the lipoxygenase in fish gill to hydroperoxides is illustrated by the response to t-butyl hydroperoxide (Figure 11). This addition of this hydroperoxide irreversibly inactivates the enzyme. This inactivation by t-butyl hydroperoxides has been shown to be similar in both fish and mammalian lipoxygenases (*40*).

This sensitivity of the enzymes to product hydroperoxides and the instability of the enzymes in general leads to certain predictions for the role of these enzymes in the temporal aspects of flavor generation and especially to the production of 'fresh' flavors. The unique pattern of flavors which result from the enzymatic activity in raw, healthy tissues will rapidly diminish as the enzymes deteriorate and autoxidative reactions predominate.

Similarly, variation between different species of fish would lead to distinct differences in flavor spectrum depending on the species alone. We have catalogued the activity of these enzymes across several species and evolutionary periods. Examination of the variation in products of the 12 and 15-lipoxygenases in fish gills across species and time illustrated several points relevant to flavor. Typical of these data are those shown by the summary figures following activity of the two enzymes in disrupted tissue stored at 0 C (Figure 12).

As typified by trout and carp, the two lipoxygenases are not equally distributed amongst fish species. Trout initially exhibited very high 12- and virtually undetectable 15-lipoxygenase activities. In Carp although the 12-lipoxygenase was most active the 15-lipoxygenase was also relatively abundant. In sturgeon, a relatively ancient species, the 15-lipoxygenase was actually the predominant enzyme. Also evident from these data is the striking instability of these enzymes. The half time of the 12-lipoxygenase at 0 C. was

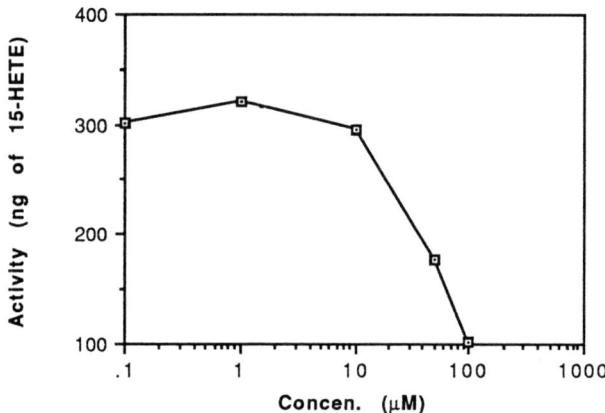

Figure 11. Effect of *t*-butyl hydroperoxide on the 15-lipoxygenase of sturgeon. Fresh homogenates of sturgeon gill were incubated for 10 minutes with the indicated concentration of hydroperoxide prior to addition of arachidonic acid. Activity was then determined by HPLC.

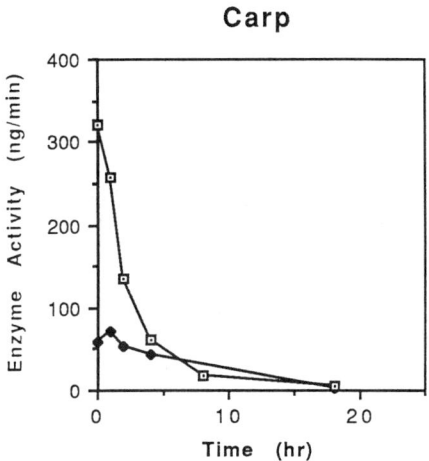

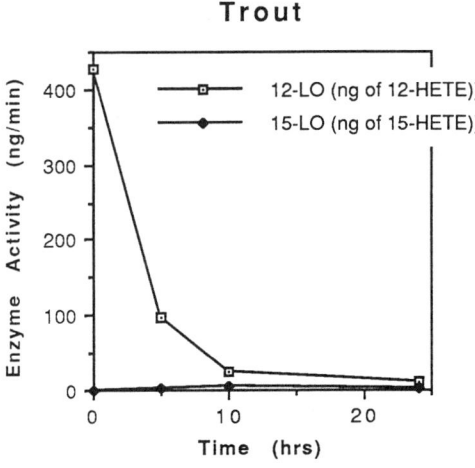

Figure 12. Lipoxygenase activity in carp and trout as a function of storage time at 0°C. Following indicated times of storage the enzyme activity towards exogenous arachidonic acid was determined by HPLC as described.

less than 3 hours. The half time of the 15-lipoxygenase was greater than 10 hours. Thus although the 12-lipoxygenase activity was predominant initially, in both carp and trout after 24 hours of storage the 15-lipoxygenase was the major enzyme activity remaining.

These observations are potentially significant to the time course of flavor development in the fresh fish tissues. Due to the differences in incidence of the enzymes, fresh volatiles are predicted to differ between species. Even more important the changing activities of the two enzymes with time will result in changing volatile patterns as the tissue ages.

As our knowledge advances especially on the basis of instability of the enzymes, and the subsequent lytic reactions, various technologies could make use of these enzyme systems in food applications. Their activity is a very sensitive index of the age and quality of the tissue. As an example, a single rapid freeze thaw cycle inactivates the enzymes. The lipoxygenases themselves or processing wastes in which they are present could be exploited as sources of marketable flavors for seafood products. As our understanding of the biogeneration of 'fresh' flavors improves even more opportunities will arise for maintaining this vital and elusive aspect of food quality.

Acknowledgments

This work is a result of research sponsored in part by NOAA, National Sea Grant College Program, Department of Commerce, under grant number NA89AA-D-SG138, project number 110-F-N through the California Sea Grant College Program and in part by the California State Resources Agency. The U.S. Government in authorized to reproduce and distribute for governmental purposes. We acknowledge the support of an International Life Sciences Future Leader award to JBG. We are also grateful for the assistance of David Osuga and Dr Dan Jones of the Facility for Advanced Instrumentation at UCD.

Literature Cited

1. Frankel, E. *Prog. Lipid Res.* **1984**, *23*, 197-223.
2. Allen, J. C.; Hamilton, R. J. In *Rancidity in Foods*; Elsevier Applied Science: London, **1989**; 23-35.
3. Hall, G.; Anderson, J. *Lebensmittel-Wissenschaft und-Technologie* **1983**, *16*, 354-61.
4. Richardson, T.; Korycka-Dahl, M. In *Developments in Dairy Chemistry 2 Lipids* Applied Science London **1983**; 114-139.
5. Schuler, P. in *Natural Antioxidants Exploited Commercially*; Hudson, B. J. F.; Elsevier Applied Science, London, **1990**; pp 99-170.
6. Badings, H. T. in *Some Aspects of Fresh Flavor in Foods*; Bessiere, Y. Thomas, A. F.; J. Wiley, New York, **1990**; pp 171-174.
7. Ingold, K. U. *Acct. Chem. Res.* **1969**, *2*, 1.

8	Berger, R. G.; Drawert, F.; Kollmannsberger, H.; Nitz, S.; Schraufstetter, B. *J. Agric. Food Chem.* **1985**, *33*, 232-236.
9	Buttery, R. G. Teranishi, R.; Flath, R. A. Sugisawwa, H.; Marcel Dekker, New York, **1981**; pp 193.
10	Josephson, D. B.; Lindsay, R. C.; Stuiber, D. A. *J. Agric. Food Chem.* **1984**, *32*, 1347.
11	Hiatt, R.; Mill, T.; Mayo, F. R. *J. Org. Chem.* **1968**, *33*, 1416.
12	Yamamoto, S.; Yokoyama, C.; Ueda; Shinjo; Kaneko, S.; Yoshimoto, T. *Adv. Prost. Thromb. Leukotr. Res.* **1986**, *16*, 17.
13	Van Os, C. P. A.; Rijke-Schilder, G. P. M.; Van Halbreek, J. V.; Vliegenthart, J. F. G. *Biochim. Biophys. Acta.* **1981**, *663*, 177-193.
14	Maas, R. L.; Brash, A. R. *Proc. Natl. Acad. Sci. USA* **1983**, *80*, 2884.
15	Bryant, R. W.; Bailey, J. M. *Prog. Lipid Res.* **1981**, *20*, 189-194.
16	Bryant, R. W.; Bailey, J. M. *Prostaglandins* **1979**, *17*, 9-18.
17	Hamberg, M. *Lipids* **1975**, *10*, 87-92.
18	Jones, R. L.; Kerry, P. J.; Poyser, N. L.; Walker, J. C.; Wilson, N. H. *Prostaglandins* **1978**, *16*, 483-489.
19	Siedow, J. N. *Ann. Rev. Plant Physiol. Plant Mol. Biol.* **1991**, *42*, 145-188.
20	Grosch, W. *Lebensm. gerichtl. Chem.* **1987**, *41*, 40.
21	Josephson, D. B.; Lindsay, R. C. In *Enzymic Generation of Volatile Aroma compounds from fresh fish*; American Chemical Society Symposium Series 317, Washington DC, **1986**; pp 200-217.
22	Kuhn, H.; Wiesner, R.; Lankin, V. Z.; Nekrasov, A.; Alder, L.; Schewe, T. *Analytical Biochemistry* **1987**, *160*, 24-34.
23	German, J. B.; Kinsella, J. E. *Biochem. Biophys. Acta.* **1986**, *877*, 290-298.
24	German, J. B.; Berger, R.; Drawert, F. *Chemistry, Microbiology and Food Technology* **1990**, *13*, 19-24.
25	Shewe, T.; Rapoport, S. M.; Kuhn, H. *Adv. Enzymology* **1986**, 191-273.
26	German, J. B.; Creveling, R. *J. Ag. Food Chem.* **1990**, *38*, 2144-2147.
27	Beckman, J. K.; Howard, M. J.; Greene, H. L. *Biochem. Biophys. Res. Commun.* **1990**, *169*, 75-80.
28	Erben, R. M.; Bors, W.; Saran, M. *Int J Radiat Biol Relat Stud Phys Chem Med* **1987**, *52*, 393-412.
29	German, J. B.; Kinsella, J. E. *Biochim. Biophys. Acta.* **1986**, *877*, 290-295.
30	Chamulitrat, W.; Mason, R. P. *J Biol Chem* **1989**, *264*, 20968-20973.
31	Cucurou, C.; Battioni, J. P.; Daniel, R.; Mansuy, D. *Biochim. Biophys. Acta.* **1991**, *1081*, 99-105.
32	German, J. B.; Kinsella, J. E. *Biochim. Biophys. Acta* **1986**, *879*, 378-386.
33	Kanner, J.; German, J. B.; Kinsella, J. E. *Crit. Rev. Food Sci. Nutr.* **1987**, *25*, 317-64.
34	Kulkarni, A. P.; Cook, D. C. *Biochem. Biophys. Res. Commun.* **1988**, *155*, 1075-81.

35 Kulkarni, A. P.; Chaudhuri, J.; Mitra, A.; Richards, I. S. *Res. Commun. Chem. Pathol. Pharmacol.* **1989**, *66*, 287-96.
36 Yamamoto, S. *Free Radic. Biol. Med.* **1991**, *10*, 149-59.
37 Macias, P.; Pinto, M. C.; Gutierrez, M. C. *Biochim Biophys Acta* **1991**, *1082*, 310-8.
38 Stone, R. A.; Kinsella, J. E. *J. Agric. Food Chem.* **1989**, *37*, 866-868.
39 Eskin, N. A. M.; Grossman, S.; Pinsky, A. *CRC Crit. Rev. Food Sci. Nutr.* **1977**, *9*, 1-40.
40 German, J. B.; Hu, M. *J. Free Radicals Biol. Med.* **1990**, *8*, 441-448.

RECEIVED March 31, 1992

Chapter 6

Relationship Between Water and Lipid Oxidation Rates

Water Activity and Glass Transition Theory

Katherine A. Nelson and Theodore P. Labuza

Department of Food Science and Nutrition, University of Minnesota, St. Paul, MN 55108

>The role of water in lipid oxidation is examined. The water activity of a system influences lipid oxidation rates. At both very high and very low water activities, lipid oxidation rates are high compared to the rate at intermediate water activities. Glass transition theory may also be used to understand lipid oxidation rates which takes place in polymeric systems. Reaction rates within the polymer should be higher when the system is above the glass transition temperature than below this temperature due to an increase in free volume associated with the system.

Lipid oxidation is an important mechanism of deterioration for many foods. Variables such as the degree of unsaturation of lipids, the availability of oxygen and light, and the presence of metal catalysts and natural antioxidants are all factors which must be considered in order to understand and predict lipid oxidation rates (*1*). As a constituent of most food systems, water plays a particularly important role in this reaction. Thus, understanding the relationship between moisture content and the rate of lipid oxidation is necessary for extending the shelf life and retaining the quality of many food products.

Water has numerous functions in food systems. It can act as a solvent to mobilize reactants, act as a substrate in chemical reactions and interact via hydrogen bonding with food components (*2*). Additionally, water can be regarded as a plasticizing agent (*3*). The role of water in lipid oxidation has been studied extensively. Water activity has been used to control lipid oxidation in susceptible food products and to explain the relationship between lipid oxidation rates and moisture content. Glass transition theory, which considers the physical state of the food system, is a new approach for describing this relationship (*4*). These theories will be discussed in terms of their applicability to understanding the relationship between water and lipid oxidation rates.

Water Activity and Lipid Oxidation

The relationship between lipid oxidation rates and water activity has been discussed by Labuza (5) and Karel and Yong (6). Water activity, the thermodynamic availability of water, is defined as

$$\frac{p}{p_o} = a_w$$

where a_w is the water activity, p is the vapor pressure of water in the food and p_o is the vapor pressure of pure water at the same temperature. A moisture sorption isotherm is shown in Figure 1 which relates moisture content and water activity. From this figure, it can be seen that most fresh or tissue foods which have moisture contents between 60 and 95% have a water activity very close to 1 (5,7). The mode of deterioration at this water activity is generally microbial or enzymatic in nature. Direct lipid oxidation (i.e. not enzyme-mediated) is not an important source of deterioration at this high water activity since other modes of degradation will deteriorate the food first. Concentration, freezing and drying are mechanisms by which the water activity of a food can be reduced. Such processes may bring the food into the intermediate moisture or low-moisture region where direct lipid oxidation becomes a more important mode of degradation. Also, during dehydration, free radicals can be formed which accelerate lipid oxidation (8). Freezing can also lead to acceleration of lipid oxidation rates through concentration of substrates and catalysts in the unfrozen portion of the system.

For most deteriorative chemical reactions such as nonenzymatic browning, as the water activity is decreased from 1, the rate of aqueous phase reactions initially increases, reaches a maximum in the 0.6 to 0.8 water activity range and then decreases again as water activity decreases and finally falls close to zero at the moisture monolayer as shown in Figure 2 (9). The monolayer moisture value is generally considered the most stable moisture for a dry food, as was first proposed by Salwin (10). At the monolayer, water is bound tightly to the food surface and thus cannot act as an aqueous phase nor is it available to mobilize reactants. Salwin (10) indicated that water may attach to sites on the food surface which would exclude oxygen from these sites, thus preventing oxidation reactions. However, modern theories of water mobility based on NMR measurements of water diffusion would dispute this idea (11). Unlike most aqueous chemical reactions, however, the rate of lipid oxidation which takes place in the oil phase is observed to increase as water activity is decreased below the monolayer (Figure 2, line b). This can be explained by considering the role of water in this reaction. Water can form a hydration sphere around metal catalysts such as Cu, Fe, Co and Cd (12). In the dry state the metal catalysts are most active. As water activity increases, the metals may hydrate which may reduce their catalytic action thus slowing the rate of lipid oxidation.

Maloney et al. (13) measured oxygen absorption in a model system containing methyl linoleate. From their results and the work of Bateman (14),

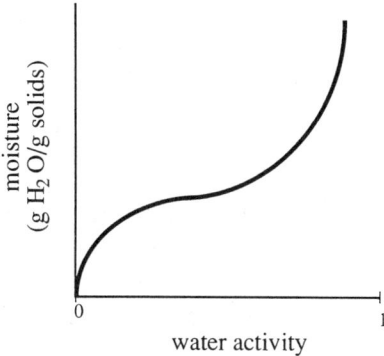

Figure 1. Moisture sorption isotherm for a food.

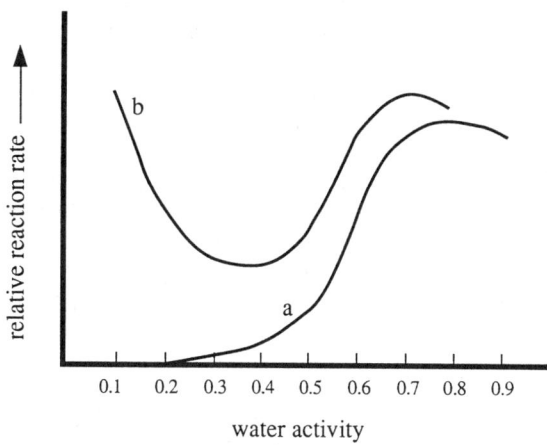

Figure 2. Relative rate of reaction as a function of water activity. a. Nonenzymatic browning b. Lipid oxidation

it was suggested that water forms hydrogen bonds with the hydroperoxides which are produced in the propagation step of free radical-mediated oxidation of unsaturated lipids. They proposed that as the amount of water/lipid phase boundary increases with increasing water activity, the polar ROOH groups move to the interface and become hydrogen bonded to the water, effectively taking them out of the reaction. This interfacial bonding was partially verified by infrared studies by Bateman (*14*), but needs further research.

Water has also been shown to decrease the free radical concentration by facilitating termination reactions as proposed by Simatos (*15*). Termination reactions involve the recombination of free radicals to form nonradical products. As the water activity increases, more water is available to mobilize the more polar radicals to the lipid/water interface which would then allow faster termination to take place.

At low water activities, insufficient water is available to perform the rate-retarding functions discussed above. Thus, the rate of lipid oxidation at low water activities has been shown to be quite high, but decreases as the water activity increases up to 0.2-0.4 as shown in Figure 2. However, between $a_w = 0.4$ and $a_w = 0.7$, the rate of oxidation of unsaturated lipids increases. This can be partially explained by the increased mobilization of the catalysts in the aqueous phase which would bring them to the lipid/water interface making them available to break down hydroperoxides into free radicals. The reduced viscosity of the aqueous phase may also allow for other materials to come to the lipid interface such that they can catalyze the reaction. It is clear that the increased rate is not due to an increased concentration of oxygen available in the aqueous phase since the reaction takes place in the lipid phase and the solubility of oxygen in organic materials is generally ten times greater than that in water (*16*). Above $a_w = 0.7$, the rate of lipid oxidation decreases, possibly as the result of the dilution of catalysts, but other unknown factors may be responsible.

A number of studies have verified the relationship between the rate of lipid oxidation and water activity shown in Figure 2. The oxidative deterioration of walnuts was found to follow this relationship (*17*). Maximum stability was found for the walnuts as measured by peroxide value at an approximate water activity of 0.30 with the rate of deterioration more rapid at both higher (0.45 and 0.54) and lower (0.26) water activities. Quaglia et al. (*18*) found that the lipid oxidation rate for freeze-dried beef measured by TBA at $a_w = 0.27$ was lower than the rate at lower water activities. Quast and Karel (*19*) found that the oxidation rate of potato chips measured by peroxide value was the slowest at $a_w = 0.4$ compared to lower water activities (Figure 3). They found the relative rate of lipid oxidation to be the highest at $a_w = 0.01$, which is below the system monolayer value. This was further verified in a study using direct oxygen absorption. Maloney et al. (*13*) found that water had an antioxidant effect on the rate of oxidation of methyl linoleate in a freeze dried model system. The rate of oxidation decreased with increasing water activity from the dry state up to a water activity of approximately 0.5 (Figure 4).

Finally, Cavaletto et al. (*20*) studied the relationship between moisture content and the oxidative stability of macadamia nuts in the range of 1.4 to

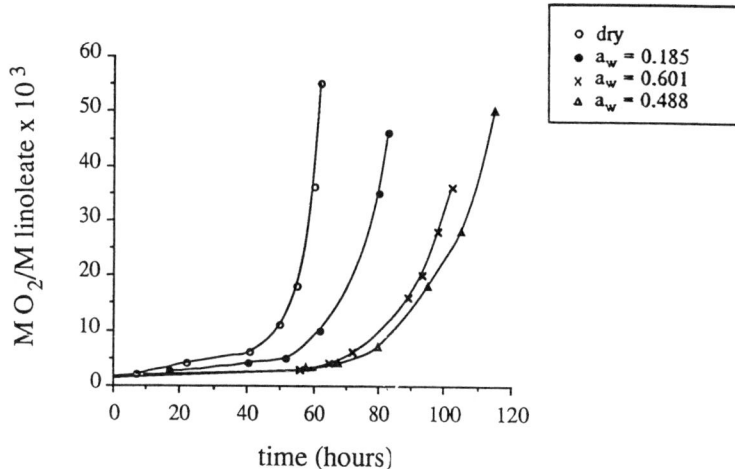

Figure 3. Oxidation of potato chips measured by peroxide value at three water activities (T=37°C). (Adapted from ref. *19*)

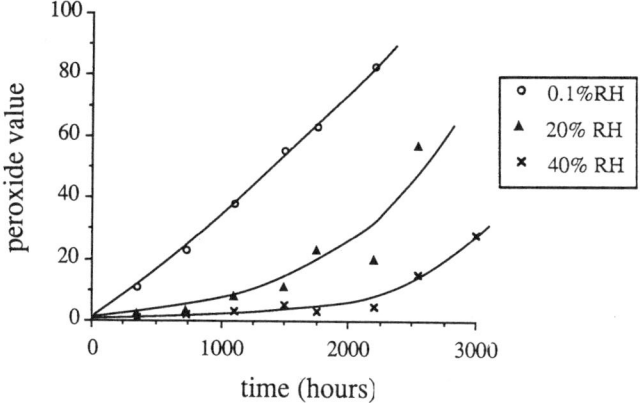

Figure 4. Oxygen absorption in a model system containing methyl linoleate. (Adapted from ref. *13*)

4.3% moisture. They measured lipid oxidation organoleptically and found the maximum stability at the lowest moisture studied with decreasing stability with increasing moisture content. It is possible that all moistures of the nuts in this study were above the monolayer value which was not measured.

In foods susceptible to lipid oxidation, control of the initial moisture content or water activity is necessary. Choosing a packaging film material which has a low water vapor permeability to prevent the gain or loss of moisture during distribution is also important. When moisture control is not possible, low and intermediate moisture foods can be protected from lipid oxidation by such methods as vacuum packaging and the addition of chelators, antioxidants or the incorporation of oxygen scavengers in the package.

Polymer Glass Transition Theory and Lipid Oxidation

Moisture content and water activity alone can not fully explain the relationship between water and the rate of free radical-induced lipid oxidation. The physical state of a food system probably also influences its susceptibility to lipid oxidation. Freeze-dried materials, for example, generally have a porous matrix. This porosity allows oxygen to diffuse through the system and react with unsaturated lipids. The physical state of a system is dependent upon several factors including the processing steps which it has undergone. Also, whether a food is in a crystalline or amorphous state can influence the rate of reactions occurring within the matrix. Scientists such as Slade and Levine (21) have suggested that an alternative approach, polymer glass transition theory, be used to understand water relationships in food systems.

Glass transition theory has been successfully used by polymer scientists to understand the physical properties of polymer systems in their amorphous state (22) and is gaining wider acceptance in the food area (4). Because many food systems such as starch, proteins and gums are polymers, it is logical that glass transition theory might be applicable to these systems as well. Glass transition theory has also been shown to be applicable to the physical state of solid systems made up of small molecular weight compounds such as sugars (23,24).

Glass transition theory states that at low temperatures and low moistures, amorphous polymers exist in their glassy state while at higher temperatures and higher moistures the polymers are in the rubbery state as shown in the state diagram in Figure 5. The glass transition temperature (T_g) is the temperature at which amorphous regions of a polymer change from a glassy state to a rubbery state.

Free volume is defined as the space in a polymer matrix which is not taken up by the polymer chains. There is less free volume in the glassy state than in the rubbery state. Levine and Slade (3) suggest that diffusion through a polymer matrix in its glassy state is limited due to the reduced free volume through which molecules can traverse and the slow polymer relaxation rates in the glassy state. This is important for lipid oxidation since oxygen must be able to diffuse to the lipid substrate in order for oxidation to take place. It is not clear, however, whether the limited void volume present in the glassy state is sufficient for preventing small molecules such as oxygen from diffusing through the matrix. Thijssen (25) and Flink and Labuza (26) found that the rate of diffusion of small molecular weight organic molecules in a porous matrix increased with decreasing size of the diffusing molecule. However, Stannett (27) showed that diffusion of oxygen in a polyvinyl acetate film was not influenced by the state of the film (i.e. rubbery or glassy) because of the small size of oxygen molecules. This is supported by Ma et al. (28) who found that oxidation of encapsulated orange oil continued to occur in the glassy state of the encapsulation matrix (Figure 6).

According to glass transition theory, one important function of water is its ability to act as a plasticizing agent. Plasticization of a matrix involves swelling of the polymer matrix when moisture is increased. The resulting increase in free volume might allow for faster diffusion of substrates in the aqueous phase which may lead to faster reaction rates. Plasticization may also increase the contact of the absorbed aqueous phase with the lipid phase. The number of catalytic sites increases such that the rate of lipid oxidation increases. Chou et al. (29) investigated the rate of lipid oxidation in a system which showed hysteresis. In hysteresis, systems on the desorption branch of the hysteresis curve have higher moistures than systems on the adsorption branch at the same water activity. They found that the rate of lipid oxidation measured by oxygen uptake was higher for a model system prepared by desorption than for the system prepared by adsorption. This increase in rate of lipid oxidation beyond that predicted by water activity can be attributed to the plasticizing effect of additional water in the desorption system. The plasticization of food polymers by water has been shown for starch (30), maltodextrins (31) and gluten (32).

Once in the rubbery state, the polymer system may undergo collapse. Collapse is observed as a shrinkage of the polymer matrix. This can occur when the viscosity of the polymer matrix is reduced such that the effect of gravity on its mass causes it to internally collapse (33). The caking and sticking of powders is the result of collapse (34). This is undesirable in many manufacturing processes where free flow is required, but can be desirable in the case of agglomeration of spray dried powders. Some theorize that upon collapse, compounds which are present within a matrix may be forced out due to the reduction in free volume (35). A study by Ma et al. (28) showed that collapse may also act to protect encapsulated products from oxidation. They found that the rate of limonene oxide formation in orange oil was slower when the maltodextrin M100 encapsulating agent was in the rubbery state compared to the reaction rate in the glassy state of this encapsulating agent, the opposite of what is expected. They attribute this decrease in oxidation rate to collapse of the maltodextrin M100 matrix which would decrease the pore size at the surface of the matrix. This would act to protect the encapsulated limonene from oxidation since the diffusion of oxygen from the surface to the interior would be restricted. In this case, the rubbery state and collapse are associated with greater stability as compared to the glassy state.

While in the rubbery state, the polymer chains may have sufficient mobility to undergo crystallization. The rate of crystallization is related to the glass transition temperature (36). When crystallization occurs, the amount of void volume in a sample decreases since the polymer chains come together and

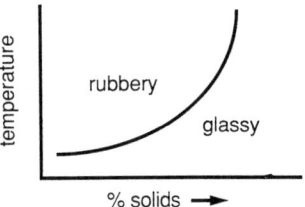

Figure 5. Theoretical state diagram for amorphous polymer system.

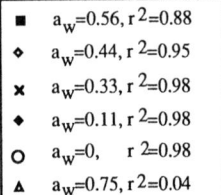

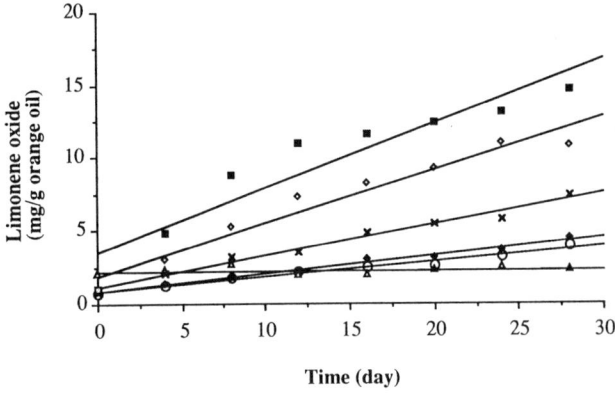

Figure 6. Oxidation of limonene in orange oil encapsulated in maltodextrin M100 at several water activities (*28*).

exclude all other components. Shimada et al. (*37*) studied the oxidation of methyl linoleate in encapsulated lactose undergoing crystallization. They found that oxidation did not occur below the crystallization temperature for the lactose matrix. Upon crystallization, however, oil was forced out of the particles and then became susceptible to oxidation since it was now on the surface of the matrix. This could also be expected upon collapse of a matrix.

Crystallization may also act to release water. In a controlled relative humidity environment, the polymer matrix will equilibrate to environmental conditions. However, when the food is in a closed container where evaporation and reequilibration are not possible, the released water may exist as a new phase on the food matrix surface. This may either increase or decrease lipid oxidation rates in this region as discussed previously.

Depending on the attribute of interest, kinetic analysis of lipid oxidation can be based on peroxide formation, oxygen consumption, hexanal formation or based on other attributes including sensory analysis (*5*). The temperature dependence of such rates have traditionally been modeled using Arrhenius kinetics (*8*) and most published data show a 1.5 to 3 fold increase in rate of oxidation for a 10°C rise in temperature (*38*). The glass transition theorists Slade and Levine (*21*) have suggested that the kinetic approach developed by Williams, Landel and Ferry (*39*) for polymers may be more appropriate for describing the rate of chemical reactions which occur in polymer systems which are in the rubbery state. This theory predicts a rate increase of 670 times for a 10°C increase above the glass transition temperature. A 110 fold increase in rate is predicted for a 10°C temperature rise between 10 and 20°C above the glass transition temperature. This has not yet been supported by experimental results.

Conclusion

The relationship between rates of lipid oxidation and moisture is complex. The amount of water, the water activity and the state of water in a food, along with other factors, must all be considered in the formulation and stability prediction of a food product which is susceptible to lipid oxidation.

Glass transition temperature should also be considered in stability prediction. Formulation to bring the food into the glassy state or to increase the glass transition temperature of the food matrix may be worthwhile. In the glassy state, the food will not be susceptible to collapse or crystallization and diffusion may be restricted which would reduce reaction rates. Controlling the temperature during freeze drying to prevent collapse can retain food product stability. Formulation techniques such as using high molecular weight polymers, adding components with high glass transition temperatures or crosslinking the polymers will all increase the glass transition temperature of the food product (*21*). However, collapse of an encapsulating matrix may act to protect components from oxidation as shown by Ma et al. (*28*). Thus, it is not yet certain whether it is always favorable to formulate to the glassy state. It is certain, however, that the control of moisture is an important factor in the control of lipid oxidation in many food systems.

Acknowlegments

Published as paper No. 19,651 of the contribution series of the Minnesota Agricultural Experimental Station based on research conducted under Project 18-78. This research was partially funded by a USDA National Needs Fellowship.

Literature Cited

1. Scott, G. *Atmospheric Oxidation and Antioxidants*; Elsevier Publishing: Amsterdam, 1965.
2. Labuza, T.P. *Food Tech.* **1980**, *4*, 36-59.
3. Levine, H.; Slade, L. In *Water and Food Quality*; Hardmman, T.M., Ed.; Elsevier Applied Science: New York, 1989; pp 71-134.
4. Noel, T.R.; Ring, S.G.; Whittam, M.A. *Trends in Food Sci. & Tech.* **1990**, 62-67.
5. Labuza, T.P. *CRC Critical Reviews in Food Technology*, **1971**, 355-405.
6. Karel, M.; Yong, S. In *Water Activity: Influences on Food Quality*; Rockland, L. B. and G.F. Stewart, Eds; Academic Press: New York, 1981; pp 511-529.
7. Labuza, T.P. *Food Tech.*, **1968**, 263-272.
8. Munday, K.A.; Edwards, M.L.; Kerkut, G.A. *J. Sci Food Agr.* **1962**, *13*, 455-458.
9. Labuza, T.P. *Sorption Isotherm: Practical Use and Measurement*; AACC Press: St. Paul, MN, 1984.
10. Salwin, H. *Food Tech.* **1959**, *13*, 594-595.
11. Nagashima, N.; Suzuki, E. In *Water Activity: Influences on Food Quality*; Rockland, L.B. and G.F. Stewart, Eds.; Academic Press: New York, **1981**; pp 247-264.
12. Labuza, T.P.; Maloney, J.F.; Karel, M. *J. Food Sci.* **1966**, *31*, 885-891.
13. Maloney, J.F.; Labuza, T.P.; Wallace, D.H; Karel, M. *J. Food Sci.* **1966**, *31*, 878-884.
14. Bateman, L. *Quarterly Rev., Chem. Soc.* **1954**, *8*, 147-167.
15. Simatos, D. *Advances in Freeze Drying*; L. Rey, Ed; Hermann Pub.: Paris, 1966.
16. Glassstone, S. *Textbook of Physical Chemistry*; Van Nostrand: New York, 1940.
17. Rockland, L.B.; Swarthout; D.M.; Johnson, R.A. *Food Tech.* **1961**, *15*, 112-116.
18. Quaglia, G.B.; Sinesio, F.; Veccia-Scavalli, D.; Avalle, V.; Scalfati, G. *Intern. J. of Food Sci. and Tech.* **1988**, *23*, 241-246.
19. Quast, D.; Karel, M. *J. Food Sci.* **1972**, *37*, 584-588.
20. Cavaletto, C.; Dela Cruz, A.; Ross, E.; Yamamoto, H.Y. *Food Tech.* **1966**, *20*, 1084-1087.
21. Slade, L.; Levine, H. *1985 Faraday Division, Royal Society of Chemistry Discussion Conference-Water Activity: A Credible Mmeasure of Technological Performance and Physiological Viability?* Cambridge.
22. Sperling, L.H. *Introduction to Physical Polymer Science;* John Wiley & Sons: New York, 1986; pp 224-295.

23. Roos, Y.; Karel, M. *Biotech. Prog.* **1990**, *6*, 159-163.
24. White, G.W.; Cakebread, S.H. *J. Fd. Technol.* **1966**, *1*, 73-82.
25. Thijssen, H.A. *J. Appl. Chem. Biotechnol.* **1971**, *21*, 372-377.
26. Flink, J.M.; Labuza, T.P. *J. Food Sci.* **1972**, *37(4)*, 617-618.
27. Stannett, V. In *Diffusion in Polymers;* Crank, J. and Park, G.S., Eds.; Academic Press: London, 1968; pp 41-73.
28. Ma, Y.; Reineccius, G. A.; Labuza, T.P. Univ. of Minnesota. Unpublished data.
29. Chou, H.E.; Acott, K.M.; Labuza, T.P. *J. Food Sci.* **1973**, *38*: 316-319.
30. Zeleznak, K.J.; Hoseney, R.C. *Cereal Chem.* **1984**, *64(2)*, 121-124.
31. Roos, Y.; Karel, M. *J. Food Sci.* **1991**. In press.
32. Hoseney, R.C.; Zeleznak, K.; Lai, C.S. *Cereal Chem.* **1986**, *3*. 285-286.
33. Simatos, D.; Karel, M. In *Food Preservation by Moisture Control;* C.C. Seow, Ed.; Elsevier Applied Science: New York, 1988; pp 1-41.
34. Wallack, D.A.; King, C.J. *Biotech. Prog.* **1988**, *4(1)*, 31-35.
35. To, E.C.; Flink, J.M. *J. Fd.Tech.* **1978**, *13*, 583-594.
36. Roos, Y.; Karel, M. *1990 AIChE Summer Meeting*, San Diego.
37. Shimada, Y.; Roos, Y.; Karel, M. *J. Ag. Fd. Chem.* **1991**. *36*, 637-641.
38. Ragnarsson, J.O.; Labuza, T.P. *Food Chem.* **1977**, *2*, 291-308.
39. Williams, M.L.; Landel, R.F.; Ferry, J.D. *J. Am. Chem. Soc.*, **1955**, *77*, 3701-3706.

RECEIVED May 26, 1992

Chapter 7

Lipid Oxidation
Effect on Meat Proteins

Arthur M. Spanier[1], James A. Miller[1], and John M. Bland[2]

[1]Food Flavor Quality Research and [2]Environmental Technology Research, Agricultural Research Service, U.S. Department of Agriculture, Southern Regional Research Center, 1100 Robert E. Lee Boulevard, New Orleans, LA 70124

A model is described for studying the effect of free radical reactions on beef proteins. While most beef flavor studies have focused on examining lipid volatiles, the contribution of peptides to flavor has been generally overlooked. The model uses artificial membranes in the form of multilamellar liposomes that are made from extracted beef lipids. The desired protein(s) is (are) encapsulated into the liposomes for investigation of the effect of free radicals on protein structure and function. A free radical oxidation generating solution is used to generate radicals through lipid peroxidation that is monitored by analyzing the production of thiobarbituric acid reactive substances. Changes in protein composition and function are assessed by capillary electrophoresis and enzyme activity measurements. Induction of free radicals leads both to inactivation and activation of enzymes and induces both loss and appearance of proteins and peptides.

Meat Flavor Quality.

Meat flavor quality is dependent upon several pre- and post- mortem factors (*1*). Perhaps some of the most important factors are those developed during the postmortem aging period (*2*) and during the cooking and subsequent storage of the product (*3-7*). For example, during the postmortem aging period and during cooking/storage meat shows significant changes in the level of endogenous chemical components such as sugars, organic acids, proteins, peptides, free amino acids, and metabolites of adenine nucleotide metabolism such as adenosine triphosphate (ATP). The chemical modifications that occur during postmortem aging and during

This chapter not subject to U.S. copyright
Published 1992 American Chemical Society

the subsequent handling of meat function as a large pool of reactive flavor compounds that can interact to form flavor notes during cooking, e.g., sugars and amino acids react during heating to form Maillard products (*8, 9*). It is, therefore, apparent that the development of flavor in meat is a complicated process that occurs continuously from the moment of slaughter, continues through cooking and storage, and ends when the food is eaten and the flavor perceived. The final quality of meat, therefore, involves several external and internal factors.

A major factor involved in the loss of flavor quality in meat products is a problem called warmed-over flavor or WOF. WOF was first recognized by Tims and Watts (*10*) who defined it as the rapid onset of rancidity in cooked meats during refrigerated storage. Several other chapters in this symposium have already dealt with WOF directly and thus, it will be covered only briefly here. Oxidized flavors are readily detectable after 48 hours in cooked meats as opposed to the more slowly developing rancidity encountered in raw meats or fatty tissues which becomes evident only after prolonged freezer storage (*11, 12*). WOF can also develop rapidly in raw meat that has been ground and exposed to air (*13-15*). It is now generally accepted that any process that involves disruption of muscle structure (e.g. cooking, grinding or restructuring) enhances the development of WOF (*15*) hereafter called MFD for meat flavor deterioration (*1, 5*).

Many of the compounds identified in meat are products of lipid oxidation and free radical reactions; these compounds play an important role in the development of the distinctive flavor character of meats such as beef, pork, and lamb (*16*). Hornstein and Crowe (*17*) and others (*18, 19*) suggested that the fat portion of meat contributed to the unique flavor that characterizes meat of one species from that of another. As lipids oxidize, they produce many secondary reaction products, such as alcohols, hydrocarbons, ketones, fatty acids, and aldehydes, each capable of supplying a different aroma, and collectively, several different aromas. (*20-24*). While the fat portion of meat is associated with species specific flavor and even though over 600 volatile compounds have been identified from cooked beef, there has not been one single compound identified to date that can be attributed to the aroma of "cooked beef" (*25*). Because of this, the lean portion of meat (the non-lipid portion) is credited with being the major contributing factor to the basic meaty flavor of muscle foods (*19*). While the lipids and lean portions of the meat are each identified as yielding a separate influence on meat flavor, no real attempts have been made to determine what effect the free radicals produced from lipid peroxidation would have on the development and deterioration of that portion of meat flavor that is associated with the lean soluble-portions. This chapter attempts to respond to this necessary area of research.

Background to the Problem

The Lipids. Yonathan and Watts (6) showed that the polyunsaturated fatty acids of the cellular lipids (the phospholipids) were involved in flavor deterioration in cooked meat. Later, Igene and colleagues (27) found that phospholipids oxidize faster than triglycerides in MFD samples. Using capillary gas chromatography, Dupuy et al. (28) demonstrated that subcutaneous fat produced about fifty volatile compounds during MFD development, whereas intramuscular fat produced more than 200 volatile compounds. There is no question that lipid oxidation of phospholipids seems to be the primary source of off-flavor notes produced during the MFD process as seen in Figure 1. However, where these lipids lie in situ, which ones are the most susceptible to oxidation, and the mechanism of catalysis remain mostly unanswered. The general consensus seems to revolve around the action of some form of iron ions, either bound or free, inducing lipid oxidation followed by the generation of a cascade of free radical reactions (29-32; see Simic, Kanner, and St. Angelo chapters this Symposium).

The increase in lipid volatiles such as hexanal and in thiobarbituric acid reactive substances (TBARS), appear to follow closely the changes in sensory descriptors such as painty (15, 33, 34, see St. Angelo chapter this Symposium). Bitter and sour are undesirable tastes that develop during storage (Figure 1) and that correlate well with meat the peptides and amino acids of the lean portion of meat rather than with meat lipids or lipid oxidation products. These two tastes are present at higher levels in fresh cooked meat than is painty and other lipid-derived flavors because of their generation via proteolysis during postmortem aging (1, 5, 15). Desirable flavors such as beefy and brothy are also lost during the storage period following cooking.

The Proteins and Enzymes. The nutritional contribution of meat proteins is well established, and significant research efforts have been directed towards understanding the effect of proteinases on meat texture and tenderization (2, 3). However, the contribution of meat proteinases and peptides to meat "tenderness", meat "flavor," and overall meat "quality" is perhaps one of the least examined areas in meat science. The process of meat tenderization during the postmortem aging period is complex and is not fully understood. Indeed, improvement in tenderness and the exact mechanisms involved in the conditioning process are still unknown. It has been emphasized that the key target of all the morphological and biochemical changes occurring in meat during the postmortem conditioning period reside with myofibrillar matrix (3, 35-38). Many of the alterations seen in muscle ultrastructure during the postmortem period have been ascribed to the action of lysosomal enzymes, especially cathepsin B and L (1, 39) and these enzymes have been implicated in the process of postmortem tenderization (3). However, the full extent of the changes occurring during aging can only be attributed to a synergistic action of lysosomal and calcium-dependent proteinases (40).

Because proteins comprise the major chemical components of beef, they represent the ideal class of compounds to study regarding their relationship to flavor quality. Increased amounts and kinds of proteinaceous components such as peptides and amino acids have been shown to occur both during postmortem aging (1) and at different cooking temperatures (5). For example, two major classes of peptides are found in beef roasts after cooking (5). One of these, the 1800 M.W. class, is composed of two subclasses as determined by reverse phase high pressure liquid chromato-graphy, i.e. one is enriched with hydrophilic residues and the other is enriched with hydrophobic residues. Hydrophilic residues are associated with desirable flavor while hydrophobic residues are associated with undesirable or off-flavors. Both residues are produced during cooking and there is little evidence that their final tissue level is due to the end-point cooking temperature; it is more likely that these flavor peptides have been generated during the postmortem conditioning stage by proteolytic activity (1, 5). As precooked meat is stored in the refrigerator, there is a decline in the presence of the hydrophilic residue with no major effect on the hydro-phobic residue. It is thought that the loss in the hydrophilic residue with storage is a function of fragmentation of the parent hydrophilic residue via free-radical mechanisms. This loss in hydrophilic peptide residue is closely related to changes observed in the flavor of the stored meat sample where one generally sees an increase in the bitter and sour notes and a decline in the cooked beef and brothy notes during storage.

The Model

Preparation of Liposomes. Because of the complex structure and biochemistry of muscle it is difficult to clearly define the mechanism(s) and targets of the change in beef flavor in whole meat systems. An *in vitro* model, which permitted the study the effect of free radicals on meat proteins, was thus developed. The model used endogenous meat lipids to prepare artificial membranes called "liposomes" into which various protein-flavor components are encapsulated or entrapped .

Cellular lipids were extracted from meat with organic solvents as per Bligh and Dyer (41). As seen in Figure 2, the organic lipid-extract was placed into a rotary evaporator flask. A thin film was prepared in the flask by evaporation of the solvent. Aqueous buffered solution containing the desired protein or proteins was added to the flask with the lipid film. The flask was agitated by vortexing for 5 minutes. The resulting suspension contained multilammellar liposomes with the encapsulated protein (42) as well as unencapsulated or entrapped material. Free and entrapped protein were separated by size exclusion chromatography using Sephadex .

The formation of free-radicals was induced by exposing the liposomes

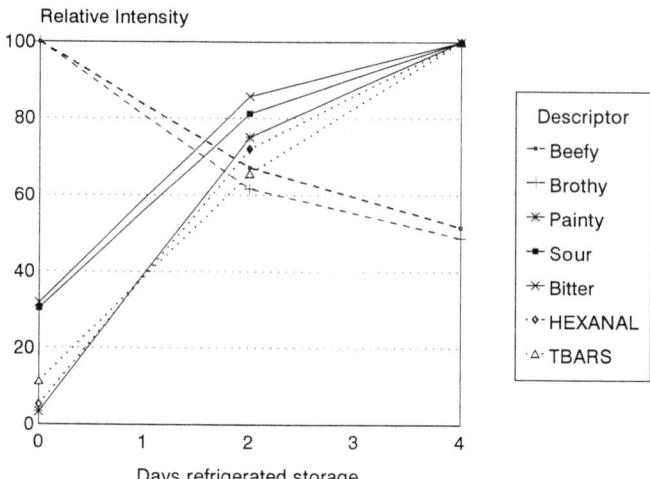

Figure 1. Effect of storage on the sensory and instrumental flavor-attributes of precooked beef patties. Sensory and instrumental analysis are performed as described previously (2, 16). Chemical analysis are presented in all capitals to separate them from sensory descriptors.

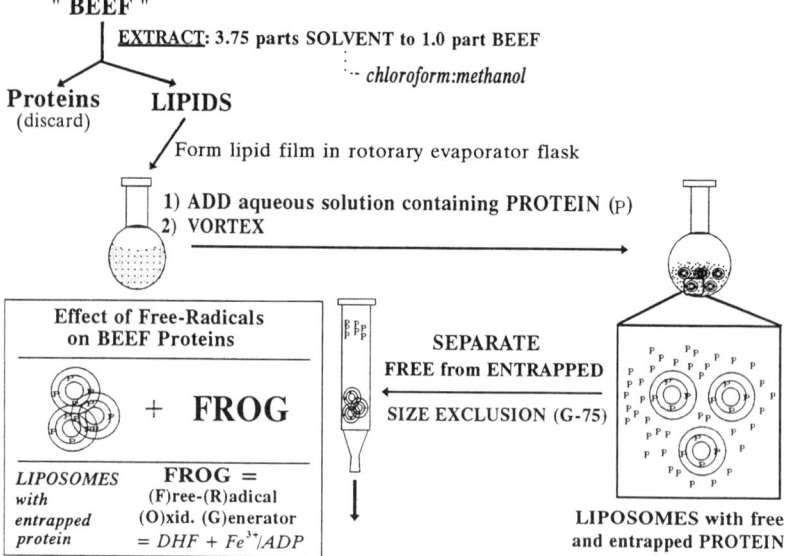

Figure 2. Diagrammatic representation of the preparation of multilamellar liposomes.

containing the entrapped protein to a free-radical oxidation-generating system which is given the acronym "FROG". Lipid oxidation and free-radical production is induced experimentally by addition of the "FROG" reaction mixture which consists of 3.3 mM dihydroxyfumarate (DHF) and a mixture of 0.1 mM iron as ferric chloride with 1.0 mM adenosine diphosphate (ADP; 43). Incubations were initiated by the addition of the iron/ADP mixture. Reactions were stopped and the all controls contained a mixture of 0.025% each of the antioxidant, propyl gallate (PG), and the chelator, ethylenediaminetetraacetic acid (EDTA). Lipid oxidation was assessed by the measurement of TBARS. Changes in protein(s) were monitored electrophoretically.

Induction of Lipid Oxidation and Free Radicals. Aerobic oxidation of DHF generates large steady state levels of superoxide anions (O_2^-) which have been suggested to generate additional active oxygen radical capable of inducing lipid peroxidation (44, 45). Once formed, superoxide (O_2^-) will lead to the generation of other reactive oxygen species such as hydrogen per-oxide (H_2O_2), hydroxyl radicals (•OH), and singlet oxygen ($^1O_2$) through the following well documented reaction scheme (see earlier chapters in this symposium).

$$2O_2^- + 2H^+ \rightarrow H_2O_2 + O_2 \qquad (1)$$

$$H_2O_2 + O_2^- \rightarrow O_2 + \bullet OH + OH^- \qquad (2)$$

$$H_2O_2 + O_2^- \rightarrow \bullet OH + OH^- + {}^1O_2 \qquad (3)$$

Reaction #2 may be catalyzed by the Fe^{3+}/ADP as follows:

$$O_2^- + Fe^{3+}\text{-ADP} \rightarrow O_2 + Fe^{2+}\text{-ADP} \qquad (4)$$

$$Fe^{2+}\text{-ADP} + H_2O_2 \rightarrow Fe^{3+}\text{-ADP} + OH^- + \bullet OH \qquad (5)$$

The cascade of free radicals produced from the reactions initiated by the "DHF-Fe^{3+}/ADP" or FROG system can effect every membrane, lipid, protein, enzyme and enzyme system in the muscle. Free radical damage to cellular membranes, particularly to that of the sarcotubular/lysosomal (39) system, can cause amino acid oxidations, malondialdehyde release from oxidized fatty acids, lipid-lipid crosslinking, fatty acid oxidation, lipid-protein crosslinking, disulfide crosslinking, protein-protein or peptide-peptide crosslinking, and protein strand scission or fragmentation, to name just a few (Figure 3).

Lipid Oxidation Induced by Free Radical Generation. Liposomes without entrapped protein were exposed to the FROG system for determination of free radical-induced lipid peroxidation. The incubation mixture

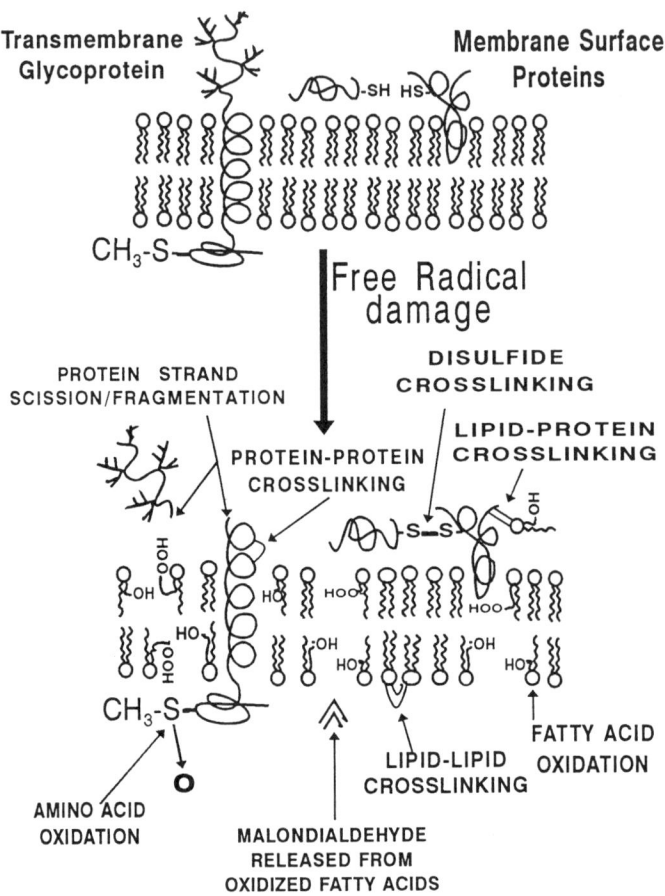

Figure 3. Free radical damage to membranes. [Adapted from Freeman, B.A.; Crapo, J.D. *Lab. Invest.* **1982** *47(5)*:412-426]

for the experimental group contained the liposomes with the FROG reaction mixture. The controls were the same as the experimental with the exception that they contained equal amounts (0.025%) of PG and EDTA. No change was seen in the background levels of TBARS in the control group (Figure 4). Neither DHF nor Fe^{3+}-ADP by themselves showed any increase in TBARS levels (not shown). On the other hand, incubation of the experi-mental group, i.e., the liposomes with the FROG system, led to a 2.7-fold increase in TBARS that reached a maximal level after 5 minutes.

The Effect

The Effect of Free Radicals on Beef Enzyme Activity. The data presented in Figure 5 show the effect of free radicals/lipid oxidation on the activity of two muscle enzymes, creatine phosphokinase (CPK) and N-acetyl-β-glucosaminidase (NAGA). In these experiments, the liposomes encapsul-ated or entrapped a 35,000xg aqueous extract of beef (top round or *semi-membranosus* muscle). Enzyme activity was analyzed after appropriate incubation with the FROG system or with the FROG system plus the anti-oxidant/chelator (PG/EDTA) mixture. CPK activity was activated by the free radicals/lipid oxidation induced via the FROG system. Maximal activity was observed between 10 and 15 minutes when activity began to decline. The decline in CPK activity is attributed to the thermal inactivation of CPK during incubation (*1*). While CPK activity was stimulated as a result of the FROG treatment, NAGA activity was significantly and continuously reduced as a result of the free radical and lipid peroxidation (Figure 5).

The effect of free radicals and lipid oxidation on the activity of a muscle proteinase, cathepsin D, was also examined (Figure 6). The data show that the FROG system with the liposomes (FR+L) and the liposomes by themselves (L only) had background levels of cathepsin D activity. Activity was measured as the mg tyrosine released per milliliter per hour from acid denatured hemoglobin as described previously (*1*). The three groups containing the muscle extract (ME) all showed the same level of catheptic activity at time 0 (the unincubated control and experimental groups). On the other hand, there was an almost 33% reduction in catheptic activity in the group incubated for 60 minutes (FR+L+ME); since this same decline is not observed in the 60 minute control group (C-60m) containing the antioxidant/chelator (PG/EDTA) mixture, the decreased activity in the FR+L+ME group is attributed to free radicals production and lipid oxidation.

The Effect of Free Radicals on a Specific Beef Peptide. Yamasaki and Maekawa (*46*) found a delicious-tasting, meaty-flavored, linear, octapeptide (Lys-Gly-Asp-Glu-Glu-Ser-Leu-Ala) in papain digests of beef. The potential impact of peptides in general and this octapeptide in

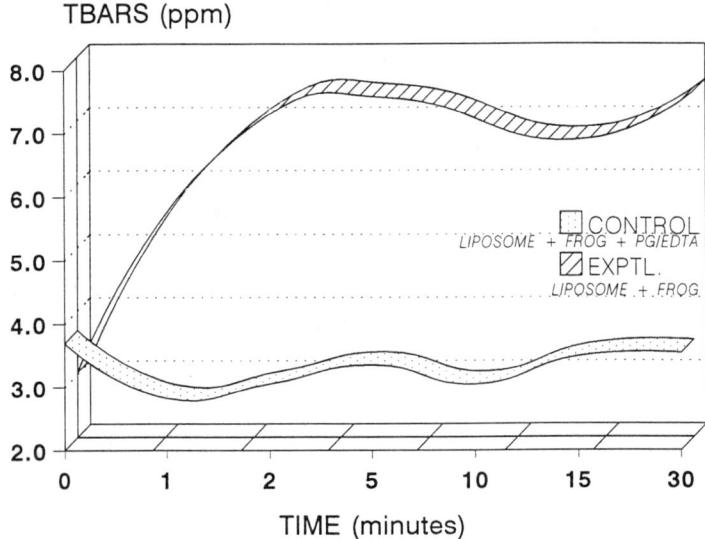

FROG: Free Radical Oxidation Generator
PG/EDTA = Propyl gallate, Ethylene diamine tetraacetic acid

Figure 4. Effect of the FROG system on the level of TBARS in an artificial liposome model without encapsulated protein(s).

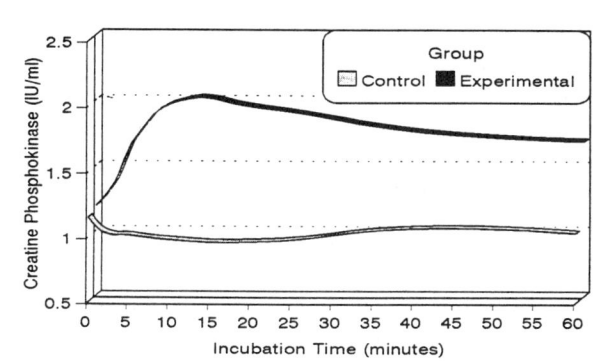

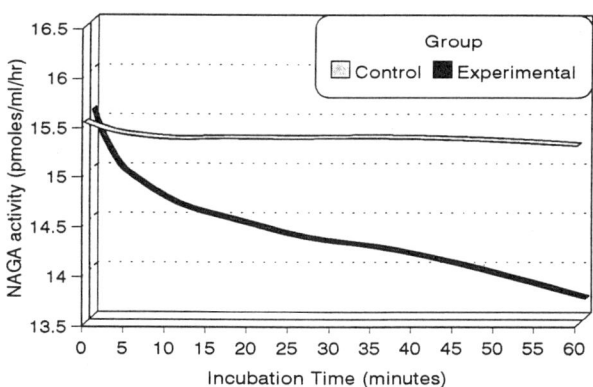

Figure 5. Effect of free radicals on the activity of creatine phosphokinase (CPK) and N-acetyl-β-glucosaminidase (NAGA) of beef extracts encapsulated into a multilamellar liposome model.

particular to beef flavor has been demonstrated by Kato et al. (4) who reported that enzymatic cleavage of the two amino-terminal amino acids from the "delicious peptide" yielded a hexapeptide residue imparting a "bitter" taste. This is similar to an earlier observation in precooked/stored meat in which it was demonstrated that there is a loss in the level of a hydrophilic/desirable flavored peptide and the cooked beef/brothy flavor notes along with a concomitant increase in the off-flavor notes bitter and sour (1). The delicious peptide, "BMP" (beefy meaty peptide) proves to be valuable tool to examine the behavior and role of peptides in meat flavor quality.

Figure 7 shows the effect of the FROG system on BMP in the absence of liposomes (top), liposomes only (bottom), and BMP entrapped within multilamellar liposomes (middle). Data in the upper graph indicate that the free radicals generated by the FROG system have no effect on BMP when lipids are absent. The bottom electrophoretic profile shows the effect of free radicals and lipid oxidation generated by the FROG system on the electrophoretic mobility of the multilamellar liposomes without entrapped protein. In this case, the electrophoretic migration of the liposomes seems to be affected not by the FROG system but rather by the incubation period since both the 60 minute control and the 60 minute experimental groups show the same shape and increased rate of migration; furthermore, no difference is observed in either the shape or the migration rate of the liposomes both with or without the antioxidant/chelator mixture. The middle electrophoretic profile shows the effect of the FROG reaction mixture on the combined BMP/liposome system. No direct effect on BMP by either free radicals nor lipid oxidation is seen in this complete system. The only effect appears to be an incubation effect on the BMP/liposome similar to that observed for the liposomes only.

The data presented in Figure 8 show the high performance capillary electrophoresis profile of beef protein extracts (35,000xg aqueous supernatants) both before and after the induction of free radicals and lipid oxidation using the FROG system. The uppermost profile is the electrophoretic pattern of unincubated control beef extract in liposomes, while the lower (middle) profile depicts the same liposome/meat-extract sample after 60 minutes incubation with the FROG system. In this case a decrease in several major proteins and peptide groups is seen such as the clusters between 2-3 minutes and between 3-4 minutes, the peaks at 4.2 and 4.4 minutes, and the peak at 5.1 minutes. The data also show the appearance of new peaks not present prior to FROG treatment such as the peaks between 8-9 minutes and between 12-13 minutes, and the change in shape and migration time of the peak between 10-11 minutes. The electrophoretic peak between 5.0 and 5.3 minutes is intriguing since it has the same initial shape and migration-time as synthetically-prepared BMP (bottom electrophoretic profile). Semiquantitative analysis of the peak areas, indicates that there is a 16.6% reduction in area (0m → 60m) suggesting that the molecular composition of this peak has been altered.

7. SPANIER ET AL. *Lipid Oxidation: Effect on Meat Proteins* 115

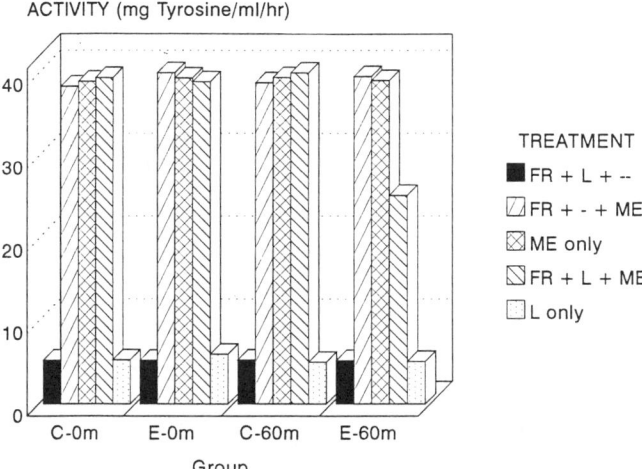

Figure 6. Effect of free radicals on cathepsin D activity. "FR" = free radical oxidation generating system described in text; "L" = multilamellar liposomes with or without entrapped beef extract; "ME" = beef *semimembranosus* muscle extract; "C" = control samples containing the antioxidant (propyl gallate; PG) and chelator (ethylene diamine tetraacetic acid; EDTA) cocktail; "E" = experimental samples without the PG/EDTA; Numbers refer to the time (minutes; m) incubated at 37°C.

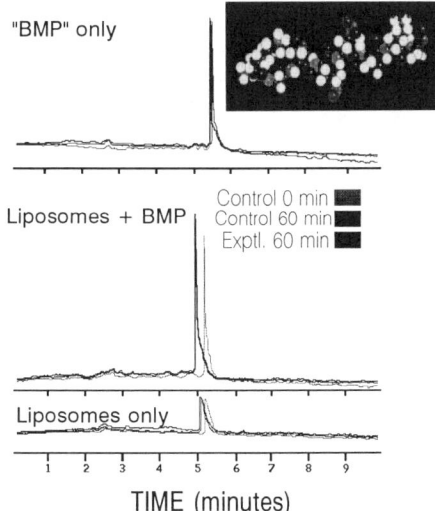

Figure 7. Capillary electrophoretograms demonstrating the effect of free radicals on beefy meaty peptide (BMP).

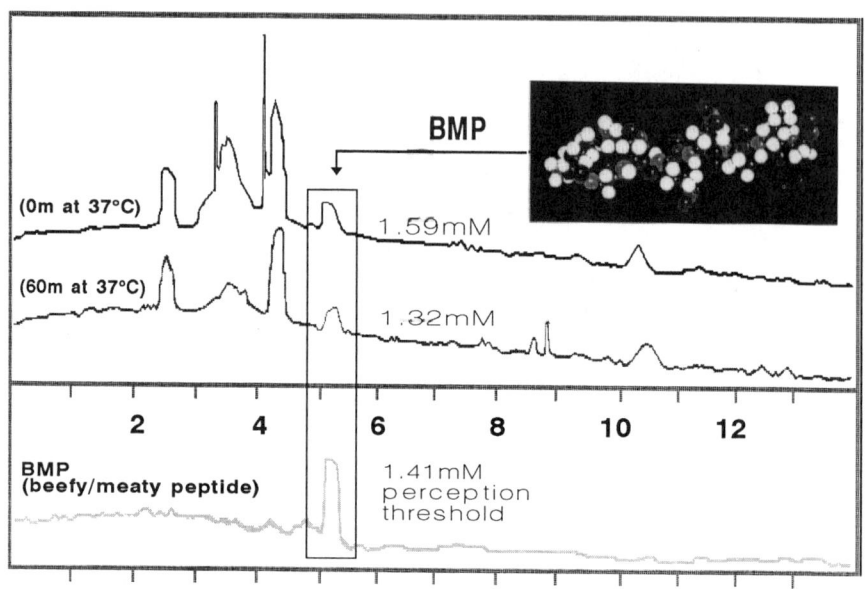

Figure 8. Capillary electrophoretograms demonstrating the effect of free radicals and lipid oxidation on *semimembranosus* muscle extracts entrapped within multilammellar liposomes.

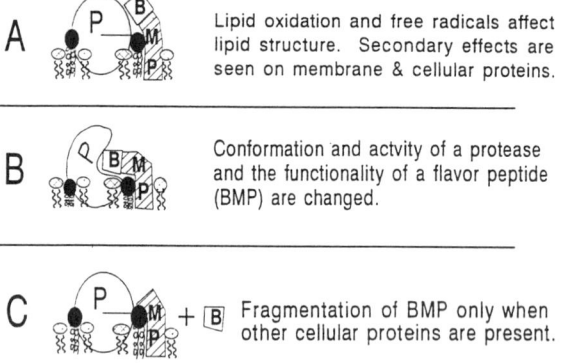

Figure 9. Postulated mechanism for the fragmentation of a beefy/meaty peptide (BMP) of beef by free radicals. Packman figure with postulate "P" = protease; Circles with squiggly lines = lipids.

Further analysis of these integrated peaks revealed that the area under the peak of the unincubated beef extract represents 44.9% of the area of the standard or approximately 2.69 μg of octapeptide. This represents a sample containing an equivalent of 1.59 mM of octapeptide which is greater than the 1.41 mM perception-threshold reported by Dr. Okai and his colleagues for this peptide (46). The measured 16.6% reduction in the peak area due to incubation with the FROG system yields a 1.32 mM concentration which is just below the threshold level for the sensory perception of BMP.

Since the BMP in the liposome-entrapped meat extract is fragmented only after incubation with the FROG system (Figure 8) and not modified after incubation of pure BMP and liposomes with the FROG system (Figure 7), we feel that the fragmentation in the presence of other cellular components is a secondary effect of free radicals and lipid oxidation (Figure 9). We suggest that the free radicals and the oxidation of muscle membrane lipids cause conformational changes to occur in other muscle proteins such as the proteinases, cathepsins B, H, and/or L. It is hypothesized that the conformation change in the protease is such that its function and/or activity against BMP is enhanced, thereby, leading to the fragmentation of BMP and thus a change in beef flavor.

Summary

An efficient and desirable model system is available for studying those components of meat which contribute to flavor quality, i.e., the liposome model is used to simulate the chemical processes occurring during storage of precooked meat. The data make it apparent that not only is there an increase in undesirable flavor (painty), flavor components (hexanal) and flavor markers (TBARS) of lipid origin (5, 48; St. Angelo et al., this volume; Figure 1) but also there is an increase in protein-derived undesirable flavors (bitter and sour) and a decrease in protein-derived desirable flavors (beefy and brothy). The decline in the desirable flavored peptide components is secondarily related to the free radicals generated through lipid peroxidation. Knowledge obtained from such model systems will allow us to gain a better understanding of the complex response(s) of flavor components to different conditioning, cooking, and cooking/storage situations. Perhaps more importantly, the knowledge generated from models such as the one described above will allow us to formulate better management methods to maintain and enhance food flavor and help us to develop better predictive, adaptive or management methods for enhancing flavor quality.

Literature Cited

1. Spanier, A.M.; McMillin, K.W.; Miller, J.A. *J. Food Sci.* **1990** *55(2)*:318-326.

2. Koohmaraie, M.; Babiker, A.S.; Schroeder, A.L.; Merkel, R.A.; T.R. Dutson, T.R. *J. Food Sci.* **1988**, *53(6)*:1638-1641.
3. Etherington, D.J.; Taylor, M.A.J.; Dansfield, E. *Meat Sci.* **1987**, *20*:1-18.
4. Kato, H.; Nishimura, T. In *Umami: A Basic Taste.*; Kawamura, Y.; Kare, M.R., Eds.; Marcel Dekker, Inc: New York, NY, **1987**, pp. 289-306.
5. Spanier, A.M.; Edwards, J.V.; Dupuy, H.P. *Food Tech.* **1988**, *42(6)*:110-118.
6. St. Angelo, A.J.; Vercellotti, J.R.; Legendre, M.G.; Vinnett, C.H.; Kuan, J.W.; James, C.; Dupuy, H.P. 1987. *J. Food Sci.* **1987**, *52*:1163-1168.
7. St. Angelo, A.J.; Vercellotti, J.R.; Dupuy, H.P.; Spanier, A.M. *Food Technol.* **1988**, *42(6)*:133-138.
8. Bailey, M.E. *Food Technol.* **1988**, *42(6)*:123-126.
9. Bailey, M.E.; Shin-Lee, S.Y.; Dupuy, H.P.; St. Angelo, A.J.; Vercellotti, J.R.; In *Warmed-over flavor of meat*; St. Angelo, A.J. and M.E. Bailey, Eds.; Academic Press: Orlando, Florida, **1987**, pp. 237-266.
10. Tims, M.J.; Watts, B.M. *Food Technol.* **1958**, *12*:240-243.
11. Pearson, A.M.; Gray, J.I. In *The Maillard Reaction in Foods and Nutrition.* Waller, G.R.; Feather, M.S., Eds.; American Chemical Society: Washington, D.C. **1983**, Amer. Chem. Soc. Symp. Ser. 215, pp. 287-300.
12. Pearson, A.M., Love, J.D.; Shorland, F.B. 1977. In *Advances in Food Research.* Chichester, C.O.; Mrak, E.M.; Stewart, G.F., Eds.; Academic Press, New York, NY, **1977**, Vol. 23; pp. 1-74.
13. Greene, B.E. *J. Food Sci.* **1969**, *34*:110-113.
14. Sato, K.; Hegarty, G.R. *J. Food Sci.* **1971**, *36*:1098-1102.
15. Spanier, A.M.; Vercellotti, J.R.; James, C., Jr. *J. Food Sci.* **1992**, *57*:IN PRESS
16. Shahidi, F.; Rubin, L.J.; D'Souza, L.A. *CRC Crit. Rev. Food Sci. Nutr.* **1986**, *24*:141-243.
17. Hornstein, I.; Crowe, P.F.; *J. Agric. and Food Chem.* **1960**, *8(6)*:494-498.
18. Sink, J.D. *J. Food Sci.* **1979** *44*:1-5.
19. Wasserman, A.E. *J. Food Sci.* **1979**, *44*:6-11.
20. Wasserman, A.E.; Talley, F. *J. Food Sci.* **1968**, *33*:219-223.
21. Forss, D.A. *Lipids.* **1972**, *13(4)*:181-258.
22. Frankel, E.N. In *Recent Advances in the Chemistry of Meat*; Bailey, A.J., Ed.; The Royal Society of Chemistry: London, England; **1984**; pp. 87-118.
23. Gasser, U.; Grosch, W. *Z. Lebensm. Unters. Forsch.* **1988**, *86*:489-494.
24. MacLeod, G.; Ames, J.M. *Flavour Fragrance J.* **1986**, *1*:91-104.
25. Baines, D.A.; J.A. Mlotkiewicz, J.A.; In *Recent Advances in the*

Chemistry of Meat; Bailey, A.J., Ed.; The Royal Society of Chemistry: London, England, **1984**, pp. 119-164.
26. Younathan, M.T.; Watts, B.M. *Food Res.* **1960**, *25*:538-543.
27. Igene, J.O.; King, J.A.; Pearson, A.M.; Gray, J.I. *J. Agric. Food Chem.* **1979**, *27*:838-842.
28. Dupuy, H.P.; Bailey, M.E.; St. Angelo, A.J.; Vercellotti, J.R.; Legendre, M.G. 1987. In *Warmed Over Flavor in Meats*; St. Angelo, A.J.; Bailey, M.E., Eds.; Academic Press: Orlando, FL, **1987**; pp. 165-191.
29. Kanner, J.; Harel, S. *Arch. Biochem. Biophys.* **1985**, *237*:314-321.
30. Kanner, J.; Kinsella, J.E. *J. Agric. Food Chem.* **1983**, *31*:370-378.
31. Ladikos, D.; Lougovois, V. *Food Chem.* **1990**, *35*:295-314.
32. Spanier, A.M. In, *Recent Developments in Food Science and Technology. Proceedings of the (postponed) 1991 Samos Flavor Conference*, Charalambous, G., ed.; Elsevier: Amsterdam; **1992** in press.
33. Larick, D.K.; Turner, B.E. *J. Food Sci.* **1990**, *55(2)*:312-317.
34. Lillard, D.A. In *Warmed-over Flavor of Meat.;* St. Angelo, A.J.; Bailey, M.E., Eds.; Academic Press: Orlando, FL., **1987**, pp. 41-67.
35. Goll, D.E.; Otsuka, Y.; Nagainis, P.A.; Shannon, J.D.; Sathe, S.K. *J. Food Biochem.* **1983**, *7(3)*:137-177.
36. Penny, I.F.; Etherington, D.J.; Reeves, J.L.; Taylor, M.A.J. *Proc. 30th Eur. Meet. Meat Res. Work, Bristol.* **1984**, pp. 133-134.
37. Ouali, A.;Garrel, N.;Obled, A.;Deval, C.; Valin, C.; Penny, I.F. 1987. *Meat Sci.* **1987**, *19(2)*:83-100.
38. Mikami, M.; Whiting, A.H.; Taylor, M.A.J.; Maciewicz, R.A.; Etherington, D.J. *Meat Sci.* **1987**, *21*:81-97.
39. Bird, J.W.C.; Schwartz, W.N.; Spanier, A.M. <u>Acta biol. med. germ.</u> **1977**, *36*:1587-1604.
40. Ouali, A. *J. Muscle Foods.* **1990**, *1(2)*:129-165.
41. Bligh, E.G. ; Dyer, W.J. *Canadian J. Biochem. Physiol.* **1959**, *37(8)*:911-917.
42. Cohen, C.M.; Weissmann, G.; Hoffstein, S.; Awasthi, Y.C.; Srivastava, S.K. *Biochem.* **1976**, *15(2)*:452-460.
43. Mak, I.T.; Misra, H.P.; Weglicki, W.B.. *J. Biol. Chem.* **1983**, *258*:13733-13737.
44. Halliwell, B. *Biochem. J.* **1977**, *163*:441-448.
45. Goldberg, B.; Stern, A. *Arch. Biochem. Biophys.* **1977**, *178*:218-225.
46. Yamasaki, Y.; Maekawa, K. *Agric. Biol. Chem.* **1978**, *42(9)*:1761-1765.
47. Tamura, M.; Nakatsuka, T.; Tada, M.; Kawasaki, Y.; Kikuchi, E.; Okai, H. *Agric. Biol. Chem.* **1989**, *53(2)*:319-325.
48. Drumm, T.D.; Spanier, A.M. *J. Agric. Food Chem.* **1991**, *39*:336-343.

RECEIVED February 19, 1992

PREVENTION OF LIPID OXIDATION

Chapter 8

Maillard Reaction Products and Lipid Oxidation

Milton E. Bailey and Ki Won Um

Department of Food Science and Human Nutrition, 21 Agriculture Building, University of Missouri, Columbia, MO 65211

An important aspect of food processing involving Maillard reaction products (MRP) is their reactivity as antioxidants in food systems, particularly meat. A brief review is given of the general pathways of the Maillard reaction and the components of MRP as antioxidants and some possible mechanisms of antioxidants in the various reaction fractions. More specific results are given concerning the use of MRP to reduce oxidation of lipids in cooked meats. Both the water-soluble low molecular weight and polymeric fractions of MRP have antioxidant potential. The intermediate constituents such as maltol, dihydroxyacetone, glyceraldehyde and reductones behave as antioxidants or antioxidant precursors in these reaction systems. Low molecular weight diffusate components, as well as MRP from glucose-histidine and other precursors for meat flavor, improve and preserve the acceptance of cooked meat during storage.

A most difficult aspect of food preservation is the control of oxidation that most natural and processed foods undergo, which results in formation of characteristic undesirable odors and flavors. The most general approach for improving flavors of oxidizable foods is through the use of additives. Antioxidants most frequently used include synthetic additives such as BHA and BHT, which are constantly under safety surveillance and investigation. These and related additives are becoming less acceptable to the consuming public and food legislators.

One solution used by manufacturers for decreasing risk to consumers and increasing food acceptability is the use of natural products as additives. One such group of antioxidants that have utility in a number of food products is that produced by the Maillard reaction. The precursor chemical constituents for these reactions can be considered natural because the reactions occur normally in the processing of many foods. The use of MRP for retarding lipid oxidation in foods has been revisited and discussed by several authors (*1-4*).

The Maillard Reaction

The Maillard reaction has been discussed by numerous authors in the past and is under constant review. Some early important reviews of the Maillard reaction were those of Hodge (*5, 6*), Anet (*7*), Reynolds (*8-10*) and Baltes (*11*). More than 20 reviews have been published on the chemistry of the Maillard reaction since 1975 (*4, 12-27*), and there have been four symposia on this important subject during the same time (*28-31*). The Maillard reaction (nonenzymatic browning) is the interaction of carbonyl compounds and amines (amino-carbonyl reaction) to produce, through a series of complex reactions, a number of flavor precursors, flavorants and antioxidants, as well as polymerized brown pigments (melanoidins).

One series of reactions involves the condensation of the carbonyl and amine through the C-1 of aldoses and the C-2 of ketoses with subsequent rearrangement to the keto or aldo sugars (Amadori or Heynes intermediates) involving reactions that are reasonably well understood. An important step in these reactions is the Amadori rearrangement of the sugar moiety to irreversibly produce ketosyl compounds that enolyze and degrade by complex reactions to produce intermediate products (*4*).

The most widely accepted mechanisms for degradation pathways of Amadori compounds by dehydration and fission were suggested by the brilliant work of Hodge in early publications (*5, 6*) and there have been few improved ideas since that time, although some new supportive data have been published (*32-35*).

As diagrammed in Figure 1, three major pathways leading to the formation of intermediate compounds in flavor, antioxidant and pigment formation arise from Amadori compounds. These are the 1-deoxyosone, the 1-deoxyreductone and the 3-deoxyosone. A fourth major leg of degradation involving intermediates is the Strecker degradation.

Compounds such as maltol, isomaltol, cyclotene, 5-hydroxy-5,6-dihydromaltol, 4-hydroxy-3(2H)-furanone and similar components arise from the 1-deoxyosone pathway by dehydration. An enol form of the 1-deoxyosone (1-deoxyreductone) is degraded by fission to produce pyruvaldehyde, diacetyl, acetaldehyde, acetic acid and reductones. These compounds have tremendous potential as precursors for antioxidants and flavor compounds, particularly when heated with amines.

The third pathway begins with a 1-2-enediol form of the Amadori compound by eliminating a hydroxyl group at C-3. Deamination yields a 3-deoxy component that becomes a reductone by eliminating another molecule of water. Further dehydration under acidic conditions occurs to yield 2-furaldehydes (*16, 32*). This type of dehydration of sugars under acetic conditions without the interactions of the amines is of extreme importance in the analytical chemistry of carbohydrates and the caramelization reactions have been studied in detail by Feather and coworkers (*33, 36-38*).

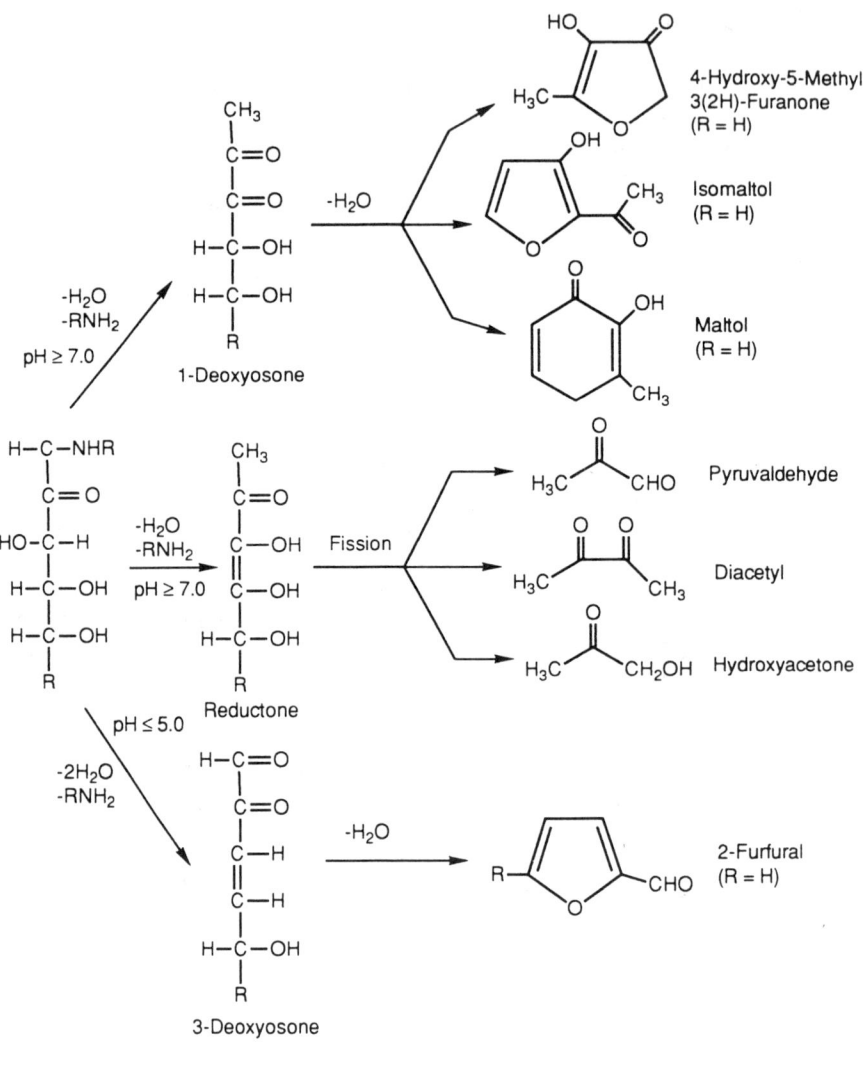

Figure 1. Deamination-dehydration of Amadori compounds to reductones and "flavor compounds" which serve as intermediates for meat flavor volatiles.

The Strecker degradation is an oxidative decarboxylation of amino acids by reactions with dicarbonyls such as glyoxal, diacetyl and perhaps 3-deoxyosones. The new Schiff base formed has one less carbon atom than the original amino acid and is hydrolyzed to an aldehyde. The Strecker aldehydes are very reactive as flavor precursors and can condense with themselves by aldol condensation or with cyclic compounds to form melanoidins that have antioxidant properties. The evolution of CO_2 during melanoidin formation has been demonstrated implicating Strecker degradation of amino acids as an important step (4).

These reactions do not, however, explain all the activities of Maillard reaction products as antioxidants and the formation of reductones by specific pathways and free radical formation through mechanisms described by Namiki and Hayashi (39), who proposed a mechanism of free-radical formation in the early stage of carbonyl-amine reaction prior to Amadori rearrangement. Analysis of hyperfine structures of ESR spectra led to the identity of radicals such as N, N^+ and pyrazine cation radicals. The radicals produced were assumed to be formed by condensation of 2 moles of 2-carbon alkylamine (enaminol) compounds produced by fragmentation of the sugar or sugar-amine with subsequent dimerization to form N, N′-disubstituted pyrazine radicals. Since the autoxidation of unsaturated fatty acids is believed to occur by free-radical mechanisms following hydroperoxide formation, there is justification for believing that components formulated by this mechanism contribute to antioxidant properties of these reactions.

MRP Antioxidants

Antioxidants can be formed at several levels during heating of carbonyl-amine mixtures, including degradation of Amadori compounds to amino reductones, to reductones, or formation of polymers with antioxidants activity. Namiki (4) recently published an excellent review of the Maillard reaction which revealed superior knowledge and interest in the development of the types of antioxidants mentioned above. His discussions are most comprehensive regarding use of MRP as antioxidants in foods.

Hodge and coworkers (40-42) were the first to demonstrate that MRP had antioxidant activity for preserving oils. Evans et al. (42) were able to demonstrate that reductones from MRP could retard oxidation of vegetable oils. Reductone is a trivial name for 3-hydroxy-2-ketopropane and consists of vicinal dicarbonyl compounds capable of enolization or enol forms forming keto groups following loss of hydrogen atoms.

$$R - C(OH) = C(OH) - C:O - R' \xrightarrow{-2H} R - C:O - C:O - C:O - R'$$

Where R and R′ can be alkyl aryl or ends of cyclizing biradicals.

Prominent examples are triose reductone, dihydroxymaleic acid, reductic acid and dihydroxy pyrogallol. Reductones as organic reducing agents were discussed by Hodge and Osman (13), who described them as having functional groups derived from conjugated enedial and carbonyl forms as above. These authors also indicated that amino analogs where hydroxyl groups were replaced by NH_2 or where carbonyl groups were replaced by -C=NR groups could behave as strong reducing agents.

Triose reductone (H-C(OH)=C(OH)-C:O-H) is formed when reducing sugars are heated in alkaline solution. Reductic acid is easily formed from uronic acids such as pectin in acid solutions above 100°C. The cyclic reductones such as reductic acid and ascorbic acid are generally more stable than their acyclic counterparts.

Crystalline amino-reductones have been prepared from hexoses and secondary amines (42), which were shown to inhibit peroxide formation in a variety of animal fats and oils. Oxidation inhibition by some of these compounds is approximately a linear function of concentration when used between 0 and 0.02%. Some of these reductones were much more effective in reducing oxidation of soybean oil and cottonseed oil than treatments with propyl gallate of the same level.

The structure of amine reductones was elucidated (43) using piperidino reductone as a model. This reductone was characterized as N-[1-methyl-1,2,3-trihydroxy-cyclopenten(2)-ylidine(4)]piperidinium beatin (N→2 or 3). The other amino-hexose reductones in all probability have analogous structures.

Ledl et al. (44) recently expanded and confirmed some of these reactions where they identified several new heterocyclic and carbocyclic compounds from reactions between sugars and secondary amines. Among products identified were β-oxypyridinium betainols, β-pyranones, pyridones and cyclopentenones.

Another important group of compounds formed by further dehydration of 2,3-enolization compounds (2,3-osones) following recondensation with amines are amino reductones, which have possible activities as antioxidants. Davidek et al. (45) summarized reactions from other workers (46) showing the degradation of D-fructoseamine in alkaline solution (Figure 2). The importance of this type of compound as an antioxidant in MRP reaction mixtures cannot be overemphasized since the presence of an amino group appears to be essential for the formation of antioxidants.

Amino reductones are possibly more effective and stable than their enediol counterparts. The amino reductones of Evans et al. (42) were excellent antioxidants because the nitrogen moiety can contribute electrophilic groups which might form chelates with metal ions such as copper and iron, which are known for their oxidative catalytic activities.

As reviewed by Bailey et al. (2), the melanoidin polymers formed from the interaction of carbonyl amines contribute to the antioxidant properties of MRP. A strong contributor to data regarding the effectiveness of melanoidins as antioxidants has been Yamaguchi et al. (47), who published information on the antioxidant activity of several melanoidin fractions using a model system of linoleic acid at pH 7.0 and peroxide values as a measure of oxidation.

The melanoidins prepared by heating glycine and xylose were separated into three molecular weight fractions on Sephadex G-100 and thin layer chromatography. The highest molecular weight fraction (4,500 daltons) had the

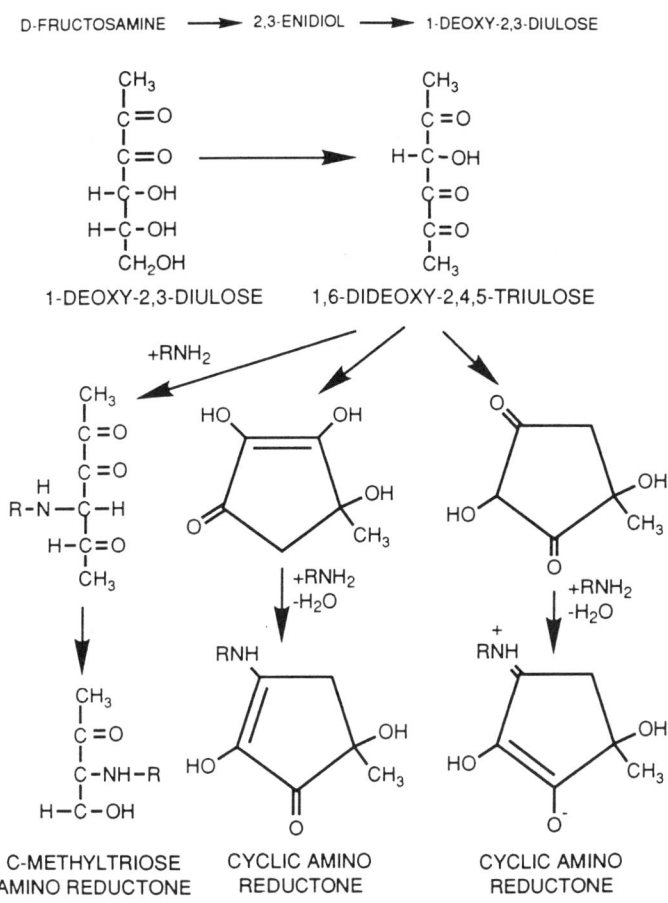

Figure 2. Deamination-dehydration of D-fructosamine to reductones. (Adapted from ref. *45*).

greatest antioxidant activity and was more effective than BHA, propyl gallate or erythorbic acid in protecting linoleic acid from oxidation. A synergistic effect was observed in the combination of the melanoidins with BHA, BHT or natural tocopherol.

Antioxidant Properties of Melanoidins

Two possible mechanisms which can account for the antioxidant properties of melanoidins are: (1) formation of free radical species, and (2) the formation of reductone (enaminol) structures. Namiki (*4*) discussed the possibility of free radical formation in pigments from past studies of the ESR signals in the initial stages of melanoidin formation (*39, 48*). The ESP spectra of glucose-lysine melanoidin contained 33-line hyperfine structures after 4 hr reaction (*39*), which is good evidence for the presence of free radicals.

One proposed structure for melanoidins (*49*) supports the presence of reductone-type forms which are considered to play antioxidant roles in metal chelation of melanoidins as well as having reducing activity. It has been suggested (*50*) that hydroxy pyridone or pyranone-like structures (Figure 3) in the melanoidin polymer may complex iron (Fe^{+++}) and reduce its catalytic activity in oxidative reactions. Some of these structures resemble maltol, which also chelates iron.

This type of structure, relative to antioxidant activity in some melanoidins, appears likely since ozone oxidation of the pigments made with xylose and glycine did not decrease antioxidant activity even though the pigments were discolored and reduced to lower molecular weight fragments (*51*).

Application of MRP as Antioxidants in Food

Maillard browning reactions have considerable utility in food processing as a natural approach to retarding oxidation in foods. Early work on the use of MRP as antioxidants was reported by German workers (*52, 53*), who studied the influence of MRP from glucose-glycine on the oxidation of margarine.

More recently, however, Lingnert (*54, 55*), Lingnert and Eriksson (*1, 56, 57*), and Lingnert et al. (*58*) have been the champions for describing data concerning the antioxidant properties of MRP in processed foods. Their most important contribution was to demonstrate that the basic amino acids (histidine, lysine and arginine) formed the most effective MRP antioxidants with sugars (*1*); they demonstrated that retentate pigments were more antioxidative than the dialysate after dialysis; that these materials contained stable antioxidant free radicals and that several fractions from histidine-glucose reaction mixtures yielded good data for agreement between EPR and antioxidative properties (*58*); and they improvised useful analytical methodologies for studying the antioxidative effect of MRP in processed foods (*1*).

Lingnert (*55*) summarized the use of MRP in reducing oxidation in processed foods other than meat. A number of reports have been published on the antioxidative properties of the Maillard reaction on bakery products, cereals and milk. The addition of glucose or amino acids to dough improved the oxidative stability of cookies during storage; heat treatment improved the stability of cereals and prevented oxidation of milk products.

Important features regarding the improvement of antioxidant properties of these foods by MRP include the normal reaction conditions regulating the Maillard reaction, such as temperature, time, water activity, pH and reagent concentration (55).

Inhibition of Warmed-Over Flavor (WOF) in Cooked Meat by MRP

The storage of precooked meat for a short time results in undesirable "old, stale, rancid and metallic" odor and flavor due to the catalytic oxidation of unsaturated fatty acids by iron. This objectionable flavor is more noticeable when refrigerated cooked meat is reheated and it is said to have warmed-over flavor (WOF). WOF is accompanied by a reduction of meaty flavor as well as an increase in warmed-over aroma, rancid flavor and metallic flavor (Figure 4). Nitrite is the most effective antioxidant used in preventing WOF, and its use results in the sale of over $15 billion worth of processed meat annually. Many highly desirable cured meat products are consumed throughout the world even though cured meat does not have fresh meat flavor, and the use of nitrite has some implications concerning cancer in experimental animals. Overall, however, nitrite is used in meat processing to prevent WOF and as a preservative.

Various aspects of WOF of meat have been discussed in many reviews of this topic (2, 3, 59-61). Many different types of antioxidants have been described as additives useful for preventing WOF (3), but MRP seem to have the greatest potential in preserving the desirable "meaty" flavor of processed meats. Various aspects of the use of MRP prepared by heating glucose and histidine have been described (62).

In more recent studies, Shin-Lee (63) demonstrated a high degree of antioxidant activity in reaction mixtures containing dihydroxyacetone (DHA) and histidine or glyceraldehyde and histidine following reaction for 2 hours at 130°C. Two-tenths molar solutions of the reaction mixtures added to cooked (70°C) ground pork at 10% (v/w) essentially prevented oxidation during storage at 4°C for 3 days. These data were confirmed by measuring volatiles by GLC analyses. Volatiles were barely detectable in samples treated with MRP made with DHA-histidine following storage for 1 day. There was a 40-fold decrease in hexanal content by this treatment compared to non-treated samples. Similar results were obtained by treating with MRP prepared with 0.2M glyceraldehyde-histidine.

Antioxidant Properties of Beef Diffusate

The water-soluble dialyzable low molecular weight ingredients of muscle are ideal precursors for forming meaty flavors and MRP antioxidants. They were first recognized as Maillard reaction meat flavor precursors by Wood and Bender (64) and demonstrated to be dialyzable by Hornstein and Crowe (65).

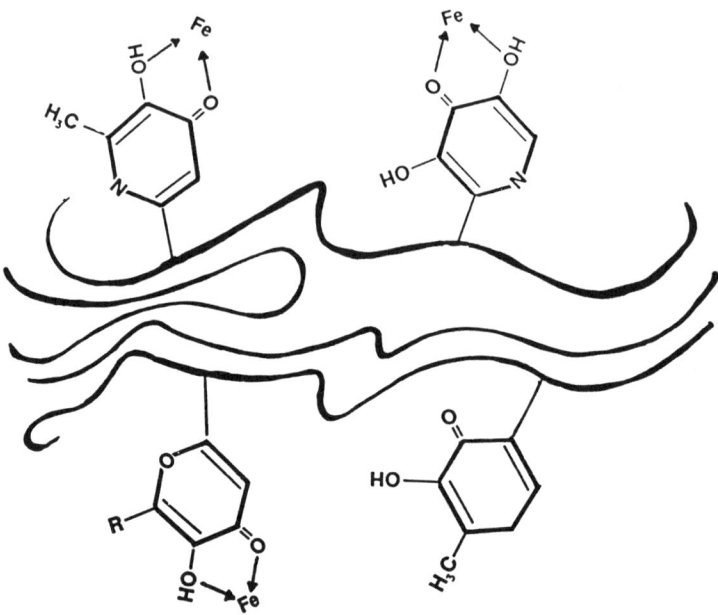

Figure 3. Hydroxypyridone and pyranone structures of melanoidins that act as antioxidants by sequestering iron. (Adapted from ref. *50*).

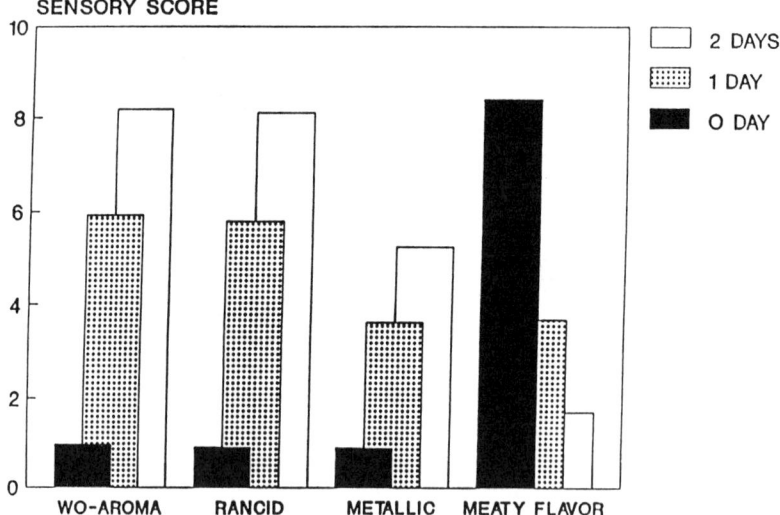

Figure 4. Increase in undesirable aroma and flavor and decrease in meaty flavor during storage of cooked ground beef (70°C) at 4°C for two days. (Adapted from ref. *63*).

Their activities in carbonyl-amine browning were demonstrated by Macy et al. (66, 67) and by Wasserman and Spinelli (68).

Shin-Lee (63) found that diffusate from beef browned extensively when heated at 180°C for 2 hours, and inhibited oxidation (TBA numbers) of cooked beef 95% during storage for 3 days at 4°C (Figure 5). When the diffusate was heated for 2 hours at 180°C and added to beef at 2.0% prior to cooking, it inhibited formation of oxidatively formed volatiles. Hexanal was inhibited 40-fold during storage of cooked beef for 3 days at 4°C. 2,3-Octanedione was inhibited 44-fold by the same treatment.

Um et al. (69) also demonstrated that diffusate prepared from beef muscle and heated at 135°C for 30 minutes and pumped into beef at 0.5-1.0% prior to cooking (70°C) inhibited warmed-over aroma and TBA values during 8 days storage at 4°C (Figure 6).

Synthetic MRP and Meat Flavor Mixture as Meat Flavor Preservative

It has previously been reported (3) that MRP prepared by heating sugars and amino acids inhibit lipid oxidation during refrigerated storage of cooked meat. MRP prepared by heating 200 mM of histidine and glucose and added to pork or beef at 0.72% offered good protection for cooked meat for 3 days, which may be a critical period for retail sales. Several different analytical procedures were discussed for measuring the stability of cooked meats during storage.

In attempts to improve the quality of products prepared with MRP as an antioxidant, Bailey et al. (70) studied the oxidation of beef roasts during storage at 4°C. MRP was prepared by heating a mixture of 0.2M glucose, 0.2M histidine and 0.2mM cystine at 100°C for 4 hours. The reaction mixture was then pumped into lean beef clod roasts at two levels (0.14 and 0.28%), with and without 0.75% NaCl and 0.38% sodium tripolyphosphate, and evaluated for warmed-over aroma, warmed-over flavor, browned meaty flavor, overall acceptability and TBA values during storage for 21 days at 4°C. The data revealed that the MRP at 0.28% injected into roasts with 0.75% NaCl and 0.38% sodium tripolyphosphate resulted in significantly ($P<0.05$) superior sensory scores and TBA values compared with the control or samples treated with MRP alone (Figure 7). Increasing the MRP concentration to 0.56% or 1.10% did not reduce WOF and warmed-over aroma significantly, but did enhance beefy flavor and reduce TBA values compared to treatment with 0.28% MRP.

A major detrimental aspect of lipid oxidation during storage of cooked fresh meat is the rapid decrease in meaty flavor soon after the meat has been cooked and cooled. In an attempt to improve flavor of packaged cooked meat during refrigerated storage at 4°C, Um et al. (69) prepared a synthetic mixture that served both as an antioxidant and as a meat flavor enhancer. The antioxidant was prepared by heating 0.2M glucose and 0.2M histidine at 150°C for 2 hours in 40% glycerol. This mixture was then used to disperse 0.5% cysteine, 0.5% thiamine, 1.0% glycine, 3.0% autolyzed yeast and 2.0% beef fat, which was then heated at 125°C for 1 hour in an oil bath. The resulting mixture was then referred to as synthetic meat flavor (SMF). When added to

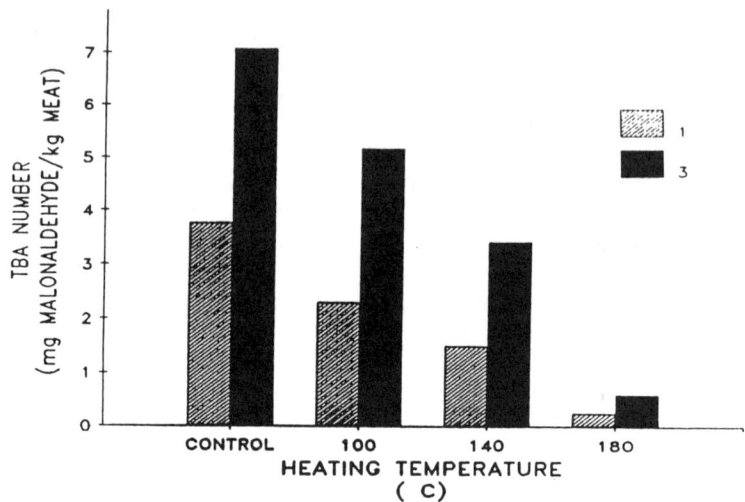

Figure 5. TBA numbers of cooked ground beef treated with 0.5% diffusate (heated at 180°C, 2 hours) and stored for 1 and 3 days at 4°C.

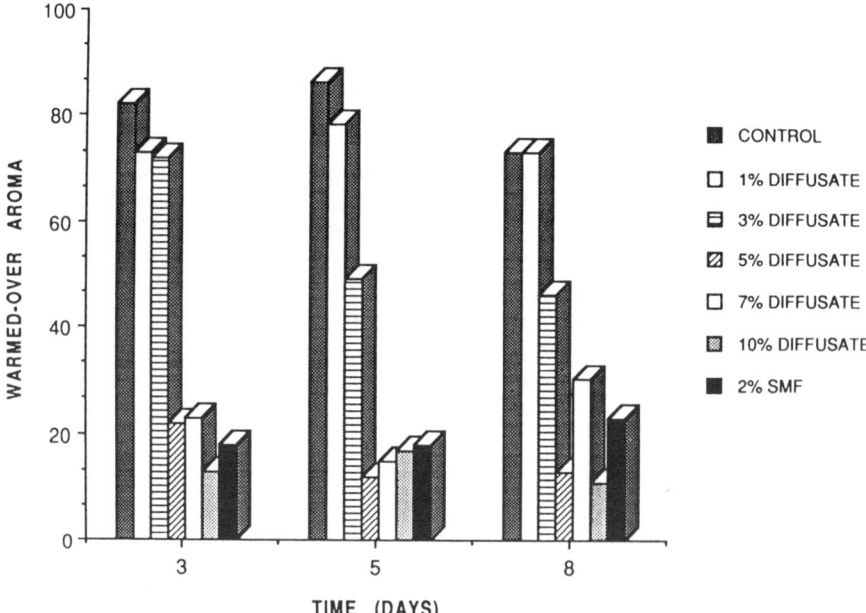

Figure 6. Warmed-over aroma of cooked ground beef treated with 0.1-1.0% diffusate antioxidant prepared by heating 1-10% solutions at 180°C for 2 hours and stored for 8 days at 4°C.

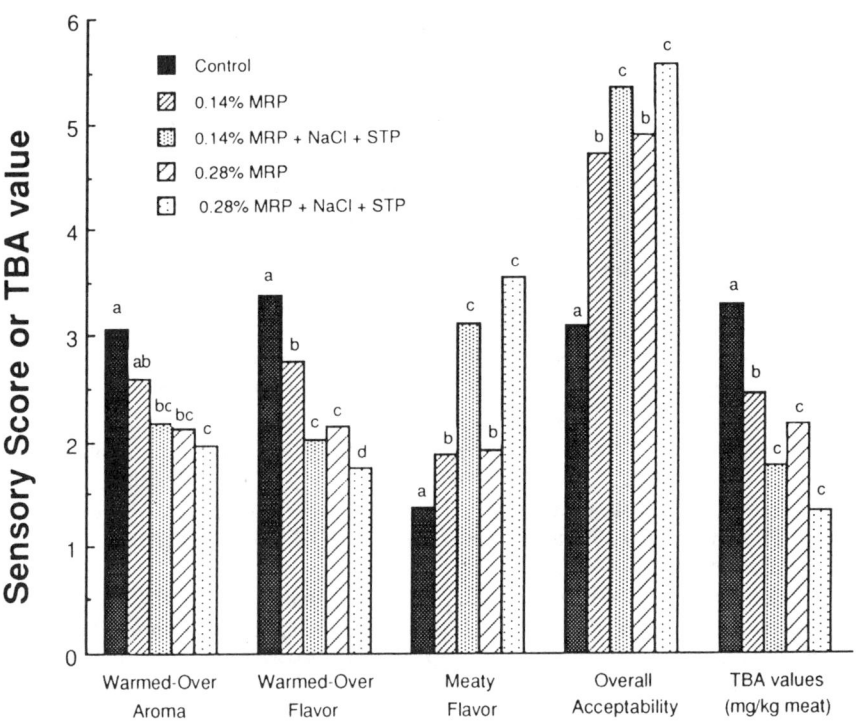

Figure 7. Mean sensory scores and TBA values for precooked beef roasts treated with MRP during storage for 21 days at 4°C. Sensory score 0=none; 10=intense. NaCl concentration was 0.75%; STP was 0.38% sodium tripolyphosphate. MRP was prepared according to Bailey et al. (70). For each attribute, different letters indicate significant differences (P<0.05).

meat prior to cooking, these ingredients were superior to all commercial samples tested where both antioxidant and meaty flavor properties were evaluated during storage of cooked meat.

Figure 8 shows the effect of NaCl (1%), phosphate (0.4% LEM-O-FOS); NaCl (1%), phosphate (0.4% LEM-O-FOS) and SMF (0.05%); and NaCl (1%) and SMF (0.05%) on the warmed-over flavor of pumped and cooked (71°C) roast beef wrapped in aluminum foil and stored for 1 week at 4°C. After 7 days storage, the SMF and the SMF with phosphate were significantly lower in WOF than samples treated with only phosphate or samples receiving only water treatment.

From Figure 9, after 1 day and 1 week storage the meaty flavor of samples treated with SMF or SMF plus LEM-O-FOS was significantly ($P<0.05$) superior to samples heated with only LEM-O-FOS or the control samples.

These results were supported by data for TBA values and volatile carbonyl compounds during storage for 1 week. Samples treated with SMF and LEM-O-FOS and those treated with SMF alone evaluated by both chemical methods were determined to have less lipid oxidation than control (NaCl samples) or samples treated with phosphate and NaCl. NaCl (1%) samples had the highest TBA values (Figure 10), while samples treated with a mixture of NaCl (1%), phosphate (0.4% LEM-O-FOS) and SMF (0.05%) had the lowest TBA values. There were no significant differences in TBA values among samples receiving the phosphate or SMF treatments alone.

These studies were followed by two consumer sensory tests involving 50 panelists each, where the consumer was asked to judge for preference for beef slices from beef roasts treated with 1% NaCl; 1% NaCl plus 0.4% LEM-O-FOS; 1% NaCl plus 0.05% SMF plus 0.4% LEM-O-FOS, or samples containing 1% NaCl plus 0.05% SMF.

Data from one study in Figure 11 show that 78% of the consumers preferred samples treated with NaCl plus SMF; 16% preferred samples treated with NaCl and phosphate (LEM-O-FOS); and 6% preferred samples treated with only NaCl. There was a decided preference for samples treated with SMF in both experiments. Eighty-four percent of a group of twelve meat processors preferred samples treated with SMF; 8% chose samples treated with phosphate; and 8% chose samples treated with only NaCl.

Conclusions

1. Reductones and free radicals are involved in the mechanisms of the antioxidant quality of Maillard reaction products (MRP) in processed food.
2. Heating the water-soluble low molecular weight diffusate from beef at 180°C for 2 hours forms MRP which prevents warmed-over flavor and enhances meaty flavor of cooked meat during storage at 4°C.
3. A reaction mixture consisting of amino acids, glucose, thiamine and yeast heated for 3 hours at 125-150°C protects desirable flavor of roast beef for 1 week at 4°C as judged by sensory and chemical analyses.

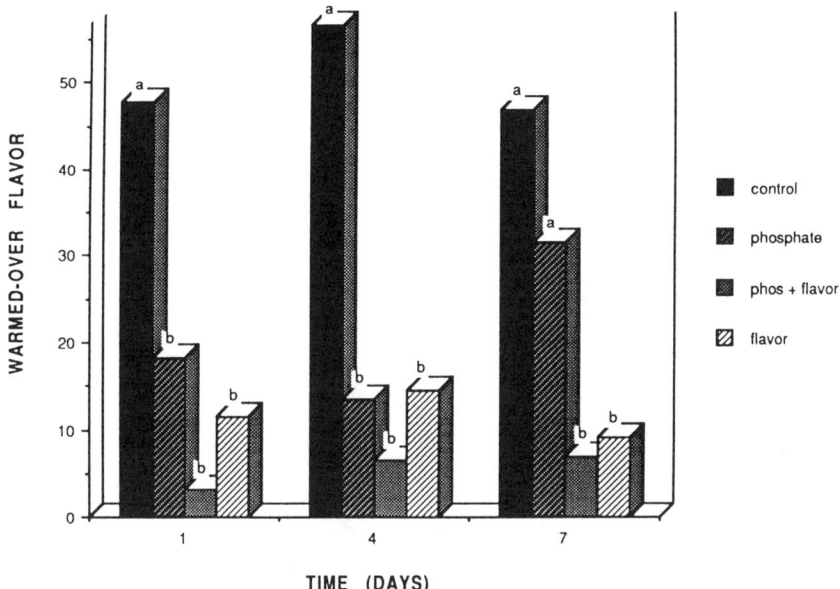

Figure 8. WOF of cooked (71°C) roast beef stored at 4°C for one week. Control = 1% NaCl; phosphate = 0.4% STP in 1% NaCl; flavor = 0.05% synthetic meat flavor in 1% NaCl. For each storage period, different letters indicate significant differences (P < 0.05).

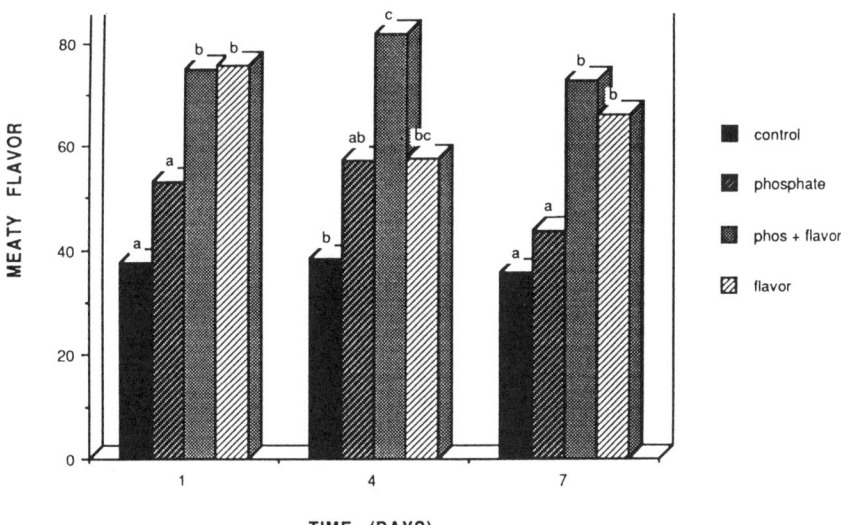

Figure 9. Meaty flavor of cooked (71°C) roast beef stored at 4°C for one week. Control = 1% NaCl; phosphate = 0.4% STP in 1% NaCl; flavor = 0.05% synthetic meat flavor in 1% NaCl. For each storage period, different letters indicate significant differences (P < 0.05).

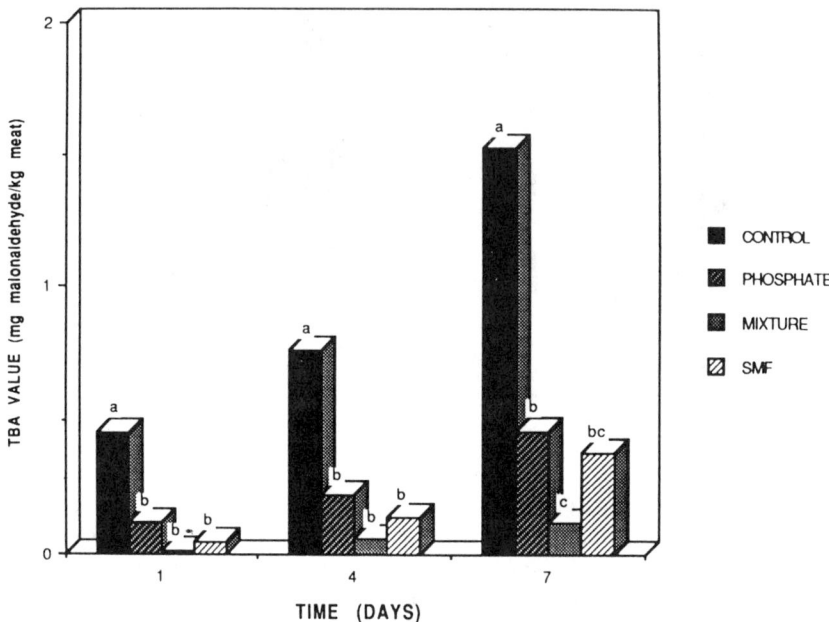

Figure 10. TBA values of cooked (71°C) roast beef stored at 4°C for one week. Control = 1% NaCl; phosphate = 0.4% STP in 1% NaCl; flavor = 0.05% synthetic meat flavor in 1% NaCl. For each storage period, different letters indicate significant differences ($P < 0.05$).

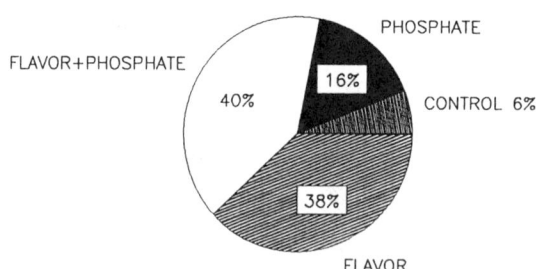

Figure 11. Consumer preference results of cooked (71°C) roast beef stored for 4 days at 4°C. Control = 1% NaCl; phosphate = 0.4% STP in 1% NaCl; flavor = 0.05% synthetic meat flavor in 1% NaCl.

Acknowledgments

The editorial and artistic efforts of Louise Noland are gratefully acknowledged.

Literature Cited

1. Lingnert, H.; Eriksson, C.E. *Prog. Food Nutr.* **1981,** *5,* 453.
2. Bailey, M.E.; Shin-Lee, S.Y.; Dupuy, H.P.; St. Angelo, A.J.; Vercellotti, J.R. In *Warmed-Over Flavor of Meat;* St. Angelo, A.J.; Bailey, M.E., Eds.; Academic Press: Orlando, FL, 1987; pp 237-266.
3. Bailey, M.E. *Food Technol.* **1988,** *42(6),* 123-126.
4. Namiki, M. *Adv. Food Res.* **1988,** *32,* 115-184.
5. Hodge, J.E. *J. Agric. Food Chem.* **1953,** *1,* 928-943.
6. Hodge, J.E. In *Chemistry and Physiology of Flavors;* Schultz, H.W.; Day, E.A.; Libbey, L.M., Eds.; AVI Publishing: Westport, CT, 1967; pp 465-491.
7. Anet, E.F.L.J.; Reynolds, T.M. *Austr. J. Chem.* **1957,** *10,* 182-192.
8. Reynolds, T.M. *Adv. Food Res.* **1963,** *12,* 1-52.
9. Reynolds, T.M. *Adv. Food Res.* **1965,** *14,* 167-283.
10. Reynolds, T.M. In *Symposium on Foods: Carbohydrates and Their Roles;* Schultz, H.W.; Cain, R.F.; Wrolstad, R.W., Eds.; AVI Publishing: Westport, CT, 1969; pp 219-252.
11. Baltes, W. *Ernahrungsumschau* **1973,** *20,* 35-38.
12. Mabrouk, A.F. In *Food Taste Chemistry;* Boudreau, J.C., Ed.; ACS Symposium Series 115; American Chemical Society: Washington, D.C., 1975; pp 205-247.
13. Hodge, J.E.; Osman, E.M. In *Principles of Food Science, Part I. Food Chemistry;* Fennema, O., Ed.; Marcel Dekker: New York, NY, 1976; pp 41-138.
14. Hurrell, R.F.; Carpenter, K.J. In *Physical, Chemical and Biological Changes in Food Caused by Thermal Processing;* Hoyem, T.; Kuale, O., Eds.; Applied Science: London, 1977; pp 168-184.
15. Paulsen, H.; Pflughaupt, K-W. In *The Carbohydrates, Chemistry and Biochemistry, 2nd Edition;* Pigman, W.; Horton, D.; Wander, J.D., Eds.; Academic Press: New York, NY, 1980, Vol. 1B; pp 881-927.
16. Mauron, J. *Prog. Food and Nutr. Sci.;* **1981,** *5(1-6),* 5-36.
17. Feeney, R.E.; Whitaker, J.R. In *Food Protein Deterioration. Mechanisms and Functionality;* Cherry, J.P., Ed.; ACS Symposium Series 206; American Chemical Society: Washington, DC, 1982; pp 201-229.
18. Hurrell, R.F. In *Food Flavours. Part A. Introduction;* Morton, I.D.; MacLeod, A.J., Eds.; Elsevier: Amsterdam, 1982; pp 399-437.
19. Feather, M.S. In *Chemical Changes in Food During Processing;* Richardson, T.; Finley, J.W., Eds.; AVI Publishing: Westport, CT, 1985; pp 289-303.
20. Nursten, H.E. In *Concentration and Drying of Foods;* MacCarthy, D., Ed.; Elsevier: London, 1985; pp 53-68.
21. Danehy, J.P. *Adv. Food Res.* **1986,** *30,* 77-138.

22. Heath, H.B.; Reineccius, G. *Flavor Chemistry and Technology;* AVI Publishing: Westport, CT, 1986.
23. Yayoujan, V.; Sporns, P. *Food Chem.* **1987**, *26*, 283-305.
24. Ledl, F.; Beck, J.; Sengl, M.; Osiander, H.; Estendorfer, S.; Severin, T.; Huber, B. In *The Maillard Reaction in Aging, Diabetes and Nutrition;* Baynes, J.W.; Monnier, V.M., Eds.; Alan R. Liss, Inc.: New York, NY, 1989; pp 23-42.
25. O'Brien, J.; Morrissey, P.A. *CRC Crit. Rev. Food Sci. Nutr.* **1989**, *28*, 210-248.
26. Eskin, N.A.M. *Biochemistry of Foods;* Academic Press: San Diego, CA, 1990.
27. Ledl, F. In *The Maillard Reaction in Food Processing, Human Nutrition and Physiology;* Finot, P.A.; Aeschbacher, H.U.; Hurrell, R.F.; Liardon, R., Eds.; Birkhäuser Verlag: Basel, 1990; pp 19-42.
28. Eriksson, C. *Maillard Reactions in Food;* Progress in Food and Nutrition Science; Oxford: 1981; Vol. 1-6.
29. Waller, G.R.; Feather, M.S. *The Maillard Reaction in Foods and Nutrition;* ACS Symposium Series 215; American Chemical Society: Washington, DC, 1983.
30. Fujimaki, M.; Namiki, M.; Kato, N. *Amino-Carbonyl Reactions in Food and Biological Systems;* Developments in Food Science 13; Elsevier: Amsterdam; 1986.
31. Finot, P.A.; Aeschacher, H.U.; Hurrell, R.F.; Liardon, R. *The Maillard Reaction in Food Processing, Human Nutrition and Physiology;* Birkhäuser Verlag: Basel, 1990.
32. Anet, E.F. *J. Adv. Carbohyd. Chem.* **1964**, *19*, 181-218.
33. Feather, M.S. *Prog. Food Nutr. Sci.* **1981**, *5*, 37-45.
34. Ledl, F. *Dev. Food Sci.* **1986**, *13*, 569-573.
35. Baltes, W.; Kunert-Kirchhoff, J.; Reese, G. In *Thermal Generation of Aromas;* Parliment, T.H.; McGarrin, R.J.; Ho, C-T., Eds.; American Chemical Society: Washington, DC, 1989; pp 143-155.
36. Feather, M.S.; Russell, K.R. *J. Org. Chem.* **1969**, *34*, 2650-2652.
37. Feather, M.S. *Tetrahedron Lett.* **1970**, 4143-4147.
38. Feather, M.S.; Harris, J.F. *Adv. Carbohyd. Chem.* **1973**, *63*, 161-224.
39. Namiki, M.; Hayashi, T. In *The Maillard Reaction in Foods and Nutrition;* Waller, G.R.; Feather, M.S., Eds.; ACS Symp. Series 215: Washington, DC, 1983; pp 21-46.
40. Hodge, J.E.; Rist, C.E. *J. Am. Chem. Sco.* **1953**, *75*, 316.
41. Hodge, J.E.; Evans, C.D. *U.S. Patent 2,806,794* **1957**.
42. Evans, C.D.; Moser, H.A.; Cooney, P.M.; Hodge, J.E. *J. Am. Oil Chem. Soc.* **1958**, *35*, 84.
43. Mills, F.D.; Hodge, J.E.; Parks, L.W. *J. Org. Chem.* **1981**, *46*, 3597.
44. Ledl, F.; Fritsch, G.; Hubl, J.; Pachmayr, O.; Severin, T. *Dev. Food Sci.* **1986**, *13*, 173-182.
45. Davidek, J.; Velisek, J.; Pokorny, J. In *Chemical Changes During Food Processing;* Developments in Food Science 21; Elsevier: Amsterdam, 1990; pp 58-168.

46. Ledl, F.; Severin, T. *Z. Lebensm. Unters. Forsch.* **1979**, *169*, 173-178.
47. Yamaguchi, N.; Koyama, Y.; Fujimaki, M. *Prog. Food Sci. Nutr.* **1981**, *5*, 429-451.
48. Wu, C.H.; Russel, G.F.; Powrie, W.D. *Dev. Food Sci.* **1986**, *13*, 135-145.
49. Kato, H; Tsuchidi; H. *Prog. Food Nutr. Sci.* **1981**, *5*, 147-156.
50. Hashiba, H. *Dev. Food Sci.* **1986**, *13*, 155-164.
51. Yamaguchi, N. *Dev. Food Sci.* **1986**, *13*, 291-299.
52. Franke, C.; Iwainsky, H. *Dtsch. Lebensm. Rundsch.* **1954**, *50*, 251-254.
53. Iwainsky, H.; Franke, C. *Dtsch. Lebensm. Rundsch.* **1956**, *52*, 129-133.
54. Lingnert, H. *J. Food Proc. Preserv.* **1980**, *4*, 219-233.
55. Lingnert, H. In *The Maillard Reaction in Food Processing, Human Nutrition and Physiology;* Finot, P.A.; Aeschbacher, H.U.; Hurrell, R.F.; Liardon, R., Eds.; Birkhäuser Verlag: Basel, 1990, pp 171-290.
56. Lingnert, H.; Eriksson, C.E. *J. Food Proc. Preserv.* **1980**, *4*, 161-172.
57. Lingnert, H.; Eriksson, C.E. *J. Food Proc. Preserv.* **1980**, *4*, 173-181.
58. Lingnert, H.; Eriksson, C.E.; Waller, G.R. In *The Maillard Reaction in Foods and Nutrition;* Waller, G.R.; Feather, M.S., Eds.; ACS Symposium Series 215; American Chemical Society: Washington, DC, 1983; pp 335-345.
59. Watts, B.M. In *Proceedings Flavor Chemistry Symposium;* Campbell Soup Co.: Camden, NJ, 1961; pp 83-96.
60. Pearson, A.M.; Love, J.D.; Shorland, F.B. *Adv. Food Res.* **1977**, *34*, 1-74.
61. Pearson, A.M.; Gray, J.I. In *The Maillard Reaction in Foods and Nutrition;* Waller, G.R.; Feather, M.S., Eds.; ACS Symposium Series 215; Washington, DC, 1983; pp 287-300.
62. Bailey, M.E.; Shin-Lee, S.Y.; Dupuy, H.P.; St. Angelo, A.J.; Vercellotti, J.R. In *Warmed-Over Flavor of Meat;* St. Angelo, A.J.; Bailey, M.E., Eds.; Academic Press: Orlando, FL, 1987; pp 237-266.
63. Shin-Lee, S.Y. *Warmed-Over Flavor and its Prevention by Maillard Reaction Products;* PhD thesis, University of Missouri, Columbia, 1988.
64. Wood, T.; Bender, A.E. *Biochem. J.* **1957**, *67*, 366-373.
65. Hornstein, I.; Crowe, P.F. *J. Agric. Food Chem.* **1960**, *8*, 494-498.
66. Macy, R.L. Jr.; Naumann, H.D.; Bailey, M.E. *J. Food Sci.* **1964**, *29* 131-141.
67. Macy, R.L. Jr.; Naumann, H.D.; Bailey, M.E. *J. Food Sci.* **1964**, *29*, 142-148.
68. Wasserman, A.E.; Spinelli, A.M. *J. Food Sci.* **1970**, *35*, 328-332.
69. Um, K.W.; Hooppaw, C.; Bailey, M.E., University of Missouri-Columbia, unpublished data.
70. Bailey, M.E.; Clarke, A.D.; Hedrick, H.B.; Shin-Lee, S.Y.; Choate, G.K. In *Proceedings 37th International Congress of Meat Science and Technology;* Kulmbach, Germany, Sept. 1-6, 1991, Vol II(5:4); pp 684-688.

RECEIVED February 19, 1992

Chapter 9

Chemical and Sensory Evaluation of Flavor in Untreated and Antioxidant-Treated Meat

Allen J. St. Angelo, Arthur M. Spanier, and Karen L. Bett

Agricultural Research Service, U.S. Department of Agriculture, Southern Regional Research Center, 1100 Robert E. Lee Boulevard, New Orleans, LA 70124

> Meat flavor deterioration primarily develops from the oxidation of polyunsaturated fatty acids catalyzed by metal ions and other free radical generating systems. As off-flavors intensify with storage, the desirable meaty flavor notes decrease. Experiments were designed to examine mechanisms that prevent these flavor changes. Ground beef patties, freshly cooked and cooked/stored for up to 5 days at 4°C, were examined by direct gas chromatography and chemical and sensory methods of analysis. Results demonstrated that combined vacuum packaging and use of chelators and antioxidants act synergistically to prevent lipid oxidation and to preserve desirable meat flavor.

In reports on the mechanism responsible for warmed-over flavor (WOF) in cooked meat (also referred to as meat flavor deterioration, or MFD), the generally accepted theory is that MFD is caused by the oxidation of membrane phospholipids, catalyzed by some ionic form of iron (1, 2). When lipids oxidize, as in the case of MFD, a mixture of aldehydes, ketones and alcohols are produced. Their concentrations increase with storage. Appreciable amounts of hexanal, one of the primary oxidation products originating from linoleic acid (3), can be observed by GC within hours after storage (4). Many other oxidation products, such as propanal, butanal, pentanal, heptanal, 2,3-octanedione, and nonanal, were readily measured upon analyses of MFD roasts (4, 5).

In addition to the increase in lipid oxidation products, which generate off-flavor notes such as painty (PTY) and cardboard (CBD), a loss of desirable beefy flavor notes, such as cooked beef/brothy (CBB) occurs (6-8). Sulfur-containing compounds that also contribute to the desirable meaty flavor will also

This chapter not subject to U.S. copyright
Published 1992 American Chemical Society

change (9). Thus, MFD appears to result from a combined chemical process whereby lipid oxidation products are evolved and desirable meaty flavor compounds not of lipid origin are diminished. According to Liu et al. (10), lipid oxidation products are found in concentrations of ppm, whereas the desirable meaty compounds are present at ppb levels. Volatile heteroatomic meat flavor principles were also confirmed at ppb levels by Vercellotti et al. (11).

Several mechanisms cause food flavors to deteriorate. Owing to the abundance of lipid oxidation products formed during the development of MFD, lipid oxidation is the most frequently cited mechanism. Over the past three decades, chelators have been used successfully to retard or inhibit iron-catalyzed lipid oxidation. However, several of these chelators do not prevent the loss of the desirable beefy flavor, CBB. On the other hand, antioxidants that function as free radical scavengers were not only successful in inhibiting WOF formation, but also appeared to retain CBB at a higher level than the chelators (12). Obviously, free radical mechanisms not only effect lipid oxidation, but are also involved in the oxidation of compounds other than lipids, such as proteins (13).

Other factors that contribute to MFD are final cooking temperature, length of refrigerated storage, and the availability of oxygen (7, 14, 15). Using combined instrumental, chemical and descriptive sensory methodologies, the MFD process was shown to be inhibited or retarded by the use of antioxidants (9) or the exclusion of oxygen (15). More recently, results from studies on combined use of chelators and free radical scavengers as antioxidants have shown a significant amount of flavor loss when the antioxidant-treated cooked meat was maintained under vacuum (16). Based on the results from the dose-response combinations obtained in those initial studies, the objectives of this paper were to determine the correlation between chemical and flavor attributes and to investigate the synergism of antioxidants and chelators in beef stored both in the presence and absence of oxygen.

Experimental Procedures

Sample Preparation. Samples consisted of USDA-Choice, top round steaks (*semimembranosus* muscle). The meat was trimmed of separable fat and then ground by two passes through each of two grinding discs with holes of 1.0 and 0.75cm diameter. Fat content ranged from 4-5%. Glass petri dishes (100mm x 20mm) were used as molds to form 85 grams of beef into patties of uniform shape. A 425g portion of the ground meat was used to make five 85 gram patties as standards. These reference standards, placed in covered petri dishes, were stored in a freezer immediately after preparation. The remaining ground meat was divided into 850g portions to make ten 85 gram patties per replicate per treatment. Each portion was mixed with either 10 ml of the additive in water or water only since water served as the additive carrier. The additives used were propyl gallate (PG) and the tetrasodium salt of ethylenediaminetetraacetic acid (EDTA). The additives were presented at 50 ppm for PG and 100 ppm for EDTA, based on the wet weight of raw ground meat. Experimental samples were allowed to stand (marinate) overnight, ca. 18 hrs, at 4°C. All patties were cooked on a Farberware grill (Model 455ND) for 7 min/side. The internal temperature of the patties was 65 ± 1°C. After the "marinated" experimental samples were cooked, they were placed in covered petri dishes, which were then placed in vacuum desiccator jars and stored in a refrigerator for a period of 3 days. Simulated vacuum packaging was accomplished by purging the desiccator jars twice with nitrogen. Final vacuum was less than 4 mm Hg. Samples stored without vacuum were placed in vacuum jars that were left open to the atmosphere. On the morning of the experiment, the cooked/refrigerated samples were rewarmed for 25 minutes at 121°C (250°F). Frozen samples were thawed and then cooked on the grill. All samples were placed in warming trays (Hobart, Model CF021) at 52°C until evaluated by the panel. Samples to be analyzed by chemical and instrumental methods were placed in the freezer until assayed.

The experiment was statistically designed as a balanced incomplete block. The blind standard was given the notation "BNO" to represent the "standard" (B) with "no vacuum" (N) and "no chelator or antioxidant" (O). The experimental groups or "treatments" were given the following notation: (1) "ENO" represented the samples in which the flavor had been allowed to deteriorate for 3 days with "no vacuum"; this sample represented the MFD sample. (2) "EVO" represented the MFD sample maintained under "vacuum" (V). (3) "ENA, ENB, or ENX" represented the MFD samples containing the "additives" (where A = Propyl Gallate, B = EDTA, and X = both PG & EDTA). (4) "EVA, EVB, or EVX" represented the MFD samples maintained with both "vacuum and additives".

Samples that were used in the 5-day correlation (chemical and flavor attributes) study, were prepared as described above, except that cooked patties were stored in covered petri dishes at 4°C until assayed. They contained no additive other than water and they were not stored under vacuum.

Sensory Analysis. Sensory profiling of the patties was performed by the method described by Meilgaard et al. (*17*). Evaluations were accomplished by a trained panel of 12 using the spectrum universal intensity scale, 0-15, as described by Meilgaard et al. (*17*). Sensory attributes included the following descriptors as originally described by Johnsen and Civille (*6*) and modified by Love (*8*): "salty", "cooked beef/brothy", "painty", "serum/raw," "browned/caramel", "cooked liver", "cardboard", "sour", "sweet", and "bitter".

Chemical Analysis. Chemical attributes such as thiobarbituric acid reactive substances (TBARS) were measured by the distillation method of Tarladgis (*18*), and were reported as mg malonaldehyde per kg sample. A standard curve was prepared with 1,1,3,3-tetraethoxypropane.

Instrumental Methodology. Instrumental attributes were determined by obtaining the volatile profile of meat samples utilizing the direct gas chromatographic (GC) method described by Dupuy et al. (*4*). This method separates the volatiles by GC after applying the sample into an external closed inlet device interfaced with a packed Tenax GC/PMPE column.

Statistical Analysis. Results were statistically analyzed using SAS (19) for both analysis of variance (ANOVA) and principal components analysis (PCA). PCA is a multivariate statistical method that is used when an empirical summary of patterns of inter-correlations among variables is desired. The purpose of PCA is to reduce a large number of variables to as small a number of factors as possible. In the present work, the variables were the experimental treatments and the sensory, chemical and instrumental attributes.

Results

Correlation of Chemical and Flavor Attributes. Cooked ground meat that was stored for 5 days was examined for hexanal content by direct GC and for desirable and off-flavor attributes by a trained sensory panel. Hexanal production and the off-flavor aromatic descriptor called painty (PTY) increased with storage (Figure 1). As noted, there was a direct correlation between the two. Other off-flavors that increased with MFD were cardboard (CBD), bitter (BTR) and sour (SOU) (data not shown). As MFD progressed during storage, the desirable flavor notes, such as cooked beef/brothy (CBB), decreased. To date, no compound has been directly correlated to off-flavors other than PTY or to the desirable flavors as hexanal has been to PTY.

Synergism of Antioxidants and Chelators. Concomitantly with the production of lipid oxidation products, proteins and peptides were shown to decompose via a free-radical mechanism (20). Consequently, the two compounds chosen to serve as antioxidants in these experiments were EDTA, which functions primarily as a strong chelator, and PG, which can chelate iron, but functions primarily as a free radical scavenger. Hopefully, EDTA, being the stronger chelator, would bind the metal and thereby allow the PG to react with the free radicals. The action of these two compounds coupled with oxygen exclusion (via vacuum) would create an atmosphere that would inhibit or retard MFD.

Chemical and Instrumental Attributes. The effect of vacuum and additives on chemical (TBARS) and instrumental (hexanal, 2,3-octanedione and nonanal) attributes of beef is shown in Figures 2 and 3. All bars were presented as the mean (± s.e.m.) result of data obtained from ANOVA with appropriate linear contrasts tested. Each graph was organized into 3 clusters with each bar representing a different treatment. The first cluster represents the blind-standard and the 3 day MFD beef pattie. The second cluster represents the effect of the different additives, i.e., A = propyl gallate, B = EDTA, X = the two additives combined. The third cluster was similar to the second except that all samples were maintained under vacuum so that the bars from left to right represented "vacuum only", "vacuum + PG", "vacuum + EDTA" and "vacuum + both PG and EDTA".

In Figure 2, the upper graph shows the relative change in the level of TBARS whereas the bottom graph and those in Figure 3 describe the GC area count of three selected products of lipid oxidation, i. e., hexanal, 2,3-octanedione, and nonanal. The GC-volatiles and the TBARS data show

144 LIPID OXIDATION IN FOOD

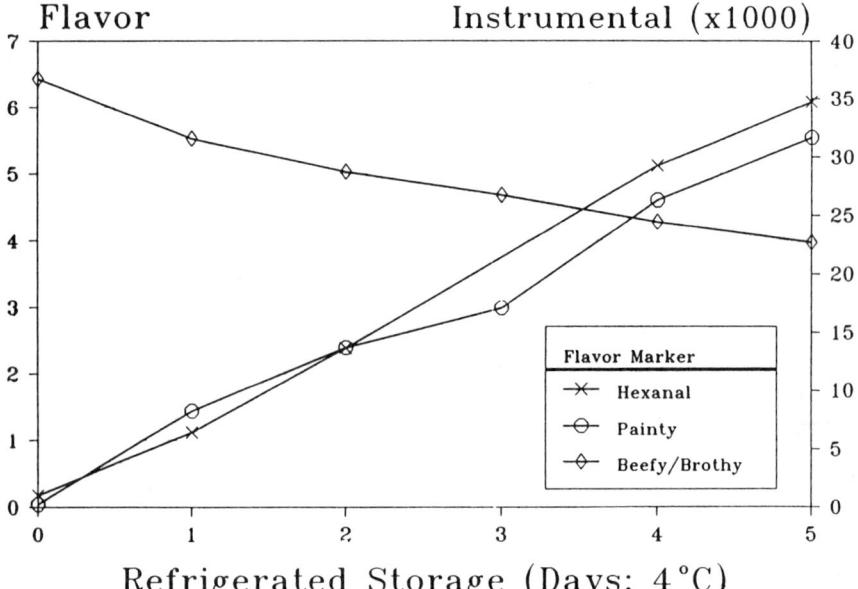

Figure 1. Cooked beef flavor changes during storage.

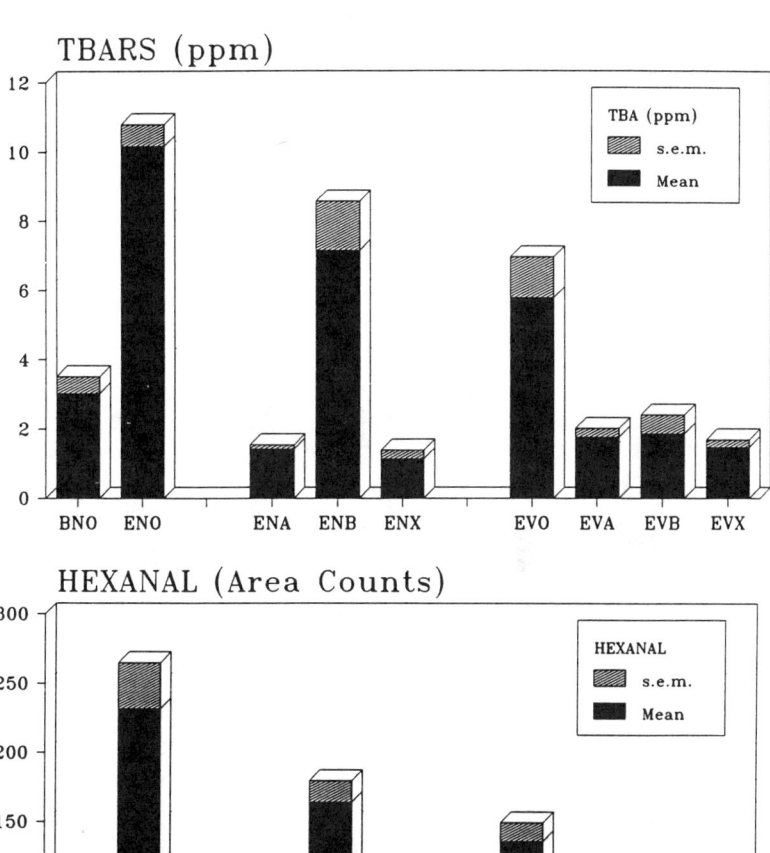

Figure 2. Effect of vacuum and additives on beef as measured by chemical (TBARS) and instrumental (hexanal) methods. Treatment codes are BNO, blind standard, no storage, no vacuum (N), and no additives (O); ENO, experimental MFD sample with no vacuum and no additives; ENA, ENB, and ENX, experimental samples with no vacuum and either 50 ppm PG (A), 100 ppm EDTA (B), or both (X) added; EVO, experimental samples with vacuum and no additives; EVA, EVB, and EVX, experimental samples with vacuum and either A, B, or X added.

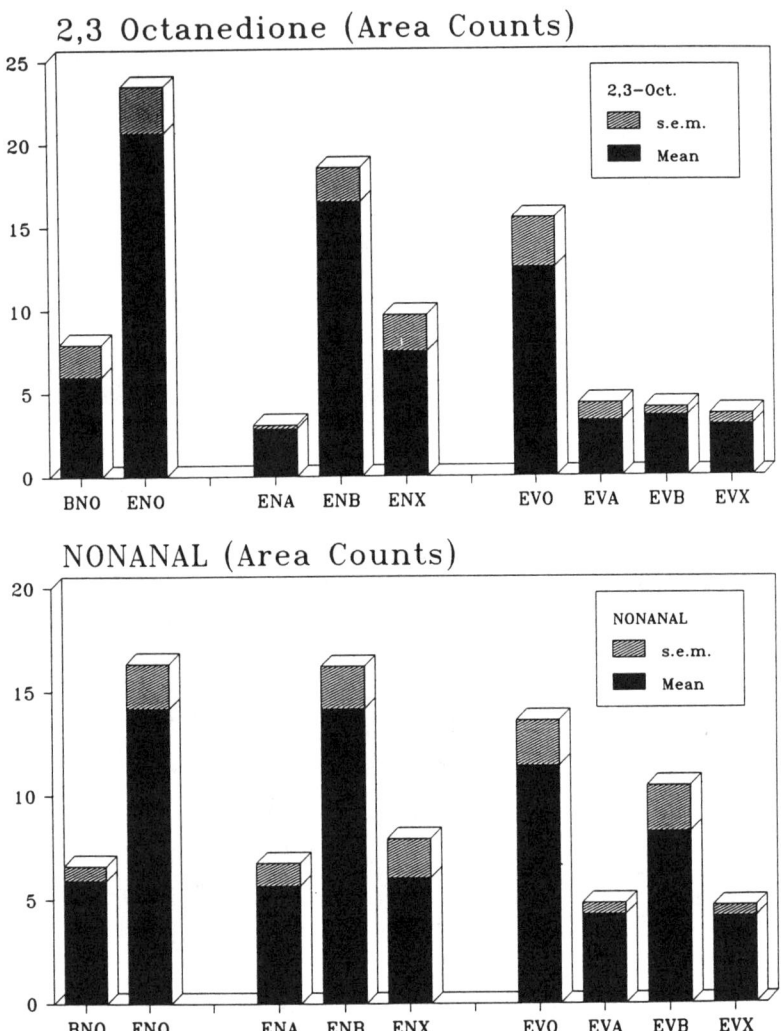

Figure 3. Effect of vacuum and additives on beef as measured by instrumental methods. Treatment codes are same as described in Figure 2.

response to each treatment that is similar to one another. For example, in each instance the 3-day MFD sample (ENO) showed a significant increase in the intensity of the attribute measured when compared to the blind standard (BNO). Vacuum alone (EVO) showed a lowering effect of the intensity of each attribute when compared to the MFD samples (ENO). In general, each additive had its own individual response, with the antioxidant, PG, being more efficient in decreasing the intensity of the chemical and instrumental attributes (compare treatments ENA to ENB). When vacuum was not used, combining PG with EDTA (ENX) reduced hexanal and 2,3-octanedione production, but not as much as PG alone (ENA). The decrease in TBARS and nonanal production were comparable. When EDTA was combined with vacuum (EVB), there was a significant reduction of lipid oxidation when compared to that without vacuum (ENB). When PG was combined with vacuum (EVA) or with EDTA plus vacuum (EVX), lipid oxidation was reduced (lower than the blind standard that had been stored frozen, BNO), and the difference was statistically significant when compared to ENA. Combining additives with vacuum, resulted in a synergistic response, which suggested that more than one type of free radical chemistry may be occurring during MFD. Furthermore, these data showed that MFD definitely involves more complex chemistry than mere lipid oxidation.

Off-Flavor Attributes. The effect of vacuum and additives on off-flavor attributes is shown in Figures 4 and 5. All four graphs illustrate the intensities of off-flavor descriptors, i.e., painty, cardboard, bitter and sour, representing two aromas, Figure 4, and two tastes, Figure 5, respectively. All bars are presented as the mean results of data from treatments as described in Figures 2 and 3.

The 3-day MFD sample (ENO) showed a significant increase in each of the four off-flavor attributes when compared to the blind standard (BNO). The response of all four descriptors to the different additives is seen in the second cluster. These data indicate that all four descriptors responded positively to the 50 ppm concentration of the antioxidant, PG (samples ENA). On the other hand, only two of the descriptors, i.e. painty and sour, showed a response to the 100 ppm of the chelator, EDTA (ENB). However, similarly shown by the chemical data in Figures 2 and 3, combining EDTA with vacuum (EVB) caused a synergistic effect for all sensory notes (compare EVB to ENB and EVO). There also was a synergistic sensory response to a mixture of PG with the chelator (ENX). In this case, the perception of the flavor intensities approached the intensity levels of the standard patties (BNO). Again, these data showed the presence of more than one mechanism involved in MFD. Samples treated with vacuum alone (EVO) had the effect of reducing the intensity of all four off-flavor descriptors of the cooked patties at levels lower than without vacuum (ENO), but higher than the standard patties (BNO). Combining the vacuum treatment with the additives indicated a synergistic response to the additives as seen by data for all four descriptors in the second cluster.

Desirable Flavor Attributes. The effect of vacuum on the desirable on-flavor attributes of beef patties is shown in Figure 6. This figure is similar to Figures 4 and 5 with the exception that it pertains to the desirable flavor

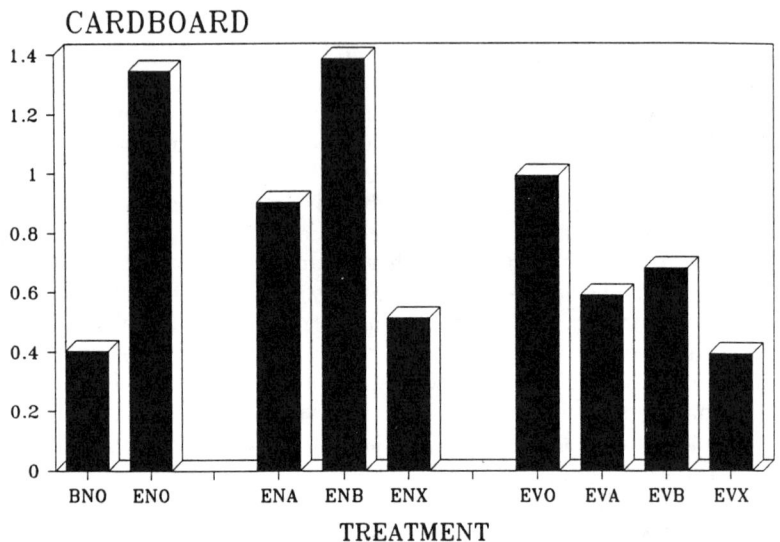
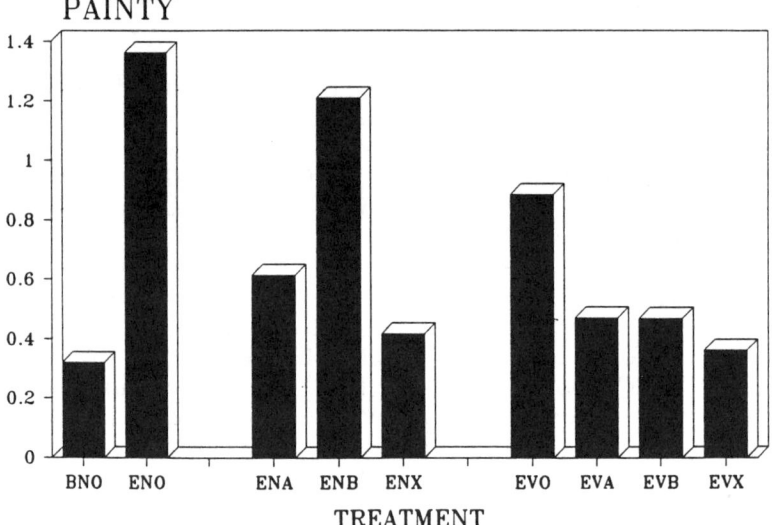

Figure 4. Effect of vacuum and additives on beef as measured by off-flavor aromas. Treatment codes are same as described in Figure 2.

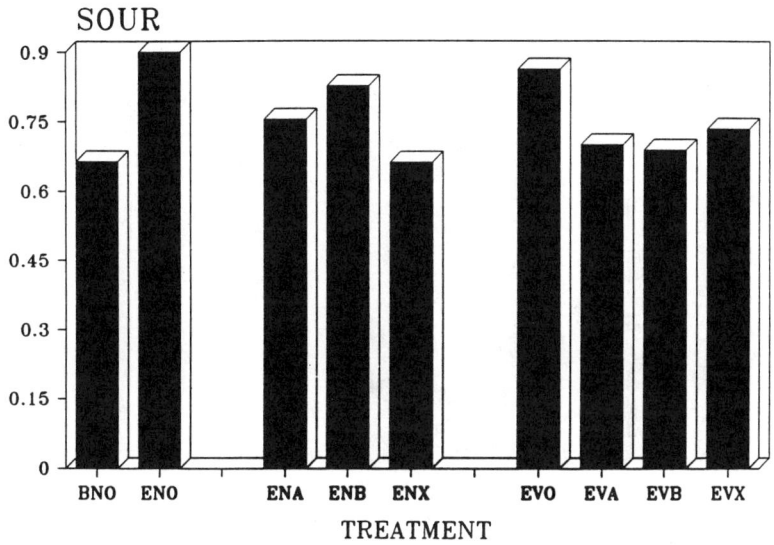
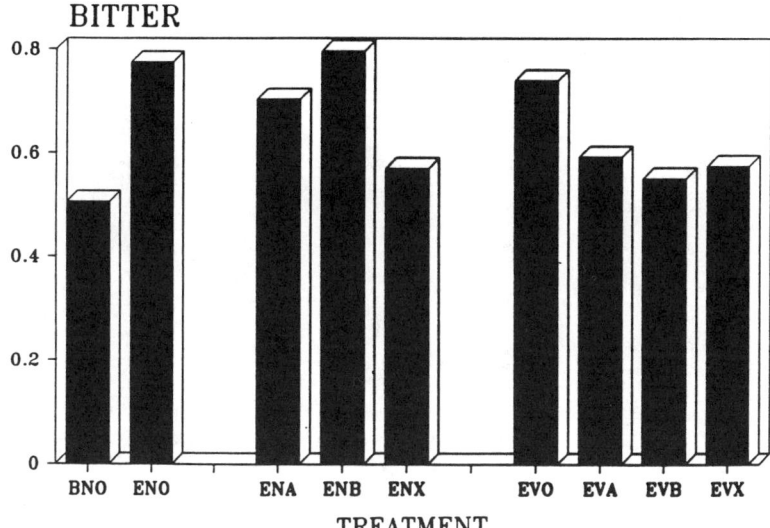

Figure 5. Effect of vacuum and additives on beef as measured by off-flavor tastes. Treatment codes are same as described in Figure 2.

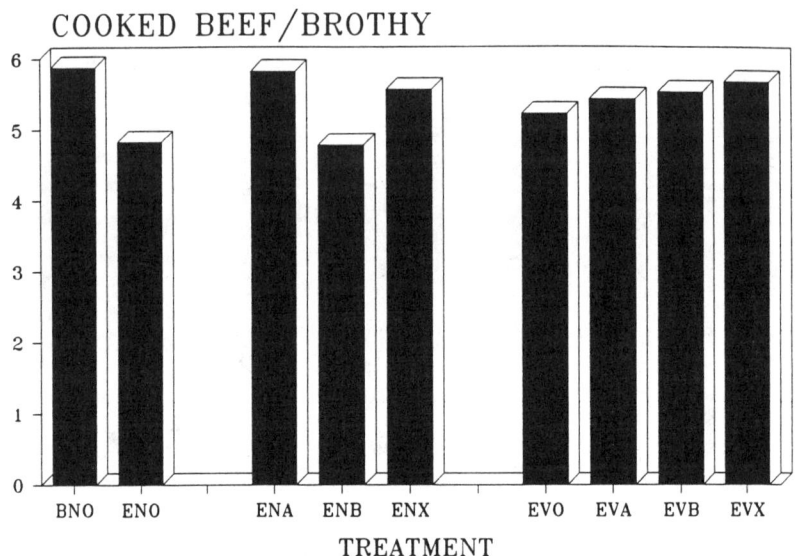
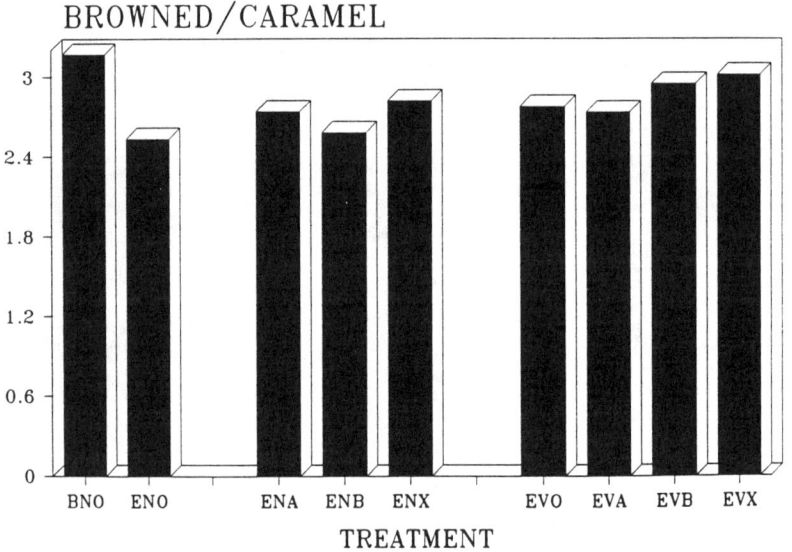

Figure 6. Effect of vacuum and additives on beef as measured by on-flavor aromas. Treatment codes are same as described in Figure 2.

attributes, cooked beef/brothy and browned/caramel (BRC). As shown by the 3-day MFD sample (ENO), the two desirable attributes decreased compared to those of the blind standard. The sample treated with no vacuum but with PG (ENA) retained the high intensity of CBB during storage for 3 days. EDTA alone did not have an effect during storage, as seen from sample ENB. Again, samples treated with EDTA plus vacuum (EVB) showed a synergistic effect. Based on the retention of high intensity values of the desirable notes, CBB and BRC, and coupled with the rentention of low intensity values for the off-flavor notes (CBD, PTY), the most successful treatment was observed in patties treated with vacuum plus both compounds, PG and EDTA, as seen in sample EVX. This point was also confirmed in Table I, which expressed significant differences between means of EVX and ENO for all flavor descriptors. CBD, PTY, SOU and BTR were significantly lower in EVX than in ENO. CBB and BRC were significantly higher in EVX than ENO. ENX and EVB treated samples had significantly different means from ENO for all 6 descriptors also. Although CBB and BRC were not as intense in the ENX and EVB samples as in the EVX samples, PTY and CBD were higher in intensity for the ENX and EVB samples than for the EVX samples.

Statistical Evaluation by Pearson Correlation Coefficients

Sensory Data. Statistical analysis of the raw sample sensory data, including 290 - 323 separate observations, revealed several significant correlations among the sensory flavor descriptors examined (Table II). Values that were greater than 0.5 were considered "very" highly correlated, while values greater than 0.3 but less than 0.5 were considered "highly" correlated. Values greater than 0.25 but less than 0.3 were correlated to a lesser degree, but were still highly significant. In all cases, the level of significance is 0.0001, which indicated statistically that these correlations were highly significant. For example, in the column listed as CBB, painty, cardboard, sour and bitter were negatively correlated with cooked beef/brothy, whereas browned/caramel and sweet had a strong positive correlation with cooked beef/brothy. In all cases, those descriptors that were negatively correlated to one descriptor, were positively correlated with each other. For example, the off-flavor descriptors, cardboard and painty, were negatively correlated to the desirable-flavor descriptor, cooked beef/brothy. Yet, these descriptors yielded the highest positive statistical correlation between themselves with a correlation of 0.80.

Chemical and Instrumental Data. Statistical analysis of chemical and instrumental attributes, including 56 separate observations, also revealed several significant correlations among the attributes examined (Table III). The intensity of these attributes increased with storage and the onset of MFD. Therefore, as the analysis indicated, they were positively correlated to one another, and the correlations were highly significant, i. e., a significance level of less than 0.0001. The correlation between TBARS and 2,3-octanedione had the highest positive correlation, 0.90. The second highest was between TBARS and hexanal, 0.79. The lowest was between hexanal and nonanal, 0.59.

Table I. Treatment Contrasts[a] (P[b]-values) of Sensory Descriptors[c]

Treatment[d] (ENO vs)	CBD	PTY	SOU	BTR	CBB	BRC
BNO	0.0001[b]	0.0001[b]	0.0130[b]	0.0029[b]	0.0001[b]	0.0001[b]
ENA	0.0019[b]	0.0001[b]	0.0156[b]	0.0927	0.0645	0.0909
ENB	0.2198	0.0437[b]	0.0557	0.2985	0.8619	0.4638
ENX	0.0001[b]	0.0001[b]	0.0035[b]	0.0118[b]	0.0015[b]	0.0307[b]
EVO	0.0053	0.0013[b]	0.1106	0.1835	0.0590	0.0698
EVA	0.0001[b]	0.0001[b]	0.0071[b]	0.0177[b]	0.0070[b]	0.1025
EVB	0.0002[b]	0.0001[b]	0.0056[b]	0.0091[b]	0.0029[b]	0.0062[b]
EVX	0.0001[b]	0.0001[b]	0.0194[b]	0.0165[b]	0.0006[b]	0.0021[b]

[a]Degrees of freedom = 1.
[b]Means were significantly different at a probability of 0.05.
[c]CBD, cardboard; PTY, painty; SOU, sour; BTR, bitter; CBB, cooked beef/brothy; BRC, browned/caramel.
[d]ENO, experimental, no vacuum, no additives; BNO, blind standard, no vacuum, no additives; ENA, experimental, no vacuum, plus PG; ENB, experimental, no vacuum, plus EDTA; ENX, experimental, no vacuum, plus PG and EDTA; EVO, experimental, plus vacuum, no additives; EVA, experimental, plus vacuum and PG; EVB, experimental, plus vacuum and EDTA; EVX, experimental, plus vacuum, PG and EDTA.

Table II. Pearson Correlation Coefficients[a] of Sensory[b] Attributes

	CBB	PTY	SER	BRC	CBD	SOU	SWT	BTR
CBB	1							
PTY	-0.5280	1						
SER	-0.0568	-0.1851	1					
BRC	0.5983	-0.2069	0.1251	1				
CBD	-0.5854	0.7965	-0.0826	-0.2866	1			
SOU	-0.3135	0.2588	-0.1977	-0.0518	0.3276	1		
SWT	0.3059	-0.2038	-0.2554	0.1389	0.2205	-0.3695	1	
BTR	-0.4377	0.3035	0.2736	-0.1404	0.3949	0.6264	-0.4754	1

[a]Abbreviations: CBB, Cooked Beef/Brothy; PTY, Painty; SER, Serum Raw; BRC, Browned/Caramel; CBD, Cardboard; SOU, Sour; SWT, Sweet; BTR, Bitter.
[b]Values >0.5 are very highly significant, >0.3 but <0.5 are highly significant, >0.25 but <0.3 are significant.

Table III. Pearson Correlation Coefficients[a] of Chemical[b] Attributes

	TBARS	HEXANAL	2,3-OCT	NONANAL
TBARS	1			
HEXANAL	0.7854	1		
2,3-OCT	0.8977	0.7438	1	
NONANAL	0.6848	0.5873	0.7604	1

[a]Values >0.5 are very highly significant, >0.3 but <0.5 are highly significant, >0.25 but <0.3 are significant.
[b]Abbreviatons: TBARS, Thiobarbituric acid reactive substances; 2,3-OCT, 2,3-Octanedione.

Sensory, Chemical and Instrumental Data. Statistical analysis of sensory, chemical and instrumental attributes, including 56 separate observations (Table IV), confirmed many of the significant correlations among the attributes that were described above. Regarding the chemical and instrumental indicators of MFD (hexanal, TBARS, 2,3-octanedione and nonanal), salty, serum, sweet, cooked beef/brothy and browned/caramel, all desirable flavor notes, were negatively correlated. Conversely, the undesirable notes (painty, cardboard, sour and bitter) were positively correlated to the MFD markers. Again, most of the correlations were highly significant (<0.0001).

Table IV. Pearson Correlation Coefficients of Chemical[a] and Sensory[b] Attributes

	TBARS	HEXANAL	2,3-OCT	NONANAL
CBB	-0.4256	-0.4492	-0.4195	-0.3203
PTY	0.5183	0.5757	0.5319	0.4086
SER	-0.0099	-0.0488	-0.0271	0.0204
BRC	-0.3091	-0.2606	0.2678	-0.2091
CKL	0.0303	0.0438	0.0089	0.0514
CBD	0.4657	0.5421	0.4894	0.3993
SOU	0.2104	0.2395	0.2305	0.1277
SWT	-0.1536	-0.1778	-0.1780	-0.1383
BTR	0.2767	0.3122	0.3394	0.2256

[a,b]Values and abbreviations are same as reported in Tables II and III.

Principal Components Solution. The principal components analysis (PCA) for all of the experimental treatments and the chemical, instrumental and sensory attributes is shown in Figure 7. The usefulness of this method for graphically representing correlations among treatments and descriptors has been documented previously (15). The principal components factors for the

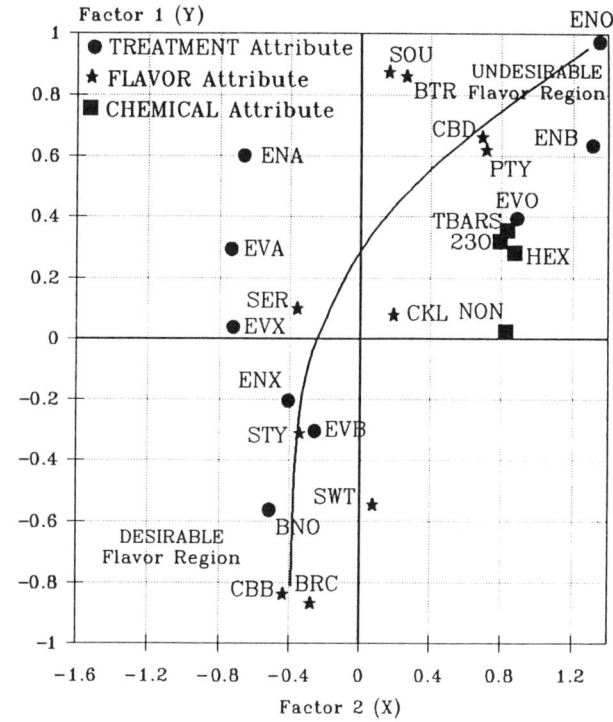

Figure 7. The principal component analysis of chemical and flavor treatments as they affect beef flavor. Codes are the same as those in Figures 2–6.

treatments are shown as circles in this figure. In general, the trend of the treatments is displayed as a curved line that runs from the most desirable flavor region in the lower left through the upper left region, and then to the region of the more undesirable flavor in the upper right portion of the grid. The blind standard beef pattie or BNO served as the anchor for the desirable flavor region, whereas the 3-day MFD or cooked/stored pattie, labeled as ENO, served as the anchor for the undesirable or off-flavor region. All treatments with vacuum or additives were located between these two anchors. The PCA of the chemical and instrumental marker compounds are shown in Figure 7 as squares and appeared in the undesirable region of the grid, although nonanal was not clustered with the other three. The PCA of the flavor attributes (presented as stars) showed a differential clustering of the variables depending upon their associated flavor response. For example, PTY, SOU, BTR, and CBD all clustered in the upper right quadrant of the grid or that area associated with the off-flavor treatments and the off-flavor chemical and instrumental markers. On the other hand, SWT, CBB and BRC clustered in the desirable-flavor region of the grid, i.e. the lower left quadrant along with that of the standard beef patties.

Discussion

When cooked meat is stored under refrigeration (4°C), a series of complex reactions begin to take place. The first lipid oxidation product observed is hexanal, whose origin is linoleic acid (*3*). Hexanal can be measured by GC methodology almost immediately. As storage time increases, so do the concentrations of other secondary reaction products from lipid oxidation (*4*), e. g., nonanal and 2,3-octanedione (see Figure 3). After three days storage, hexanal has a ten-fold increase from 2000 area counts to 20,000 (Figure 1). As these oxidation products are produced, many off-flavors begin to appear. Painty, one of the aromatics described as off-flavor, has been described as oily, linseed-like, and is associated with the rancid oil aroma (*6, 8*). A highly significant correlation (P=0.0027) was reported between the formation of PTY and hexanal (*13*). Results shown in Figure 1 also confirm those findings. No such correlation has been made with other off-flavor compounds, e.g., cardboard, that form during MFD.

With the development of off-flavor compounds during the MFD process, the loss of desirable flavor notes has also been confirmed (*6, 8*). One such aromatic note, beefy/brothy, is shown in Figure 1. However, there is evidence that many of the desirable beefy notes are indeed heterocyclic compounds that contain oxygen, nitrogen or sulfur. For example, according to Hsu et al. (*21*) and MacLeod and Seyyedain-Ardebili (*22*), 575 compounds were mentioned in the literature as components identified in the volatiles of cooked beef. Many of these compounds contained oxygen, nitrogen or sulfur. Additional heterocyclic compounds were reported by Shahidi, et al (*23*), many of which exist at the ppb level in meat (*10, 11*). Sulfur compounds degrade during MFD (*9, 24*). Therefore, it is reasonable to assume that aromatics such as cooked beef/brothy and cardboard probably have their origin from sources other than lipids.

Secondary antioxidants, such as chelators, have been used to retard or inhibit MFD over the past three decades. Judging from much of the literature, chelators appeared in most cases to be effective in preventing lipid oxidation. In many cases the experimental samples were judged by preference panels or by a line scale that asked the panelists to place a mark on a 10 cm scale where they think the flavor of the sample should lie. In those instances, 0 would show slight intensity and 10 would indicate strong. Other early methods of judging meat flavor quality were described by Melton et al. (25), who stated that descriptive terms for WOF were needed, as were accompanying standards to train panelists to evaluate WOF. Since compounds originating from oxidation of lipids are formed in the ppm level and the heterocyclic compounds are present in ppb levels, the obvious flavor detected in MFD samples would be those originating from fat oxidation. Hence, early concepts indicated that MFD (or WOF) was indeed a lipid oxidation problem that could be solved with chelators. While scientists agree that lipid oxidation is a vital part of MFD, they also recognize that it is not the whole picture. Using an analytical descriptive sensory panel, a set of descriptors developed by intense training to develop tasting skills, samples treated with chelators as antioxidants had lower CBB intensities than samples treated with antioxidants that functioned as free radical scavengers (12). Hence, we concluded that MFD was more complex than lipid oxidation and perhaps the free radical scavenger-type antioxidants were inhibiting protein degradation as well as lipid oxidation. These speculations were confirmed by Spanier et al (20), who showed that protein functionality was altered by free radical reactions. Furthermore, the free radical induced changes in protein functionality correlated well with the loss of desirable flavor such as that imparted by the peptide named "BMP".

The chemical and instrumental markers of lipid oxidation as affected by the various treatments (see Figures 2 and 3) clearly indicate that there are at least two separate mechanisms involving MFD. When the antioxidant, PG, was used without vacuum, the concentration of markers was much lower than when the chelator, EDTA, was used alone. By adding vacuum to PG, the concentrations were still low, and for hexanal and nonanal, the concentrations were even lower. However, when the vacuum was coupled to EDTA, lipid oxidation markers decreased significantly, which indicated a strong synergistic effect. It should also be noted from Figure 4, ENB and ENA, that without vacuum, CBD was not affected by the addition of the chelator, EDTA, whereas CBD was affected when PG was added. Painty, however, was decreased in both samples. Obviously, MFD is a multimechanistic system.

According to data presented in Figures 4-6, the most effective means of lowering the intensity of off-flavor aromatics (painty and cardboard) is using an antioxidant that functions as a free radical scavenger along with a chelator in the presence of a vacuum. With this combination, a synergistic effect was produced and the off-flavor aromatics were kept to a minimum. The use of PG also kept cardboard (probably resulting from a non-lipid source) to a lower intensity than did EDTA, both with or without vacuum. The off-flavor tastes, bitter and sour, did not seem to be affected in the same manner. In this case, EDTA and vacuum seemed to be the better system, and there was a synergistic effect. In regard to the desirable aromatics, cooked beef/brothy and browned caramel, the three combination system produced patties with the highest intensities.

In Figure 7, the 3-day experimental sample (ENO) clustered in the top right quadrant of the grid, whereas the blind frozen standard (BNO) appeared in the lower left quadrant. Using this multivariate statistical message, all of the undesirable flavor notes, i. e., PTY, CBD, BTR and SOU) were in the same quadrant as ENO, or the undesirable flavor region of the grid. The desirable flavor notes, i. e., CBB, BRC and SWT, were in the quadrant with the standard, or the desirable flavor region. Similar results were obtained in an earlier study in which we examined the dose-response of beef patties to the antioxidant, PG, and the chelator, EDTA, with or without vacuum (*16*). In the earlier study, the lower right quadrant contained an area that depicted the range occupied by the standard patties. This area was called the desirable flavor quadrant. However, the grid region occupied by the 3-day MFD patties extended across the entire upper portion of the grid. When the MFD patties were maintained under vacuum, the coordinates occupied by this group approached that of the standard and were located near the zero intercept. Furthermore, an extremely interesting distribution was seen for the patties containing both additives, i.e., PG and EDTA. In this case, the sample containing the 25 ppm mixture of additives clustered in the center of the left quadrant. As the level of the additives was increased from 25 to 50 and then to 100 ppm, the grid location of the treatment approached that of the desirable-flavored standard patties. Combining the vacuum and additive treatments, a synergistic effect, i.e., both a downward trend of vacuum and the dose-response of the additives, was observed.

Our data showed that chelator by itself (ENB) and vacuum by itself (EVO) had the least effects in maintaining the original desirable flavor, as indicated by their clustering in the upper right quadrant. The next method of least protection of desirable flavor was antioxidant by itself (ENA) and the antioxidant in combination with vacuum (EVA). These treatments clustered in the upper left quadrant of the grid. The most effective method of retaining desirable flavor (or retarding off-flavor) was the combined use of either chelator and antioxidants plus vacuum (EVX), chelator plus antioxidant (ENX), and chelator plus vacuum (EVB). Thus, by adding vacuum to the experimental samples, the trend was for the samples to move toward the desirable region. Finally, without vacuum, adding PG (ENA) was more effective than EDTA (ENB).

Conclusion

Our data, obtained by the use of several experimental tools such as vacuum, antioxidants, chelators, etc., demonstrated that meat flavor deterioration is a complex process of oxidative and free-radical reactions that change the chemical composition, and consequently the flavor, of meat. According to combinations of these tools coupled with statistical methodologies, chemical markers of off-flavor clustered in the off-flavor region of a grid that contained averages of principal component scores for each treatment. Off-flavor sensory attributes, i.e., painty, cardboard, sour and bitter, clustered in the undesirable-flavor region of the grid, whereas the on-flavor notes, i. e., sweet, cooked beef/brothy and browned/caramel, clustered in the desirable-flavor region of the grid. A chemical/oxygen-free environment was shown to prevent metal-catalyzed lipid oxidation and free radical reactions that cause the degradation of desirable beefy flavor components. Thus, cooked/stored beef patties still tasted like cooked/fresh beef patties.

Finally, total understanding of flavor quality and the process of flavor changes will require experimental tools such as vacuum, antioxidants, chelators and an analytical descriptive sensory panel coupled to sophisticated statistical techniques such as multivariate principal component solutions to assess or summarize the intercorrelations among all the variables. Only then can predictive or adaptive controls be implemented.

Acknowledgement

The authors wish to thank Dr. B. T. Vinyard for assistance with statistical evaluations, Mr. C. James, Jr. and Mrs. C. H. Vinnett for technical assistance, members of the meat sensory panel for evaluation of samples and employees of the sensory laboratory for assistance in preparing and serving the samples.

Literature Cited

1. Pearson, A. M.; Gray, J. E. In *The Maillard Reaction in Foods and Nutrition*; Waller, G. R.; Feather, M. S., Eds.; ACS Symposium Series No. 215; American Chemical Society: Washington, DC, 1983; pp 287-300.
2. Love, J. In *Warmed-Over Flavor of Meat*; St. Angelo, A. J; Bailey, M. E., Eds.; Academic Press Inc.: Orlando, FL, 1987; pp 19-39.
3. St. Angelo, A. J.; Legendre, M. G.; Dupuy, H. P. *Lipids* **1980**, *15*, 45-49.
4. Dupuy, H. P.; Bailey, M. E.; St. Angelo, A. J.; Vercellotti, J. R.; Legendre, M. G. In *Warmed-Over Flavor of Meat*; St. Angelo, A. J; Bailey, M. E., Eds.; Academic Press Inc.: Orlando, FL, 1987; pp 165-191.
5. Bailey, M. E.; Dupuy, H. P.; Legendre, M.G. In *The Analysis and Control of Less Desirable Flavors in Foods and Beverages*; Charalambous, G., Ed.; Academic Press Inc.: Orlando, FL, 1980; pp 31-52.
6. Johnsen, P. B; Civille, G. V. *J. Sensory Studies.* **1986**, *1*, 99-104.
7. St. Angelo, A. J.; Vercellotti, J. R.; Legendre, M. G.; Vinnett, C. H.; Kuan, J. W.; James, Jr., C.; Dupuy, H. P. *J. Food Sci.* **1987**, *52*, 1163-1168.
8. Love, *Food Technol.* **1988**, *42(6)*, 140-143.
9. Drum, T. D.; Spanier, A. M. *J. Agric. Food Chem.* **1991**, *39*, 336-343.
10. Liu, R.H.; Legendre, M. G.; Kuan, J. W.; St. Angelo, A. J.; Vercellotti, J. R. In *Warmed-Over Flavor of Meat*; St. Angelo, A. J; Bailey, M. E., Eds.; Academic Press Inc.: Orlando, FL, 1987; pp 193-236.
11. Vercellotti, J. R.; Kuan, J. W.; Liu, R. H.; Legendre, M. G.; St. Angelo, A. J.; Dupuy, H. P. *J. Agric. Food Chem.* **1987**, *35*, 1030-1035.

12. St.Angelo, A. J.; Crippen, K. L.; Dupuy, H. P.; James, Jr., C. *J. Food Sci.* **1990**, *55*, 1501-1505 & 1539.
13. Spanier, A. M.; Edwards, J. V.; Dupuy, H. D. *J. Food Technol.* **1988**, *42(6)*, 110-118.
14. Spanier, A. M.; McMillin, K. W.; Miller, J. A. *J. Food Sci.* **1990**, *55*, 318-322 & 326.
15. Spanier, A. M.; Vercellotti, J. R.; James, Jr., C. *J. Food Sci.*, in press.
16. Spanier, A. M.; St. Angelo, A. J. *1989 Annual Meeting, IFT.* Abstract No. 574.
17. Meilgaard, M.; Civille, G. V.; Carr, B. T. *Sensory Evaluation Techniques*; CRC Press, Inc.: Boca Raton, FL, 1987; Vol. 2, pp 1-23.
18. Tarladgis, B. G.; Watts, B. M.;Younathan, M. T.; Dugan, Jr., L. *J. Am. Oil Chem. Soc.* **1960**, *37*, 44-48.
19. SAS. *Statistical Analysis System User's Guide: Statistics.* SAS Institute, Inc.: Cary, NC, 1986.
20. Spanier, A. M.; Miller, J. A.: Bland, J. M. In *Lipid Oxidation in Foods*; St. Angelo, A. J., Ed.; ACS Symposium Series; American Chemical Society: Washington, DC, in press.
21. Hsu, C. M.; Peterson, R. J.; Jin, Q. Z.; Ho, C. T.; Chang, S. S. *J. Food Sci.* **1982**, *47*, 2068-2069 &2071.
22. MacLeod, G.; Seyyedain-Ardebilli, M. *CRC Crit. Rev. Food Sci. and Nutr.* **1981**, *14*, 309-437.
23. Shahidi, F.; Rubin, L. J.; D'Souza, L. A. *ibid*, **1986**, *24*, 141-243.
24. St. Angelo, A. J.; Vercellotti, J. R.; Dupuy, H. P.; Spanier, A. M. *Food Technol.* **1988**, *42(6)*, 133-138.
25. Melton, S. L.; Davidson, P. M.; Mount, J. R. In *Warmed-Over Flavor of Meat*; St. Angelo, A. J; Bailey, M. E., Eds.; Academic Press Inc.: Orlando, FL, 1987; pp 141-164.

RECEIVED February 19, 1992

Chapter 10

Prevention of Lipid Oxidation in Muscle Foods by Nitrite and Nitrite-Free Compositions

Fereidoon Shahidi

Department of Biochemistry, Memorial University of Newfoundland, St. John's, Newfoundland A1B 3X9, Canada

Lipid oxidation is a major cause of meat flavor deterioration (MFD). It is also responsible for changes in color, texture and nutritional value as well as wholesomeness of muscle food quality. In cured meat products, however, nitrite by virtue of its strong antioxidant properties inhibits autoxidation of meat lipids. Autoxidation of meat products may also be prevented by the use of food antioxidants and/or chelators. Studies on the application of natural antioxidant/ extracts and curing adjuncts in prevention of MFD and oxidative rancidity in nitrite-free cured meats have been carried out. The cumulative effects of nitrite-free curing ingredients on the prevention of lipid antioxidation in meat systems are evaluated and compared with the effectiveness of nitrite-curing systems.

Lipid oxidation is a major cause of quality deterioration of both raw and cooked muscle foods (1). Fat portion of meats, particularly their phospholipid components, undergo lipid oxidation/degradation (2) and produce a large number of volatile compounds. While hydroperoxides, the primary products of lipid oxidation, are odorless and tasteless, their degradation leads to the formation of a large number of secondary oxidation products such as aldehydes, acids, alkanes, alkenes, ketones, alcohols, esters, epoxy compounds, polymers, etc. The latter classes of compounds are flavor-active and possess low threshold values, thus are responsible for the development of warmed-over flavor (WOF), also referred to as meat flavor deterioration (MFD), and rancidity. The degree of unsaturation of the fatty acid constituents of meat lipids primarily dictates the rate of MFD. Thus, meat flavor deterioration and rancidity develops faster and further in the relatively unsaturated pork fat than in the harder beef or mutton fats, faster again in the very soft chicken fat and fastest of all in fish lipids, Figure 1 (3).

The C18 fatty acids are the major lipid fatty acid components or meats and their relative oxidation rates at 25°C are as follows (4):

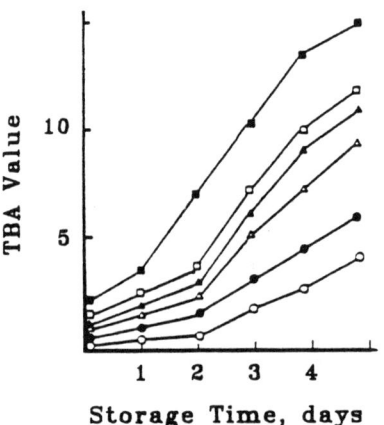

Figure 1. Effect of metmyoglobin on lipid oxidation in cooked water-extracted muscle fibres from various species stored at 4°C. ■, fish; □, turkey; ▲, chicken; △, pork; ●, beef; and ○, lamb. Adapted from Ref. 31.

ACID	NAME	RELATIVE RATES
C18:0	Stearic acid	1
C18:1 n-9	Oleic acid	100
C18:2 n-6	Linoleic acid	1200
C18:3 n-3	Linolenic acid	2500

Thus, as the degree of unsaturation of lipid fatty acids increases, their susceptibility to oxidation and flavor deterioration also increases. Although raw meat has little flavor, heat processing results in creation of the pleasant aroma of cooked, roasted and fried products. Non-volatile precursors of heat processed meat flavor are water-soluble substances which include amino acids, peptides, reducing sugars and vitamins. Upon heat treatment, free amino acids such as cysteine, produced from the action of proteolytic enzymes during the post-mortem period, react with reducing sugars, products of glycolysis, and vitamins, such as thiamin. Often, products of one reaction become precursors for others. Interaction of these volatiles with lipid-derived products may produce desirable flavors. However, progression of oxidation may mask the natural flavor of heat-processed meats which eventually leads to MFD.

Lipids in muscle tissues vary from one species to another and therefore are responsible for species-specific flavor notes in cooked meats. Thus, eliminating the lipid-derived flavor effects should reveal the natural flavor of meat itself. For simplicity one may assume that the natural flavor of meats from all species without being influenced by their lipid components, is the same.

Major proposed initiators of lipid peroxidation include singlet oxygen, superoxide radical, hydroperoxyl radical, hydroxylose radical, cryptohydroxyl radical, perferryl radical, ferryl radical, oxygen-bridged di-iron complex and phyrin cation radical (1). Presence of oxygen, heat, light and metal ions as well as their organic complexes, biological catalysts and photochemical pigments are important factors influencing the keeping quality of cooked meat. Therefore, packaging, processing and storage conditions have a definite effect on the rate of progression of MFD.

Nitrite Curing of Meat

The origin of meat curing is lost in antiquity. Nitrates, with nitrite impurities, were unknowingly used for curing of meats for many centuries (5). The first significant scientific investigations of curing, however, began in the 19th century. Polenske (6) reported the presence of nitrite in cured meats and used pickling solutions for curing. Lehmann (7) and Kisskalt (8) reported that nitrite rather than nitrate conferred the characteristic color to cured products. Subsequently, Haldane (9) proposed that the reaction of hemoproteins with nitric oxide derived from nitrite was the chemical basis for the color of the cured meats and this was later confirmed by Hoagland (10). Understanding of the process and identification of nitrite led to the use of nitrite, as such, in the cure (11). Thus, the regulated use of nitrite in the curing of meat has been practiced since 1920's. Sodium chloride is always incorporated with nitrite in the curing system as well as ascorbates or erythorbates which are used as cure accelerators, amongst their other beneficial effects. Furthermore, curing adjuncts such as sugars, phosphates/polyphosphates, and spices and herbs may be included in the curing mixture.

Nitrite is the key ingredient of the cure due to its multifunctional role. It is responsible for producing the characteristic pink color in cooked-cured products and contributes to the typical flavor associated with cured meats and this may also be attributed to its antioxidant activity (2,12,13). Reaction of nitrite with biomolecules in the meat matrix plays an essential role in preventing MFD and thus extends the shelf-life of cured meat products. In addition, nitrite inhibits the formation of the deadly neurotoxins of *Clostridium botulinum*.

A particular concern with the use of nitrite for the curing of meat has been the formation of carcinogenic N-nitrosamines in some cured products, under certain heat-processing conditions, or in the stomach (14,15). Such concerns have led to efforts to develop alternatives to the use of nitrite in meat curing. These alternatives are composite mixtures which possess the cumulative effects of nitrite.

Antioxidant and Flavor Effects of Nitrite

The relationship of nitrite to cured meat flavor was first described by Brooks *et al.* (16). These authors concluded that the characteristic flavor of bacon was primarily due to the action of nitrite and that satisfactory bacon could be made with only sodium chloride and sodium nitrite. They also presumed that an adequate cured flavor could be obtained with a relatively low, 10 mg/kg, concentration of nitrite. Later results by Simon *et al.* (17) and MacDougall *et al.* (18), however, showed that higher taste panel scores were consistently obtained as the nitrite addition level increased. A linear relationship between the panel scores and logarithm of the nitrite concentration was apparent.

The role of nitrite in modifying cured-meat flavor (13,19,20) and in suppressing lipid oxidation and MFD in cooked meats (2,12,22) is well documented. Sato and Hegarty (23) reported that while nitrite at 2000 ppm eliminated lipid oxidation in cooked beef, at 50 ppm it was capable of suppressing its development. Results of Younathan and Watts (24) and Hadden *et al.* (13) indicated lower TBA values for cured as compared to uncured meats. Similar conclusions were reached by Fooladi *et al.* (22) who also compared the TBA values of cured and uncured beef, pork and chicken.

Meat flavor deterioration, therefore, does not develop in cured meat. Nitrite has been shown to have a strong antioxidant effect in cooked pork (Figure 2). Its effectiveness is particularly evident at a 156 ppm level of addition, where the formation of malondialdehyde (MDA) and other 2-thiobarbituric acid reactive substances (TBARS) are effectively inhibited. In the presence of a reductant, such as sodium ascorbate, a synergistic effect is observed and still lower TBARS values were obtained. The antioxidant activity of nitrite in cured meat products may arise from any or a combination of the effects related to (a) stabilization of heme pigments (25-27), (b) stabilization of membrane lipids (28,29), (c) chelation of free metal ions and peroxidation catalysts (20,30,31), and (d) formation of nitrosated heme pigments possessing antioxidant effects (30-34).

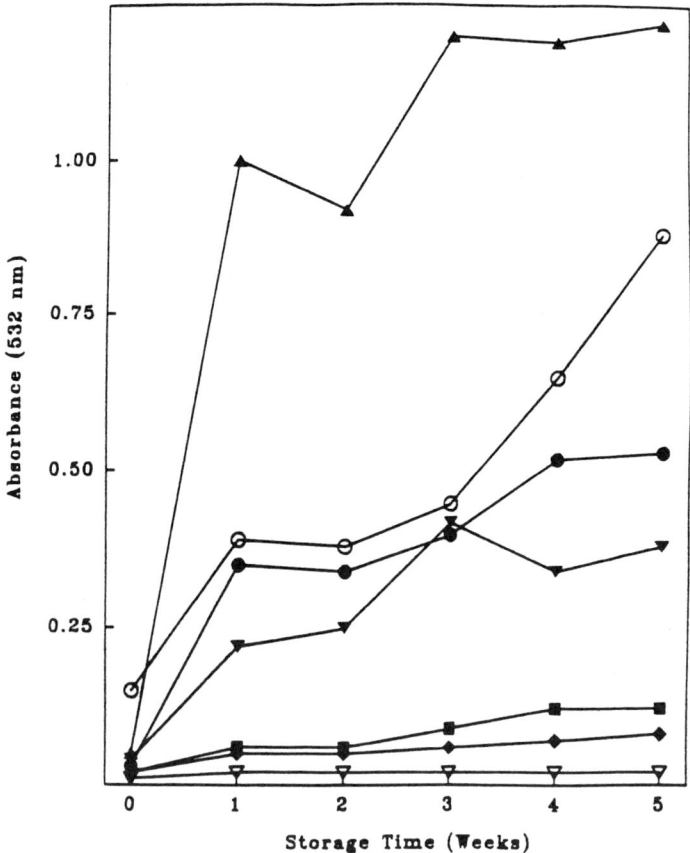

Figure 2. Antioxidant effect of sodium nitrite and cooked cured-meat pigment (CCMP), as reflected in the absorbance of TBA-MA chromogen at 532 nm, over a 5-week storage at 4°C. Symbols are: ▲, 12 ppm CCMP; ○, 25 ppm $NaNO_2$; ●, 50 ppm $NaNO_2$; ▼, 12 ppm CCMP + 550 ppm sodium ascorbate; ■, 25 ppm $NaNO_2$ + 550 ppm sodium ascorbate; ◆, 12 ppm CCMP + 550 ppm sodium ascorbate + 3000 ppm sodium tripolyphosphate; and ▽, 150 ppm $NaNO_2$.

Heme proteins and inorganic metal ions have been implicated as pro-oxidants in meat lipid oxidation (*2,23,30,35*). During cooking of untreated meat, hemoproteins are rapidly oxidized and iron-containing products and ferrous/ferric ions so produced act as active catalysts of lipid oxidation, thereby giving rise to a high content of TBARS. In cured meats, nitrite reacts with Fe^{2+} and rapid depletion of nitrite under these conditions occurs (*3,30,36,37*). The strong reaction of nitrite in heated Fe^{2+} systems may be of major significance in suppression of MFD (Table I). Shahidi and Hong (*31*) recently reported that addition of polyphosphates such as sodium tripolyphosphate (STPP) or disodium salt of ethylenediaminetetraacetic acid (Na_2EDTA) to meat systems containing pro-oxidants such as iron and copper ions or heme pigments resulted in a substantial decrease in their content of TBARS. Thus, chelating properties of nitrite may potentially be duplicated by the action of other ingredients (Table II).

Table I. Effect of nitrite and nitric-oxide myoglobin, MbNO, on lipid oxidation (TBA numbers) in cooked water-extracted pork muscle system (WF) catalysed by metals and myoglobin, Mb [a]

Treatment	$NaNO_2$, ppm			
	0	50	100	200
WF (control)	0.72 ± 0.05	0.31 ± 0.03	0.08 ± 0.02	trace
WF + MbNO	0.63 ± 0.03	0.25 ± 0.04	0.05 ± 0.02	trace
WF + Fe^{2+}	10.26 ± 0.69	4.57 ± 0.21	2.72 ± 0.11	0.07 ± 0.01
WF + Fe^{2+} + MbNO	7.21 ± 0.23	2.18 ± 0.09	1.65 ± 0.10	0.05 ± 0.02
WF + Cu^{2+}	7.63 ± 0.18	3.36 ± 0.14	2.14 ± 0.08	0.06 ± 0.02
WF + Cu^{2+} + MbNO	4.89 ± 0.21	2.03 ± 0.05	1.21 ± 0.02	0.06 ± 0.01
WF + Mb	3.43 ± 0.11	1.24 ± 0.07	0.68 ± 0.04	0.07 ± 0.01
WF + Mb + MbNO	2.82 ± 0.13	0.92 ± 0.06	0.51 ± 0.05	0.06 ± 0.02

[a] Fe^{2+}, Cu^{2+} and Mb were added at 5 ppm and MbNO was added at 10 ppm. Adapted from Ref. 30.

The nitrite used in meat curing reacts rapidly with meat components and after one to two days of storage only 50% of it remains free (*38,39*). Using ^{15}N labelling studies, Cassens et al. (*39*) have demonstrated that about 1-3% of the nitrite was bound by lipids as was found in extracts prepared by the Folch (*40*) method. Thus, stabilization of meat lipids by nitrite may be, at least partially, due to direct coupling of nitric oxide with lipid radicals. However, most of the added nitrite to meat is present in the protein-bound form and as nitrosothiol, nitrite/nitrate and gaseous nitrogen compounds as well as nitric oxide heme complex (*32,41*).

Kanner (*32*), Kanner et al. (*41*), Morrissey and Tichivangana (*30*) and Shahidi et al. (*34*) have clearly demonstrated that some nitrosylheme compounds possess antioxidant effects. Thus, the cooked cured-meat pigment (CCMP) was found to act as a weak antioxidant in meat model systems (*42*). However, CCMP together with ascorbates may form potent antioxidant combinations with evidence of strong synergism between the components (Figure 2). In the presence of

Table II. Effect of Na₂EDTA and STPP on lipid oxidation (TBA numbers) in cooked ground pork catalyzed by metals and heme compounds after 1 day storage at 4°C [a]

Treatment	No chelator	Na$_2$EDTA (500 ppm)	STPP (3000 ppm)
Control	2.90-3.15	0.07	0.22
Fe^{2+}	4.80	0.09	0.27
Fe^{3+}	4.60	0.08	0.37
Cu^{+}	3.55	0.35	0.32
Cu^{2+}	3.20	0.31	0.38
Mb	4.20	0.07	0.28
Hb	4.33	0.10	0.30
Hm	4.35	0.07	0.33
CCMP	0.42	0.30	0.30

[a] Na$_2$EDTA, disodium salt of ethylenediaminetetraacetic acid; STPP, sodium tripolyphosphate; Mb, myoglobin; Hb, hemoglobin; Hm, hemin; CCMP, cooked cured meat pigment. Adapted from Ref. 31.

polyphosphates, this synergistic effect is further accentuated. Both nitric oxide-myoglobin and CCMP have their iron atoms in the ferrous oxidation state and their coordination sites are occupied. Suggestions have been made that nitrosated iron porphyrin compounds act in the early stages of the reaction to neutralize substrate-free radicals and thus inhibit lipid oxidation (41).

S-Nitrosocysteine (RSNO) is regarded as a potent antioxidant thought to be generated during the meat curing process (33). The inhibitory effect of RSNO on lipid oxidation in a cooked turkey meat product was reported by Kanner and Juven (33). The inhibitory effect of RSNO and nitrite at equimolar concentrations was similar. Furthermore, it was demonstrated that at room temperature, or upon cooking, RSNO dissociates to form heme-NO. Only 1-3 mg of nitrite is sufficient for cured color production in meats. Furthermore, the concentration of RSNO in meat, on a molar basis, may arguably be much smaller than that of the added nitrite. Thus comparison of the antioxidant activity of nitrite with RSNO at equimolar concentrations may not be realistic. Furthermore, the antioxidant effect of nitrite in cured meats depends also on the formation of nitric oxide which interacts with metals especially those of iron. Nitrite has also been demonstrated to be effective as an antioxidant against phosphatidyl ethanolamine, the major phospholipid responsible for MFD.

It has been reported that while addition of nitrite in the presence of metmyoglobin in water-extracted mackerel meat resulted in an increase in the TBA values, in the presence of ascorbate, a relative decrease in the TBA values was evident (43). Nitrite curing of mackerel meat in the presence of ascorbate and its subsequent storage for 6 days did not increase TBA values. From these and other experiments, the above authors concluded that nitric oxide ferrohemochromogen formed in cured mackerel possessed antioxidant activity, which is partly attributable to its function as a metal chelator.

Volatiles of Nitrite-Cured Meats

So far, nearly 1000 compounds have been identified in the volatile constituents of cooked meats and poultry as it has recently been reviewed (44,45). These were representative of most classes of organic compounds (Table III). Moreover, many of these substances may be regarded as unimportant to flavor of meat and some may have been artifacts. Only a limited number of these compounds, mainly sulfur-containing substances, possessed meaty flavor notes (45).

Table III. Chemical composition of volatile constituents of cured and uncured pork [a]

Chemical Class	Uncured Pork	Cured Pork
Aldehydes	35	29
Alcohols and Phenols	33	10
Carboxylic Acids	5	20
Esters	20	9
Ethers	6	—
Furans	29	5
Hydrocarbons	45	4
Ketones	38	12
Lactones	6	—
Oxazol(in)es	4	—
Pyrazines	36	—
Pyridines	5	—
Pyrroles	9	1
Thiazol(in)es	5	—
Thiophenes	11	3
Other Nitrogen Compounds	6	2
Other Sulfur Compounds	20	30
Halogenated Compounds	4	1
Miscellaneous Compounds	1	11
Total	314	137

[a]Adapted from Ref. 45.

Furthermore, the number and/or concentration of volatile components in cured meats was drastically reduced as compared to their uncured counterparts. Thus, only 137 volatiles were noticed in cured pork flavor as compared to 314 in uncured pork (Table III). Of the volatile compounds of cooked, untreated meats, lipid-derived products constituted a large proportion. The spectrum of secondary products of lipid oxidation will of course depend on the fatty acid composition of their lipid constituents which also varies from one species to another. Nonetheless, hexanal (Figure 3) was always a major oxidation product and its content depended on the amount of linoleic acid present in meats.

In cured meats, nitrite by virtue of its antioxidant properties retards the breakdown of unsaturated fatty acids and formation of secondary oxidation products. Thus, Cross and Ziegler (46) noticed that formation of higher aldehydes such as hexanal was effectively suppressed by nitrite curing of meats. Other chemical changes responsible for the flavor attributes of cured meats have not clearly been understood. Cross and Ziegler (46) also reported that passage of volatiles of uncured beef and chicken through a solution of 2,4-dinitrophenylhydrazine converted their carbonyl compounds to non-volatile derivatives; the resultant volatiles after stripping of their carbonyl compounds, had an aroma similar to that of cured ham.

This observation indicated that the chemical nature of cured beef, pork and chicken was essentially the same. Elimination of lipid oxidation, either by curing or by stripping of carbonyl compounds from volatiles of untreated meats, displayed a major effect on flavor perception and volatile constituents of meats. Our own results lend support to the above findings, however, qualitative differences due to the possible presence of less active flavor components can not be ruled out. Nonetheless, gas chromatographic analyses of volatiles of cured meats revealed a much simpler spectrum than their uncured counterparts with drastic suppression in the content of major aldehydes (Table IV) which are known to have very low threshold values and to be responsible for MFD (Table V). Based on these findings, we proposed that any other agent, or combination that prevented lipid oxidation would, in principle, duplicate the antioxidant role of nitrite in the curing process, thus preventing MFD. This is in line with findings of other researchers and its validity was confirmed by preliminary organoleptic evaluations. However, mutton was not included in these studies.

Table IV. Effect of curing on the relative concentration of major aldehydes in pork flavor volatiles [a]

Aldehyde	Relative Concentration	
	Uncured	Cured
Hexanal	100	7.0
Pentanal	31.3	0.5
Heptanal	3.8	<0.5
Octanal	3.6	<0.5
2-Octenal	2.6	—
Nonanal	8.8	0.5
2-Nonenal	1.0	—
Decanal	1.1	—
2-Undecenal	1.4	0.5
2,4-Decadienal	1.1	—

[a] Adapted from Ref. 45.

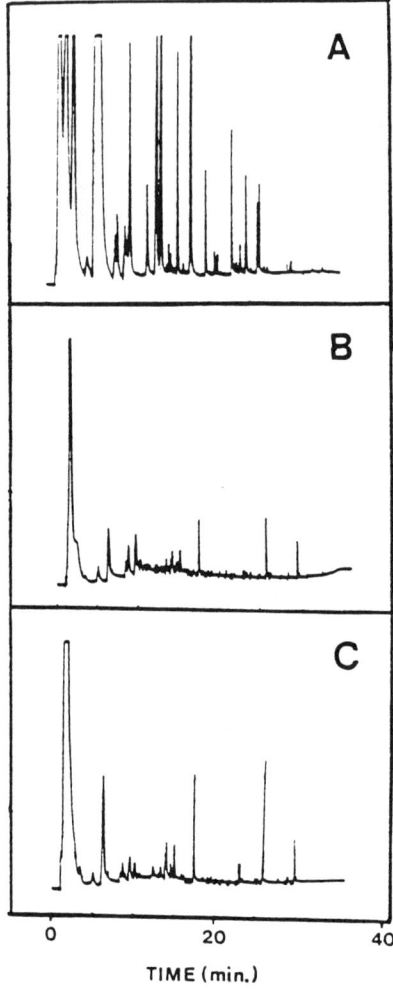

Figure 3. Gas chromatograms of volatiles of cooked meats with A: no additive, B: sodium nitrite (150 ppm) and sodium ascorbate (550 ppm), and C: sodium tripolyphosphate (3000 ppm), sodium ascorbate (550 ppm) and BHA (30 ppm).

Table V. Odor thresholds (mg/mL air) of some volatile lipid oxidation products [a]

Carbon chain length	n-Alkanal	t-2-Alkenal	t,t-2,4-Alkadienal
6	43	480	—
7	260	250	57
8	7.8	47	12
9	45	3.6	0.4

[a]Adapted from Ref. 79.

A simplistic view, attempting to present a unifying theory of the origin of the basic flavor of meat, species differentiation, and meat flavor deterioration, as described previously (46,47), has been summarized in Figure 4. It postulates that meat on cooking acquires its characteristic species flavor which is caused by volatile carbonyl compounds formed by oxidation of its lipid components, primarily phospholipids. Further oxidation during storage of cooked meat results in its flavor deterioration. Curing with nitrite suppresses the formation of oxidation products (Table IV). Hence it might be justified, as a first step, to assume that flavor of nitrite-cured meats is actually the basic natural flavor of meat from different species without being influenced by overtones derived from oxidation of their lipid components. Thus, the scheme presented in Figure 4 is in line with these views of general understanding of meat flavor volatiles as we presently know. Further support for this view has recently been provided by Ramarathnam et al. (48). However, the postulation does not easily explain the fact that intensity of cured meat flavor is proportional to the logarithm of nitrite concentrations, or the apparent persistence of the characteristic "mutton" flavor even after nitrite curing of sheep meat (49). Nonetheless it is the view of the author that the former effects are due to taste effects related to the role of Na^+ ion and the latter observations are due to effects exerted by the branched fatty acids present in the samples.

Curing Adjuncts and their Effect on Lipid Oxidation

Salt is basic to all curing mixtures and is always added to cured meats. By its dehydrating effect, salt inhibits bacterial growth and subsequent microbial spoilage. Generally 2-3% salt is added to hams and bacon preparations. Salt generally acts as a pro-oxidant (Table VI) and thus has undesirable effects on flavor and appearance. To overcome the harsh, dry and saltiness of products, formulations generally contain sugars as well.

The addition of sugar to cures is primarily for flavor. They retain the moisture and moderate the flavor of cooked meats by interacting with the free amino acids and amino groups of proteins in Maillard reactions. Their preservative effect in retarding bacterial growth in cured meats is minimal.

Ascorbic acid, erythorbic (isoascorbic) acid, or their salts, are used as curing adjuncts. It is reported that they aid the curing process by accelerating the reduction of nitrite and help to stabilize the cured color of the products. Ascorbic acid

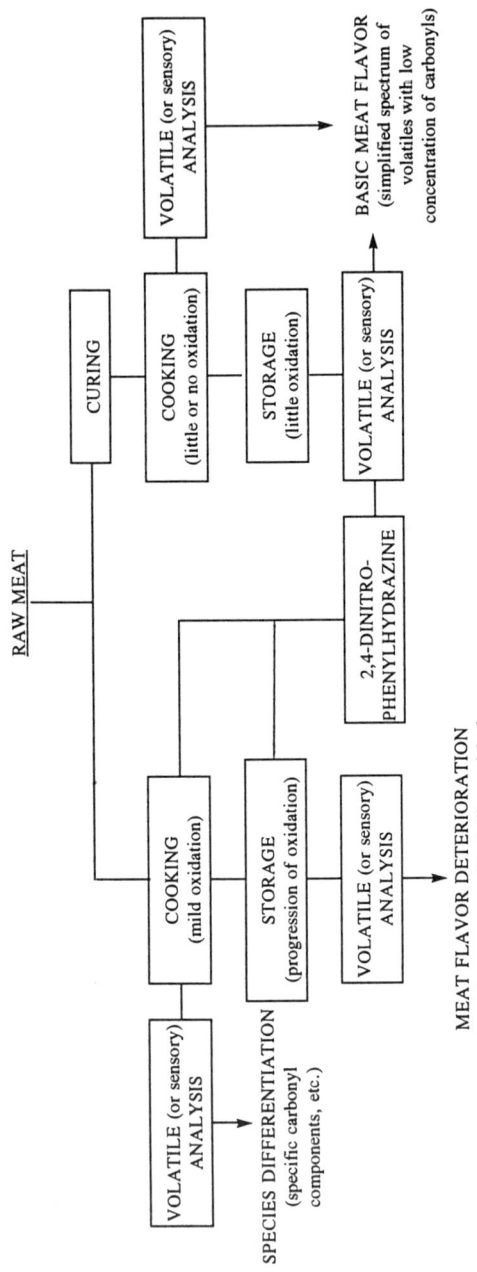

Figure 4. Consequence of cooking, curing, and storage on flavor of cooked meats and development of meat flavor deterioration.

Table VI. Inhibition (mean %) of lipid oxidation in cooked ground pork by selected additives [a]

Additive (Concentration), ppm	Inhibition, %
NaCl (2%)	-34
AA (500 ppm)	6
EA (500 ppm)	54
SA (550 ppm)	52
AP (2000 ppm)	94
STPP (3000 ppm)	92
SPP (3000 ppm)	93
SHMP (3000 ppm)	56
Na$_2$EDTA (500 ppm)	93
BHA (30 ppm)	95
BHT (30 ppm)	59
PG (30 ppm)	79
TBHQ (30 ppm)	95
NaNO$_2$ (200 ppm)	93

[a] AA, ascorbic acid; EA, erythobic acid; SA, sodium ascorbate; AP, ascorbyl palmitate; STPP, sodium tripolyphosphate; SPP, sodium pryophosphate; SHMP, sodium hexametaphosphate; Na$_2$EDTA, disodium salt of ethylenediaminetetraacetic acid; BHA, butylated hydroxyanisole; BHT, butylated hydroxytoluene; PG, propyl gallate; TBHQ, tertbutylhydroquinone. Adapted from Ref. 34.

inhibits lipid oxidation by keeping the heme pigment in its catalytically inactive form. Ascorbyl palmitate and the C16 acetal of ascorbic acid give rise to strong antioxidant effects in cooked meats. This activity may be due to their enhanced solubility in the fat portion of meat as compared to ascorbic acid itself (42).

Ascorbic acid has been shown to be useful not only in developing and maintaining cured meat color but also in improving its odor and flavor. However, presence of substances such as non-heme iron, tocopherols, citric acid and amino acids which are naturally present in meats may change the role of ascorbic acid from an antioxidant to a pro-oxidant (27). It has been concluded that ascorbic acid retards autoxidation by upsetting the balance between Fe^{2+} and Fe^{3+} or by an oxygen scavenging effect (50).

Phosphates and polyphosphates are generally added to the cure to increase water holding capacity and to improve texture and the yield of finished products. They are also known to influence color, coagulation, emulsification and microbial growth of meats. Sodium tripolyphosphate (STPP) is the most widely used of all the phosphates in meat curing; however, sodium acid pyrophosphate (SAPP) is also used in sausages.

Polyphosphates have been shown to retard lipid oxidation (Figure 5) perhaps by sequestering metal catalysts, particularly those of iron. A strong synergism has also been noted between polyphosphates and asorbates. The TBA values of meats treated with combinations of polyphosphates and sodium ascorbate or with either ascorbyl acetal or ascorbyl palmitate are fairly close to those of nitrite-cured meat

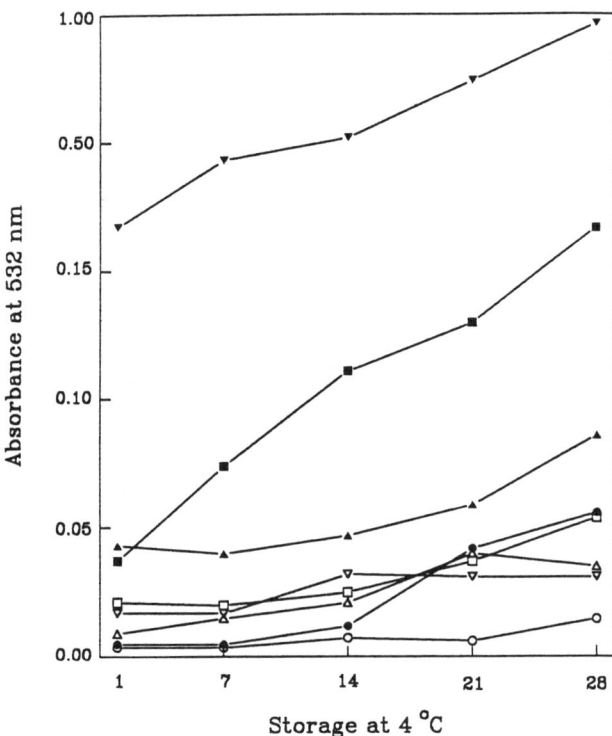

Figure 5. Synergistic effect of polyphosphates (3000 ppm) with sodium ascorbate (550 ppm) in cooked meat systems. Symbols (full) are ▼, sodium hexametaphosphate; ■, sodium tripolyphosphate; ▲, sodium pyrophosphate; ●, sodium acid pyrophosphate. Corresponding open symbols are for polyphosphate counterparts with sodium ascorbate.

in model systems. At 0.4% SAPP effectively retarded lipid oxidation in chicken nuggets. Long-chain polyphosphates are the best sequestering agents for light metal ions such as calcium and magnesium compared with short-chain polyphosphates for iron and copper ions. As pH increases, the chelating ability of long-chain polyphosphates also increases, while the opposite is true for short-chain polyphosphates (51).

In some cured products, spices and herbs may be added in conjunction with the cure to impart a desired taste and flavor to the product. Rosemary, sage, thyme, marjoram and oregano are among the herbs and clove, ginger and mace are among the spices which possess strong antioxidant effects (52). Antioxidant activity of herbs and spices is due to the presence of natural inhibitors of lipid oxidation. The antioxidant activity of spices is generally derived from a diverse group of phenolic-based compounds. For instance, clove contains 1.26% gallic acid and 3.03% eugenol, both of which are known to be strong antioxidants (53) at relatively low levels (54).

In emulsified meat products, protein extenders such as soy protein isolate/concentrate and other plant proteins may often be incorporated. Many of these protein extenders are known to contain a relatively high concentration of phenolic compounds. Rhee et al. (55) and Ziprin et al. (56) have reported that addition of protein meals from glandless cottonseed or their aqueous or methanolic extracts are able to retard lipid oxidation. Our own results for deheated mustard flour (DMF) indicated that it possessed strong antioxidant effects at 1.5-2.0% addition to meats. Its aqueous and methanolic extracts were less effective (Figure 6).

Table VI summarizes the relative antioxidant/pro-oxidant activity of meat curing adjuncts as reflected in the % inhibition of TBARS production in treated meats.

Nitrite-Free Curing Systems and their Role in MFD Prevention and Flavor Retention

The present trend in food processing is to avoid use of additives implicated as potential health hazards (57). A particular concern with the use of nitrite for the curing of meat has been the formation of N-nitrosamines, such as N-nitrosodimethylamine and N-nitrosopyrrolidine, which are known carcinogens. Research on the elimination of the use of nitrites has concentrated on formulating multi-component alternatives, as it was early recognized that the multi-functional properties of nitrite could not be duplicated by a single compound. Sweet (58) formulated nitrite-free curing mixtures consisting of a colorant, an antioxidant/sequestrant and an antimicrobial agent. He reported that phosphates and polyphosphates together with low levels of an antioxidant such as TBHQ were effective in retarding lipid oxidation. Our own approach has been a similar one. However, we have concentrated our efforts in taking advantage of various properties and effects of curing adjuncts and to use natural ingredients, wherever possible. In addition to the above ingredients, CCMP, added for its coloring effects, was found to accentuate the antioxidant activity of the ingredients used for flavor preservation in nitrite-free curing compositions.

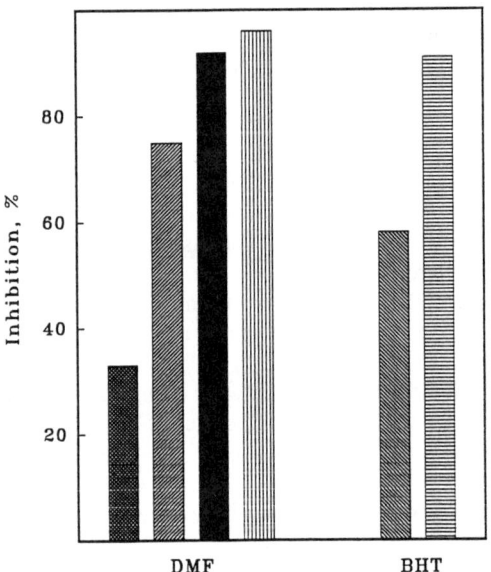

Figure 6. Average inhibition of oxidation (malondialdehyde production) of cooked meat by deheated mustard flour (DMF) at 0.5%, ▨; 1%, ▨; 1.5%, ■; and 2.0%, ▥, as compared with butylated hydroxytoluene (BHT) at 30 ppm, ▧; and 200 ppm, ▤, level of addition stored over a 5-week period at 4 °C.

The nitrite-free curing mixtures under discussion consisted of the natural cooked cured-meat pigment prepared from hemoglobin, an antioxidant/sequestrant or curing adjuncts possessing antioxidant/sequestering ability, and an antimicrobial agent or low-dose gamma irradiation. Only replacement of nitrite for its antioxidant activity and prevention of MFD will be addressed here.

Inhibition of lipid oxidation calculated from the relationship

$$\text{Inhibition } (\%) = (1 - \frac{\text{Response in the presence of additive}}{\text{Response in the absence of additive}}) \times 100$$

showed that sodium ascorbate, together with STPP effectively retarded lipid oxidation in treated meats (Table VII). Thus production of MDA, hexanal, 2-pentylfuran, nonanal, decenal and 2,4-decadienal was greatly suppressed in the presence of the above combination. Addition of 30 ppm of BHA or TBHQ nearly inhibited the oxidation of meat lipids. However, the cooked cured-meat pigment exerted a similar effect when used in combination with an ascorbate/ polyphosphate system. The effect of the above combination was comparable to that of nitrite (Table VII).

Formation of many other combinations for retarding lipid oxidation has also been attempted. Among these, use of α-tocopherol, mixed tocopherols or citric acid with ascorbates has proven to be most useful (F. Shahidi; unpublished results). Addition of these ingredients, due to their natural, rather than synthetic origin might prove advantageous. Furthermore, presence of naturally-occurring antioxidants in binders and protein extenders such as those of soybean provides an added advantage to the use of these proteins in emulsified meat formulations (59). Presence of phenolic acids and flavonoids in soybean as well as cottonseed flours has shown to enhance the keeping quality of the meats containing these flours (56,60-63). Spices used in emulsified meat products also play an important role in prevention of MFD (64). Thus, many non-meat ingredients, at appropriate concentrations, may present viable alternatives to nitrite as far as control of lipid oxidation is concerned.

Many cured meat products are smoked in order to improve their flavor characteristics. Effectiveness of smoke as an antioxidant in processed meats is well documented. The antioxidant activity of smoke is mainly due to simple phenols, derived from thermal decomposition of phenolic acids and lignin (63). Major phenols of wood smoke include guaiacol, 4-methylguaiacol, syringol and phenol (66). Our own work, in progress, has shown that the keeping quality of emulsified meats could be improved by the smoking process. Use of smoke phenolics in the nitrite-free combinations also proved beneficial.

Assessment of lipid oxidation in nitrite and nitrite-free cured meats

Lipid oxidation in muscle foods is conveniently monitored by the use of the classical TBA test. This method is simple and has been shown to correlate well with sensory data. There are also some limitations and problems associated with the use of the TBA method. In most cases, TBA numbers of meat tend to increase over the storage period and reach a maximum value and then start to decline. This decline

Table VII. Inhibition (%) of formation of malondialdehyde (MDA) and major volatile oxidation products in meats [a]

Additive(s)	Inhibition, %					
	MDA	Hexanal	2-Pentylfuran	Nonanal	Decenal	2,4-Decadienal
SA	48	50	80	21	—	63
STPP	70	62	40	34	—	57
STPP + SA	92	88	88	61	18	64
STPP + SA + BHA	99	98	96	91	96	96
STPP + SA + TBHQ	99	98	98	92	96	96
STPP + SA + CCMP	99	98	97	91	90	90
STPP + SA + TBHQ + CCMP	99	99	98	92	96	96
NaNO$_2$ + SA	98	99	98	95	95	95

[a] Additives are: SA, sodium ascorbate at 550 ppm; STPP, sodium tripolyphosphate at 3000 ppm; BHA, butylated hydroxyanisole at 30 ppm; TBHQ, tert-butylhydroquinone at 30 ppm; CCMP, cooked cured-meat pigment at 12 ppm; NaNO$_2$, sodium nitrite at 156 ppm.

may be due to further reaction of malondialdehyde with meat constituents such as amino groups, etc. Formation of adducts has been shown to occur between malondialdehyde and the amino groups of amino acids and DNA molecules (67-69). Thus, one may not know accurately which side of the hill is being dealt with. However, during the early stages of MFD no such problems should be encountered (70).

In nitrite-cured meats, evaluation of TBA values has been suggested (71) to include addition of sulfanilamide to the mixture prior to distillation according to the method of Tarladgis et al. (72). MDA at high temperatures, reached during distillation, reacts with nitrite. This reaction leads to underestimation of its content in meat products. Sulfanilamide, when added, would intercept any free nitrite present in the meat by formation of a diazonium salt. Thus a better estimation of MDA is possible. However, we have noticed that addition of sulfanilamide poses a new problem when the residual nitrite has been decipated in the system. Sulfanilamide (S) was found to interact with malondialdehyde (M) giving rise to the formation of an adduct (SMS) involving two molecules of sulfanilamide and one molecule of MDA.

Formation of SMS was positively proved by fluorescent studies (73). However, it has been reported (74) that such a complex upon addition of the TBA (T) reagent might decompose. We have noticed that this is not necessarily true. Our studies in model systems have revealed that SMT is one of the products of such possible disproportation (75). Thus, addition of sulfanilamide is not always beneficial.

It might be suggested that the use of an extraction rather than a distillation procedure would resolve the problem. In such cases, nitrite, if present, would also be extracted into the mixture and one may not be necessarily sure that nitrosation of malondialdehyde is prevented during the color development with the TBA reagent at boiling temperatures, unless the mixture is allowed to stand for 12-14 h for color formation at room temperature. However, this hypothesis has to be tested. Nonetheless, the extraction method also suffers from its own complications (76). We have noticed that the cooked cured-meat pigment, in nitrite-cured or in pigment-treated meats is extracted into the assay solution, thus giving rise to artificially large TBA values (76). Dispersion of fat particles and proteins may also result in the turbidity of the extracted solutions and thus overestimation of TBA values. Therefore, TBA methodology although simple and adequate for a rough estimate of the oxidative state of cured meats, it is not without its complications and drawbacks when accurate testing is required.

We have previously shown that hexanal may be used as an indicator of lipid oxidation in muscle foods (79). The correlation coefficient between hexanal and sensory data was 0.981. The larger the hexanal content, the lower was the acceptability of the meat. Furthermore, hexanal was found to give a more sensitive measure of the oxidation state of cooked meats in the early stages of storage. While the TBA values after 2 days of storage for a variety of treated meat samples were similar, their hexanal content was different. These results have been supported by recent findings of Stoick et al. (78) and Drumm and Spanier (70).

Conclusions

While nitrite is unique in conferring multi-functional properties to cured meat, its antioxidant effects in retarding meat flavor deterioration and warmed-over flavor is not unique. Both natural and synthetic antioxidants, as well as adjuncts generally used in the curing process are capable of duplicating the antioxygenic role of nitrite. While the antioxidant activity of nitrite plays an important role in revealing the delicate flavour of cured meats by eliminating the production of carbonyl overtones, interaction of nitrite with meat components may produce minute quantities of flavor-active volatiles which might have important effects on the flavor of cured meats.

Acknowledgements

Financial support from the Natural Sciences and Engineering Research Council (NSERC) of Canada is gratefully acknowledged.

Literature Cited

1. Asghar, A., Gray, J.I., Buckley, D.J., Pearson, A.M. and Booren, A.M. *Food Technol.* **1988.** *42*(6):102-108.
2. Pearson, A.M., Love, J.D. and Shorland, F.B. *Adv. Food Res.* **1977.** *23*:1-74.
3. Tichivangana, J.Z. and Morrissey, P.A. *Meat Sci.* **1985.** *15*:107-116.
4. Belitz, H.-D.; Grosch, W. In: *Food Chemistry*, Springer-Verlag, Berlin, p. 155.
5. Binkerd, E.F. and Kolari, O.E. *Food Cosmet. Toxicol.* **1975.** *13*:655-661.
6. Polenske, E. *Arb. a. d. K. Gsndktsamte* **1891.** 7:471-474.
7. Lehmann, K.B. *Sher. Phys.-Med. Ges Würzb.* **1899.** *4*:57-59.
8. Kisskalt, K. *Arch. Hyg. Bakt.* **1899.** *35*:11-18.
9. Haldane, J. *J. Hyg. Camb.* **1901.** *1*:115-117.
10. Hoagland, R. **1908.** USDA Bureau of Animal Industry. 25th Annual Report. p. 301. Washington, D.C.
11. Doran, G.F. **1917.** US Patent 1,212,614.
12. Fox, J.B. Jr. *J. Agric. Food Chem.* **1966.** *14*:207-210.
13. Hadden, J.P., Ockerman, H.W., Cahill, V.R., Parrett, N.A. and Borton, R.J. *J. Food Sci.* **1975.** *40*:626-630.
14. Gray, J.I. *J. Milk Food Technol.* **1976.** *39*:686-692.
15. Gray, J.I. and Randall, C.J. *J. Food Prot.* **1979.** *42*:168-179.
16. Brooks, J., Haines, R.B., Moran, T. and Pace, J. **1940.** Food Investigation Special Report No. 49, London, HMSO.
17. Simon, S., Ellis, D.E., MacDonald, B.D., Miller, D.G., Waldman, R.C. and Westerberg, D.O. *J. Food Sci.* **1973.** *38*:919-923.
18. MacDougall, D.B., Mottram, D.S. and Rhodes, D.N. *J. Sci. Food Agric.* **1975.** *26*:1743-1754.
19. Mottram, D.S. and Rhodes, D.N. *Proc. Int. Symp. Nitrite in Meat Products.* **1974.** Zeist, Wageningen. pp. 161-171.

20. MacDonald, B., Gray, J.I., Kakuda, Y. and Lee, M. *J. Food Sci.* **1980.** *45*:889-892.
21. Bailey, M.E. and Swain, J.W. *Proc. Meat Ind. Res. Conf. American Meat Inst. Foundation*, Chicago. **1973.** pp. 29-45.
22. Fooladi, M.H., Pearson, A.M., Coleman, T.H., and Merkel, R.A. *Food Chem.* **1979.** *4*:283-292.
23. Sato, K. and Hegarty, G.R. *J. Food Sci.* **1971.** *36*:1098-1102.
24. Younathan, M.T. and Watts, B.M. *Food Res.* **1959.** *24*:728-734.
25. Zipser, M.W., Kwon, T-W. and Watts, B.M. *J. Agric. Food Chem.* **1964.** *12*:105-109.
26. Waters, W.A. *J. Am. Oil Chem. Soc.* **1971.** *48*:427-433.
27. Igene, J.O., Yamauchi, K., Pearson, A.M., Gray, J.I. and Aust, S.D. *Food Chem.* **1985.** *18*:1-18.
28. Goutefongea, R., Cassens, R.G. and Woolford, G. *J. Food Sci.* **1977.** *42*:1637-1641.
29. Zubillaga, M.P., Maerker, G. and Foglia, T.A. *J. Amer. Oil Chem. Soc.* **1984.** *61*:772-776.
30. Morrissey, P.A. and Tichivangana, J.Z. *Meat Sci.* **1985.** *14*:175-190.
31. Shahidi, F. and Hong, C. *Food Chem.* **1991.** *42*:339-346.
32. Kanner, J. *J. Amer. Oil Chem. Soc.* **1979.** *56*:74-75.
33. Kanner, J. and Juven, B.J. *J. Food Sci.* **1980.** *45*:1105-1112.
34. Shahidi, F., Rubin, L.J., and Wood, D.F. *Meat Sci.* **1988.** *22*:73-80.
35. Love, J.D. *Food Technol.* **1983.** *37*(7):117-120,129.
36. Love, J.D. and Pearson, A.M. *J. Food Sci.* **1974.** *22*:1032-1034.
37. Igene, J.O., King, J.A., Pearson, A.M. and Gray, J.I. *J. Agric. Food Chem.* **1979.** *27*:838-842.
38. Sebranek, J.G., Cassens, R.G., Hoekstra, W.G., Winder, W.C., Podebradsky, E.W. and Kielsmeier, E.W. *J. Food Sci.* **1973.** *38*:1220-1223.
39. Cassens, R.G., Greaser, M.L., Ito, T. and Lee, M. *Food Technol.* **1979.** *33*(7):46-57.
40. Folch, J., Lees, M. and Sloane-Stanley, G.H. *J. Biol. Chem.* **1957.** *226*:497-509.
41. Kanner, J., Ben-Gera, I. and Berman, S. *Lipids* **1980.** *15*:944-948.
42. Shahidi, F., Rubin, L.J. and Wood, D.F. *J. Food Sci.* **1987.** *52*:564-567.
43. Ohshima, T., Wada, S. and Koizumi, C. *Nippon Suisan Dakkaishi* **1988.** *54*(12):2165-2169.
44. Shahidi, F., Rubin, L.J. and D'Souza, L.A. *CRC Crit. Rev. Food Sci. Nutr.* **1986.** *24*:141-243.
45. Shahidi, F. In: *Flavor Chemistry - Trends and Developments*. Teranishi, R., Buttery, R. and Shahidi, F. (eds.) American Chemical Society, Washington, D.C. **1989.** pp. 188-201.
46. Cross, C.K. and Ziegler, P. *J. Food Sci.* **1965.** *30*:610-614.
47. Rubin, L.J.; Shahidi, F. In: *Proceedings of the 34th International Congress of Meat Science and Technology*. **1988.** August 29-2, Brisbane, Australia, pp. 295-301.
48. Ramarathnam, N.; Rubin, L.J.; Diosady, L.L. *J. Agric. Food Chem.* **1991.** *39*: 1838-1847.

49. Young, D.A. **1991**. Personal Communication.
50. Decker, E.A.; Hultin, H.O. *J. Food Sci.* **1990.** *55*:947-950, 953.
51. Barbut, S.; Draper, H.H. and Hadley, M. *J. Food Protec.* **1989.** *52*:55-58.
52. Shahidi, F.; Wanasundara, P.K.J.P.D. *CRC Crit. Rev. Food Sci.* **1992.** *32*:67-103.
53. Kramer, R.B. *J. Amer. Oil Chem. Soc.* **1985.** *62*:111-113.
54. Al-Jalay, B., Blank, G., McConnell, B. and Al-Khayat, M. *J. Food Protec.* **1987.** *50*:25-27.
55. Rhee, K.S., Smith, G.C. and Rhee, K.C. *J. Food Sci.* **1983.** *48*:351-352,359.
56. Ziprin, Y.A., Rhee, K.S., Carpenter, Z.L., Hostetler, R.L., Terrel, R.N. and Rhee, K.C. *J. Food Sci.* **1981.** *46*:58-61.
57. Bailey, M.E. *Food Tehcnol.* **1988.** *42*(6):123-126.
58. Sweet, C.W. **1975.** US Patent 3,899,600.
59. Hayes, R.E., Bookwalter, G.N. and Bagley, E.B. *J. Food Sci.* **1977.** *42*:1527-1532.
60. Pratt, D.E. *J. Food Sci.* **1972.** *37*:322-323.
61. Rhee, K.S., Ziprin, Y.A. and Rhee, K.C. *J. Food Sci.* **1979.** *44*:1132-1135.
62. Rhee, K.S., Ziprin, Y.A. and Rhee, K.C. *J. Food Sci.* **1981.** *46*:75-77.
63. Pratt, D.E., DiPietro, C., Porter, W.L. and Giffee, J.W. *J. Food Sci.* **1982.** *47*:24-25,35.
64. Chipault, J.R., Mizuno, G.R. and Lundberg, W.O. *Food Technol.* **1956.** *10*(5):209-211.
65. Maga, J.A. *CRC Crit. Rev. Food Sci. Nutr.* **1978.** *10*:323-372.
66. Maga, J.A. *Food Rev. Inter.* **1987.** *3*:139-183.
67. Crawford, D.L., Yu, T.C. and Sinnhuber, R.O. *J. Agric. Food Chem.* **1966.** *14*:182-184.
68. Chio, K.S. and Tappel, A.L. *Biochem.* **1969.** 8:2821-2827.
69. Reiss, U. and Tappel, A.L. *Lipids* **1973.** 8:199-202.
70. Drumm, T.D.; Spanier, A.M. *J. Agric. Food Chem* **1991.** *39*:336-343.
71. Zipser, M.W. and Watts, B.M. *Food Technol.* **1962.** *16*:102-104.
72. Tarladgis, B.G., Watts, B.M., Younathan, M.T. and Dugan, L.R. Jr. *J. Amer. Oil Chem. Soc.* **1960.** *37*:44-48.
73. Shahidi, F., Pegg, R.B., and Harris, R. *J. Muscle Foods* **1991.** 2:1-9.
74. Kolodziejska, I., Skonieczny, S., and Rubin, L.J. *J. Food Sci.* **1990.** *55*:925-928, 946.
75. Shahidi, F. and Pegg, R.B. **1990.** Can. Inst. Food Sci. Technol. Annual Meeting. Abstract No. 31. Saskatoon, SK. June 3-6.
76. Shahidi, F. and Hong, C. *Food Biochem.* **1991.** *15*:97-105.
77. Shahidi, F., Yun, J., Rubin, L.J. and Diosady, L.L. *Can. Inst. Food Sci. Technol. J.* **1987.** *20*:104-106.
78. Stoick, S.M., Gray, J.I., Booren, A.M., and Buckley, D.J. *J. Food Sci.* **1991.** *56*:597-600.
79. Hall, G.; Andersson, J. *Lebensm.-Wiss. U.-Technol.* **1983.** *16*:354-361.

RECEIVED February 19, 1992

Chapter 11

Lipid Oxidation of Seafood During Storage

George J. Flick, Jr., Gi-Pyo Hong, and Geoffrey M. Knobl

Department of Food Science and Technology, Virginia Polytechnic Institute and State University, Blacksburg, VA 24061-0418

The degradation of texture, flavor, and odor of stored seafood is attributed to the oxidation of unsaturated lipids. Processing operations as salting, cooking, and mincing promote oxidation while smoking, dehydration, and freezing retarded oxidation. The rate and degree of lipid degradation in frozen fish is dependent upon the fish species and muscle type, dark or white. Lipid oxidation proceeds in the following decreasing order: skin, dark muscle, and white muscle. Lipid oxidation within a given species will vary with season and location within the tissue. Metal ions affect oxidation in the following decreasing order: Fe^{2+}, hemin, Cu^{2+}, and Fe^{3+}. Oxidation can be reduced through the use of single or combined antioxidants. However, vacuum packaging has a greater reduction on oxidation than the presence of additives.

Fish lipids are characterized by a high degree of unsaturation in the form of multiple double bonds in the fatty acids and ready susceptibility to attack by molecular oxygen *(1)*. The chemical composition of the lipid content of fish flesh differs from that of most other naturally-occurring oils and fats in: (1) having a greater variety of lipid compounds, (2) possessing larger quantities of fatty acids with chain lengths exceeding 18 carbons, (3) containing a much greater proportion of highly unsaturated fatty acids, and (4) having polyunsaturates, primarily at the ω3 rather than ω6 position. The amounts and character of the fatty acids vary with the different organs and parts of the fish.

Lipid oxidation in fish is one of the more important factors responsible for quality loss in refrigerated and frozen storage. Peroxidation of lipids and development of rancidity in frozen herring is well documented *(2)*. Lipid oxidation is a rather complex process whereby unsaturated fatty acids react with molecular oxygen via a free radical chain mechanism, and form fatty acyl hydroperoxides, generally called peroxides, or primary products of the oxidation *(3)*. The free-radical chain mechanism has been generally accepted as the only process involved in autoxidation. The autoxidation of seafoods occurs in three stages: initiation, propagation, and termination. Hydroperoxides are the major initial reaction products of fatty acids and O_2. Once the reaction has been initiated, the hydroperoxides formed are converted to free radicals, which in turn can accelerate the rate of lipid oxidation.

Development of off-flavors are one of the greatest sensory effects of lipid oxidation. Loss of n-3 fatty acids during oxidation could be significant, but can be minimized by judicious use of antioxidants, controlled exposure to oxygen, and lower temperatures. Increased oxidation at lower humidities may be attributed to concentration of prooxidants, such as metal ions or hemoglobin.

Autoxidation involves the chemical breakdown of fat in the presence of oxygen. This process is self-generating and difficult to control. It continues even at freezing temperatures, especially with high-fat fish, and is not especially affected by antioxidants. Fish oils are converted to ketones, aldehydes, and hydroxyacids. The reactions are enhanced by iron and copper ions, so that red muscle readily becomes rancid, especially in species as tuna, swordfish, bluefish, and mackerel. Red muscle appears as a thin brownish-grey layer next to the larger portion of edible flesh. The amount of red muscle and its location within the flesh tissue varies with fish species.

Lipid peroxidation was traditionally attributed to nonenzymatic reactions *(4)*. However, McDonald et al. *(5)* and Shewfelt et al. *(6)* isolated a membrane fraction from fish skeletal muscle that catalyzed the peroxidation of its lipid component in the presence of NADH and iron. The reaction was enhanced by the presence of ADP or ATP. The NADH requirement was interpreted as providing sufficient evidence that the reaction was enzymatic in nature. Additionally, further evidence for the enzymatic nature of the reaction was provided when heating destroyed the ability of the microsomal fraction to carry out the lipid peroxidation. Slabyj and Hultin *(7)* reported that enzymatic peroxidation by light and dark muscle microsomes of herring (*Clupea harengus*) required ATP or ADP, NADH, and Fe. The optimal pH for the reaction of both types of microsomes was between 6 and 7. The energy of activation for the light and dark muscle microsomes was similar. The light muscle microsomes, however, lost activity faster than the dark muscle microsomes when exposed to 35°C.

Free radicals and peroxides are produced during lipid oxidation, which destroies vitamins A, C, and E. The nutritional value of proteins is lowered by reactions with peroxides, which also destroy pigments, bleach foods, produce toxins, and/or carcinogens, and cause off-flavors and odors. The toxicity of oxidized fish lipids is attributed to the toxicity of the hydroperoxides which cause death when ingested either as concentrates or as pure compounds. Although the exact mechanism has not been completely defined, it appears that free radicals arising from decomposition of the hydroperoxides may be responsible.

The discoloration of fish muscle is partially caused by Maillard reactions of ribose with amino groups. Fujimoto *(8)* and Fujimoto et al. *(9)* believed that reactions of lipid oxidation products are directly involved in browning processes. The primary and secondary amino groups react with oxidized lipid components.

Factors Affecting Lipid Oxidation

The nature, proportion, and degree of unsaturation of fatty acids present in a food will indicate the approximate susceptibility of that product to oxidative deterioration. The higher the proportion and degree of unsaturation of the fatty acids, the more labile the food is to oxidation *(10)*. Distribution of fat in the body; i.e., contact of fat in meat with an aqueous solution containing accelerators or inhibitors of rancidity, and the orientation of unsaturated fatty acids at an interface are also important factors in oxidative reaction in fish tissue. Presence or absence of other chemical compounds in the tissue which may act as accelerators or inhibitors of rancidity reactions, and external factors, such as heat, light, and ultraviolet rays, which tend to change the equilibrium of tissue compounds, play a significant role in lipid oxidation in fish tissue *(11)*.

A linkage between phospholipid hydrolysis and lipid peroxidation in low fat fish muscle during frozen storage was reported by Han and Liston (1987). The authors hypothesized that phospholipid hydrolysis was dependent on lipid peroxidation in fish muscle during frozen storage. However, they did not provide a relationship between lipid peroxidation and PLA_2 (phospholipase A_2). In a subsequent study, (Han and Liston, 1988), the relationship between PLA_2 and lipid peroxidation was tested by analyzing lipid peroxidation and hydrolysis indicators after lipid peroxidation inhibitors (EDTA and KCN) or promoters ($FeCl_3$) at different concentrations were introduced into rainbow trout (*Salmo gairdnerii*) muscle mince prepared for frozen storage at -10°C. EDTA had no significant effect on either hydrolysis or peroxidation but KCN caused a significant reduction in phospholipid hydrolysis as measured by lysophosphatidylcholine. The addition of ferric ion increased both lipid peroxidation and lipid hydrolysis, providing support that both of the processes were connected.

Water Content. Water acts as an antioxidant at very low levels by decreasing the catalytic activity of metal catalysts, by promoting recombination of free radicals, and by promoting nonenzymatic browning, which causes production of active antioxidants. Hydration of lipid hydroperoxides, or their concentration at lipid-water interfaces, also changes the mechanism of hydroperoxide decomposition and reduces the rate of free-radical formation *(14,15)*. At high water activities, oxidation is accelerated by increased mobilization of components that are made nonreactive at low water activities by being trapped or encapsulated within a matrix of nonreactive food components.

Lipid oxidation actually may be enhanced in dry foods. This would also lead to browning reactions, decreased protein quality, and bleaching of carotenoids.

Lipid oxidation is maximized at low a_w and high temperatures. Carbonyl products of lipid oxidation can react with the four most limiting amino acids (cystine, methionine, tryptophane, lysine) resulting in protein quality loss due to Maillard reactions.

Temperature. Temperature can affect lipid oxidation or fat stability in many ways. Generally speaking, the rate of chemical reaction is directly related to temperature. Ke et al. *(16)* showed that the rate of oxidation of mackerel lipids during frozen storage is dependent upon the temperature of storage. The rate of formation of peroxides in the skin and the dark muscle of mackerel was significantly lowered at -40°C storage than at -15°C.

Freezing. Freezing slows enzyme activity and inhibits microorganism growth but does not decrease lipid oxidation. Postmortem lipolysis can occur in chilled and frozen fish, and the major products are free fatty acids and glycerol *(17)*. Deng *(18)* showed that in long periods of ice storage, lipase and phospholipase are still active in chilled and frozen mullet. The storage life of fish is limited by the development of rancidity and other off-flavors resulting from a high unsaturated fatty acid content and high activity of the lipolytic enzymes, which leads to accumulation of free fatty acids below 0°C. Takama et al. *(19)* and Shono and Toyomizu *(20)* have reported that lipid oxidation occurs more rapidly in tissues containing free fatty acids, and there is evidence that, in some fatty species, the oxidation in cold storage is enhanced by long-term storage *(20,21)*.

Tsukuda *(22)* studied changes in the lipids of skipjack tuna during frozen storage at -10°, -20° and -30°C for 80-140 storage days. After 80 days at 0 -20°C, the free fatty acid content increased from 151 mg/100g to 1,070 mg/100g (about 30% of total lipid) in the dark muscle, and from 79 mg/100 g to 156 mg/100 g (about 20% of total lipid) in ordinary muscle due to hydrolysis of phospholipids and triglycerides. Hardy et al. *(23)* confirmed that lipolysis is the major change occurring during the frozen storage of cod. They observed that oxidation was extremely slow and occurred primarily in the phospholipid fraction. The acceptability of the fish was reduced through carbonyl oxidation, primarily through the oxidation of hept-cis-4-enal.

Fresh water whitefish (*Coregonus clupeiformis*) were stored for 16 weeks at -10°C wrapped in Saran *(24)*. An organoleptic analysis indicated that rancidity had occurred during the storage period. A chemical analysis revealed that the peroxide value (POV) rose from 13.3 (meq. thiosulphate/kg lipid) for frozen muscle with no storage to a maximum of 68.3 for 12 weeks of storage. Thereafter, the POV decreased to 26.2 after 16 weeks of storage. Hiltz et al. *(25)* reported oxidative rancidity as measured by TBA values in frozen storage at -10°C for 8 weeks in both silver hake (*Merluccius bilinearis*) fillets and mince. Maximum values of 3-5μmole malonaldehyde/100 g were recorded in some minced materials after 4-7 weeks of storage which was twice the rate in fillets. Holding the fish for up to 6 days in refrigerated sea water (RSW) at 0 - 1°C before processing extended the storage life during storage at -10°C. Mince prepared from winter (March) and summer (August) fish showed little or no difference in deterioration rates. Mince stored at -26°C for up to 6 months had negligible deterioration.

The effects of double freezing and subsequent long-term refrozen storage on inshore male capelin (*Mallotus villosus*) was studied by Botta and Shaw *(26)*. Whole capelin were stored at -23°C for 2 months and 6 months prior to thawing, beheading, and refreezing. Although the quality of the twice-frozen product was in both cases inferior to a once-frozen sample, it was still acceptable after 2 years of storage. As expected, the 2-month sample was superior to the 6-month sample. The POV of both samples reached a maximum of 18 and 14 (meq $Na_2S_2O_2/100g$ fat) for the 6- and 2-month samples respectively after 16 months of storage. The values continued to decrease during the 24-month storage period. The increase in peroxide values was not indicative of appreciable rancidity development. The rancidity (if any) probably occurred at subthreshold levels since it was not detected by the taste panels.

Siu and Draper *(27)* reported that fresh fish samples yielded lower malonaldehyde (MA) levels than the frozen samples. The toughened texture, poor flavor, and unappealing odor of poorly stored frozen seafood, has been attributed to the binding of oxidized unsaturated lipids to proteins, a process by which insoluble lipid-protein complexes are formed *(11)*.

The rate of oxidation and rancidity in frozen fatty fish has been retarded by glazing with a thin layer of ice. Jadhav and Magar *(28)* showed that the storage life of frozen oil in sardines glazed with water was 2-5 months at -20°C, depending upon its oil content. Mackerel became rancid after 4 months of storage at -20°C with a water glaze. Pomfret can be stored at -18°C with a water glaze for 3 months.

Tomas and Anon *(29)* froze salmon muscle slowly and rapidly at -25°C to determine whether the cellular damage to muscle fibers caused by slow freezing resulted in increased lipid oxidation. The fish were stored at -5°C for 47 days and periodically analyzed for oxidative changes. TBA values increased slightly but significantly with storage time, but freezing rate had no effect.

Reddy et al. (30) studied the changes in fatty acids and sensory quality of freshwater prawns stored at -18°C. The results showed that the lipids contained 23% saturated, 46% monounsaturated, and 31% polyunsaturated fatty acids. The fatty acids, especially the unsaturated ones, decreased during frozen storage for 6 months at -18°C, regardless of the packaging procedure employed. No objectionable rancid flavor was detected during the 6 months of frozen storage. The susceptibility of stored seafood products to autoxidation at low temperature may be related to the highly unsaturated long chain fatty acids present in these products *(27)*.

Changes in lipids and protein of marine catfish (*Tachysuru dussumieri*) were followed over a period of 300 days at -20°C by Srikar et al. *(31)*. Phospholipids decreased and free fatty acids (FFA) and peroxide (POV) increased (Table I). A linear relationship between the decrease in phospholipid and increase in free fatty acids was observed, suggesting that FFA were formed by hydrolysis of phospholipid. Both peroxide values and free fatty acids increased during the frozen storage. Peroxide values increased from an initial value of 0.34 mmoles to 36.67 mmoles O_2 per kg lipid after 300 days of storage, indicating oxidative deterioration. Similar observations have been made in oil sardine *(32)* and in seer fish *(33)* during frozen storage. Free fatty acids increased from 1.6 to 11.4% during the same period. During frozen storage, the proportion of saturated fatty acids remained unaltered; whereas monounsaturated acids decreased and polyunsaturated acids increased. Only marginal changes were observed in other fatty acids.

Table I. Changes in Lipids of Marine Catfish During Storage at -20°C

Storage (days)	Phospholipids 100g Flesh	%TL	POV[a]	FFA 100g Flesh	%TL
0	.655	36.4	0.34	28.8	1.6
30	.501	32.9	5.3	47.8	3.2
60	.425	28.0	11.0	73.3	4.2
90	.360	20.2	19.0	116.0	6.5
150	.296	17.7	23.0	119.0	7.1
180	.363	20.9	25.0	120.0	7.3
210	.339	20.1	26.0	173.0	10.3
300	.245	14.9	37.0	81.8	11.1

[a]mmole of O_2/kg fat

Cooking. Analysis of cooked fish showed that lipolysis also occurred at higher temperatures (34), although cooking is the most common method used to avoid enzymatic deterioration during frozen storage. Lipid oxidation and flavor deterioration of herring and cod filets during cold storage were reduced by precooking of the raw herring (35,36). The extent of lipid oxidation in cooked meaty appears to be related to the intensity of heat treatment. In a survey of the malonaldehyde (MA) content of retail meats and fish, it was reported that 38% of all fresh meat samples tested had MA contents less than 1 μg/g whereas 60% of the cooked products were in the range 1-6 μg/g (27).

Sen and Bhandary (37) showed that cooking sardine fish causes a significant decrease in oxidation rate of fatty components. Raw sardines stored at refrigerated temperatures became rancid in 2-3 days; however, cooked sardines became rancid in 6 days. Protection against oxidative rancidity observed in cooked fish may be ascribed to: 1) destruction of lipoxidase; 2) formation of water-soluble antioxidants; and 3) destruction of heme compounds (11).

Minced carp (*Cyprius carpio*) tissue was cooked by baking and deep-fat frying and stored at -18°C for time periods up to 8 weeks (38). Free fatty acids increased during frozen storage of all samples, however, there was no significant change in the composition of the fatty acids during storage (Table II). Thiobarbituric acid values were higher in the cooked samples compared to the raw samples but there was no major difference between method of cooking (Table III). The carbonyl content of the samples fluctuated during storage but no specific trend occurred.

Aubourg et al. (39) studied the effect of cooking on albacore lipids, first held on ice and then frozen and stored for 6 months at -18°C, to determine whether oxidative rancidity and free fatty acid development is affected by the duration of previous ice storage. The muscle tissue was divided into three portions reported in the commercial literature as back muscle, ventral muscle, and belly flap muscle. The fish were held on ice for 10 days and divided into two groups prior to processing. One group was frozen at -40°C and stored at

Table II. Changes in Total Lipid Content Free Fatty Acids in Raw and Cooked Minced Carp During Storage at -18°C

Storage	Raw		Baked		Deep-Fried	
(week)	TLC[a]	FFA[b]	TLC	FFA	TLC	FFA
0	5.44	229	5.85	224	11.31	207
2	5.23	259	5.78	233	9.75	216
4	5.04	258	5.84	213	11.23	213
6	5.28	281	6.50	252	12.40	232
8	5.44	336	6.18	241	11.16	215

[a]TLC = Total Lipid Content; [b]FFA = Free Fatty Acids

Table III. Changes in Thiobarbituric Acid to Carbonyl Values (CV) in Minced Raw and Cooked Minced Carp During Storage at -18°C

Storage	Raw		Baked		Deep-Fried	
(week)	TBA	CV	TBA	CV	TBA	CV
0	0.44	18.92	2.11	17.54	0.94	22.99
2	0.41	47.50	0.96	24.76	0.83	8.48
4	0.42	38.14	0.93	16.67	0.67	30.76
6	0.49	61.60	0.96	50.72	0.84	59.95
8	0.44	40.74	1.11	32.73	1.06	40.22

TBA = μmoles MA/100g Wet Sample; CV = μmoles/100g Wet Sample

-18°C for 6 months. The second group was cooked with steam at 102-103°C until a final backbone temperature of 65°C. The fish were cooled at room temperature (14°C) for about 5 hours and frozen and stored as previously described. There were no major differences in composition either between tissues or after cooking (Table IV). The lipid class contents showed significant differences between the three tissue types but not when compared on a flesh composition. The higher values obtained in the cooked samples probably were a reflection of water loss during the thermal processing procedure.

Enzymes. Yamaguchi and Toyomizu *(40)* studied lipid oxidation during cold storage of fish and found that the TBA/kg skin of fatty fish was always higher than that of lean fish, however, the TBA/10g of lipid in the skin of lean fish was frequently higher than that of fatty fish. The authors reported that the prooxidant was a dioxygenase which had an optimum temperature at 25°C and a pH value of 13 and required Co^{2+} as a stabilizer and contained Fe^{3+} in the molecule. German and Kinsella *(41)* reported that skin enzymes may constitute a significant source of initiating radicals leading to oxidation of fish lipids.

Table IV. Stability of Frozen Steam Cooked Tuna (muscle tissue) Stored at -18°C for Six Months

	Back		Ventral		Belly-Flap	
	Raw	Cooked	Raw	Cooked	Raw	Cooked
Total Lipid %	5.80	4.90	4.30	3.90	18.50	14.00
Iodine Value	163.00	165.00	164.00	163.00	164.00	165.00
POV (meq/kg)	8.00	7.60	7.70	7.00	10.30	9.00
FFA	5.30	6.80	7.00	7.40	2.40	3.70
(flesh %)	.30	.30	.30	.29	.44	.52
Phospholipid	5.70	6.90	7.70	8.50	1.90	2.30
(flesh %)	.33	.34	.33	.33	.35	.32

Cho et al. *(42)* determined the involvement of enzymes in the oxidation of fish and fishery products by investigating the effects of pre-cooking on the oxidative stability of lipids in salted and dried sardines, *Sardinops melanosticta*. Lipid oxidation was observed to proceed very rapidly even at -5°C. Oxygen absorbed into salted and dried sardines from the early stage of autoxidation was consumed readily (Figure 1), not only for hydroperoxide formation but also for secondary oxidation product formation. However, lipid oxidation was found to be remarkably retarded by pre-cooking the raw sardines. Therefore, it was recognized that enzymatic oxidation is an important factor for the oxidative deterioration of salted and dried sardines.

Species, Parts, and Season. There are tremendous variations in the fat content of various fish species. Even within a single fish itself, there is a difference in the speed with which different portions undergo rancidity. Seasonal variations in susceptibility to rancidity have also been demonstrated. Yamaguchi et al. *(43)* determined lipid oxidation (TBA values) on various tissues of round fish ordinary and dark muscle and the skin after storage at 5°C for 0-3 days or at -5°C for 35 days and on various minced tissues which had been prepared prior to storage at 5°C for 7 days or at -5°C for 14 days. Lipid oxidation in the fresh and frozen round fishes proceeded in the following order of skin > dark muscle > ordinary muscle. Preferential lipid oxidation was reported in the skins of both fatty and lean fish. In the minced tissues, however, the most pronounced lipid oxidation was observed in the dark muscles instead of the skins. The authors concluded that the preferential lipid oxidation in the skin of round fish might be partially due to the aerobic condition of the skin in comparison with the anaerobic condition of other tissues.

Oxidative rancidity in the skin and muscle lipid of oil sardine was also studied by Nair et al. *(44)*. The lipid content of this fish varies with season, from as low as 2-3% in April-May to 16-20% in November-December. When the fish was stored at -18°C, skin lipids showed an increase of about 10% in the total lipid content after 4 weeks, then remained at about this level until the end of 37 weeks. There was a significant increase in free fatty acids in muscle lipids, primarily due to the breakdown of phospholipids. They reported that peroxide values reached a maximum for both skin and muscle lipids after 4 weeks of storage, then decreased. However, the peroxide value of the skin lipids was five times higher than that of the muscle lipids at that stage. TBA (2-thiobarbituric acid test) values reached a maximum after 22 weeks, at which point the TBA of the skin was nine times higher than that of the muscle lipids.

Ke et al. *(16)* studied differential lipid oxidation in various parts of frozen mackerel. They reported an unusual difference in reaction rates between skin and muscle lipids, as observed by the difference in induction period and in the overall accumulation of oxidation products. At 60°C, the rate of oxidation of mackerel skin lipids was significantly higher than the rate of oxidation of meat lipids. The results suggest that polyunsaturated fatty acids are definitely oxidized faster than monoenes in the skin fats of frozen mackerel, a process which has parallel in mackerel oil autoxidized at 60°C. The oxidation *in vitro* shows that the faster oxidation of a skin lipid is probably not due to a greater surface/mass ratio of the skin and that there may be some fat-soluble substance in the mackerel skin lipids which catalyzes their unusual lipid oxidation.

Yamaguchi and Toyomizu *(45)* suggested that the reason lipid oxidation proceeds preferentially in the skin during cold storage of fish in the round is that the lipids buried in the skin connective tissue move to the surface from the inner layer by the disintegration of the connective tissue, thereby accelerating the oxidation process.

Ingredients. Lipid deterioration in salted (10%, 15%, and 20% NaCl) sea urchin, *Strongylocentrotus nudus*, during storage at 5°C was studied by Ohshima et al. *(46)*. TBA values (Figure 2) and contents of VBN, carbonyl, and oxo-acid were generally inversely proportional to NaCl content. The triglyceride content of the gonads, a principal component of non-polar lipids, decreased during storage while the free fatty acids increased.

The effects of salt (NaCl) on the oxidation and hydrolysis of lipids in salted sardine, *Sardinops melanostictus*, filets during storage was studied by Takiguchi *(47)*. Salted sardine filets with NaCl contents of 0.39%, 2.83%, 5.36% and 9.19% were prepared and stored at 5°, -3°, -20°, and -35°C. At -3°, -20°, and -35°C, lipid oxidation of the samples with NaCl contents of 2.83%, 5.36%, and 9.19% commenced immediately after storage, while that of the 0.39% began 5 -30 days after storage, depending on the temperature. The rates of lipid oxidation was observed to increase with increased NaCl content and the storage temperature (Figure 3).

Processing Technology. Takiguchi *(48,49)* studied lipid oxidation in boiled and dried (25°C) anchovies, *Engraulis japonicus*, having different lipid contents. Although both lean and fatty anchovies had identical initial moisture contents, the dehydration rate occurred slower in the fatty fish (Figure 4). An analysis of the fish indicated that both the rate and degree of lipid oxidation during drying differed (Figure 5). A microscopic observation of the lipid distribution in fatty anchovy meat showed that the fat migrated to the inner portion of the muscular tissue in an early stage of drying. It was, therefore, concluded that the

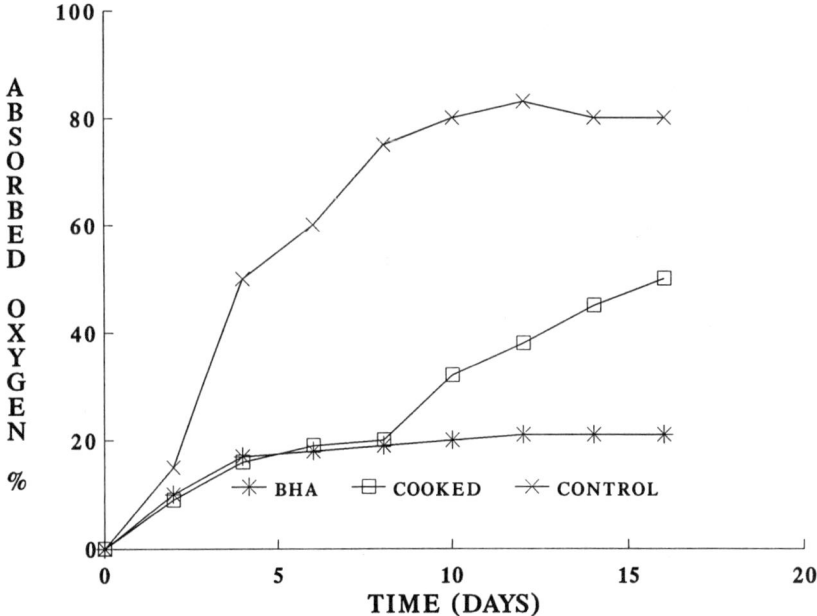

Figure 1. Oxygen absorption (percent) by cooked and salted-dried sardines during storage at -5°C.

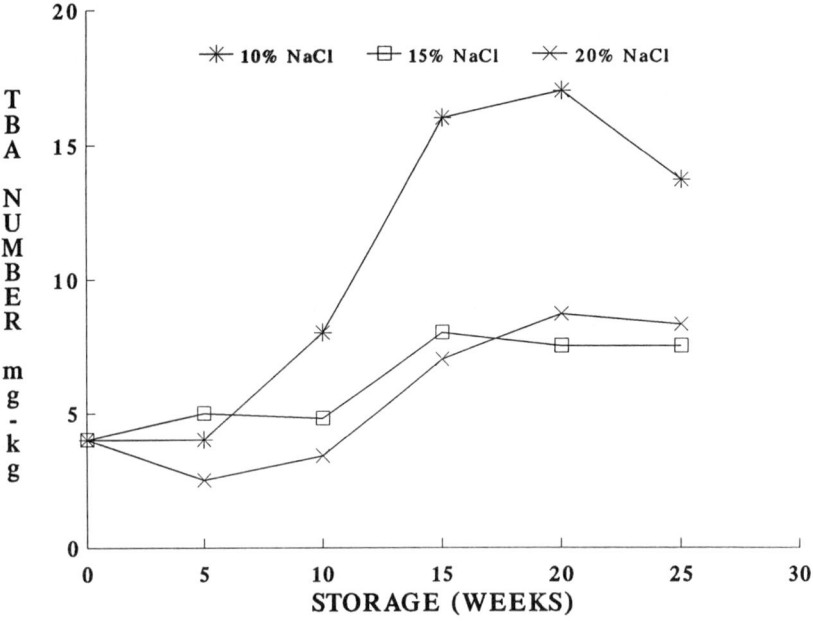

Figure 2. TBA numbers in salted sea urchins during storage at -5°C.

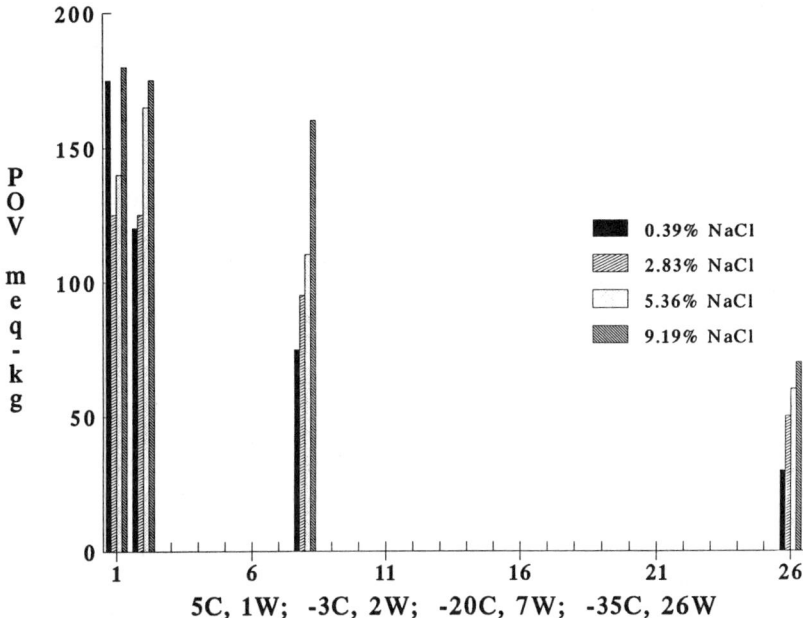

Figure 3. Peroxide values in salted sardines for 1 week at 5°C, 2 weeks at -3°C, 7 weeks at -20°C, and 26 weeks at -35°C.

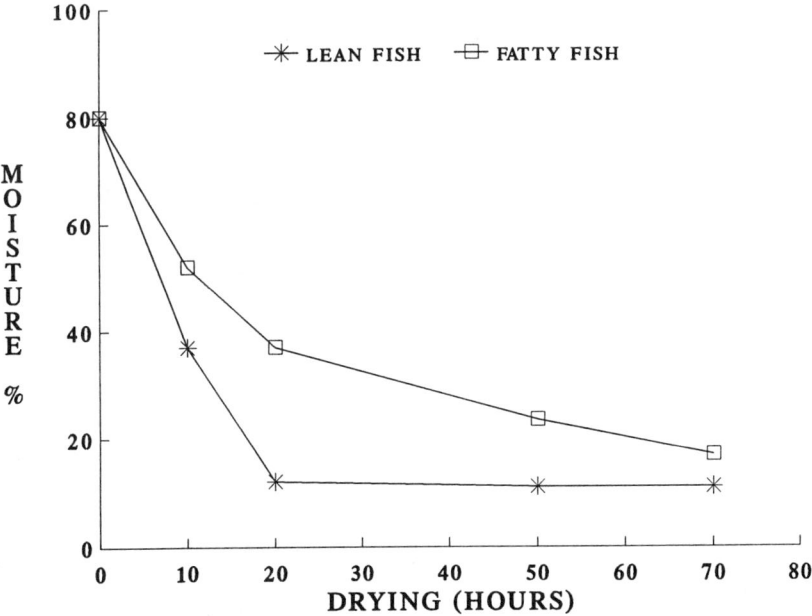

Figure 4. Percent moisture loss in dried lean and fatty anchovies during drying.

triglyceride fraction distributed in the surface portion of the fish was oxidized in an early stage of drying and that the oxidized triglyceride fraction penetrated into the inner portion of the muscular tissue to come into contact with the polar lipid fraction and participated in its oxidation. A microscopic observation on the lipid distribution in the lean meat showed that the migration of the depot fat to the inner portion of the muscular tissue did not occur during drying. This could be one of the reasons why lean fish are generally more stable to oxidation.

Takiguchi (49) investigated lipid oxidation in salted-dried anchovies, salted-dried split anchovies, and boiled-dried anchovies during drying at 30°C for 20 hrs and storage at 20°C for 50 days. The POV increased at a greater rate and to a higher degree in the salted-dry anchovies while values increased at a lower rate in the boiled-dried anchovy product (Figure 6). POVs followed a similar trend in all three products during storage. All values increased rapidly during the first 10 days of storage and decreased at a slower rate during the next 10 storage days (Figure 7). The slow increase in lipid degradation in the boiled-dried anchovy appeared to be due to the inactivation of an endogenous enzyme system involved in lipid hydrolysis occurring during boiling prior to drying. It is also possible that the heating process resulted in the formation of a water-soluble antioxidant or a significant destruction of heme compounds.

The deterioration of mechanically separated fish flesh as measured by the extent of lipid oxidation was studied by Lee and Toledo (50) using fresh mullet (*Mugil spp.*). The parameters studied included contact with iron surfaces, mechanical stress applied to the muscle during deboning, temperature of deboning drum, washing of deboned fish flesh, and post-deboning treatments including addition of antibiotic, sparging with nitrogen gas, cooking, freezing, and thawing. TBA values were used as an index of lipid oxidation. The degree of stress applied during deboning did not cause any significant change in TBA values. However, increased temperature of the deboning drum resulted in an adverse effect which became significant with increased storage (Table V). The

Table V. Effect of Drum Temperature on Lipid Oxidation in Deboned Mullet.

Temperature	TBA Value	
	3 hour	24 hour
Room Temperature (25°C)	1.58	12.22
Refrigeration (3°C)	1.52	7.82

contact of flesh with iron parts of the deboner resulted in an increase in TBA value within the short time span involved in the deboning process (Table VI). Washing improved product quality when the product was frozen for an extended period, but appeared to have no major advantage when stored under refrigeration and immediately processed (Figure 8). A marked rise in TBA value occurred during thawing with the effect being more pronounced with thawing time. The effect of thawing also became more pronounced with increased

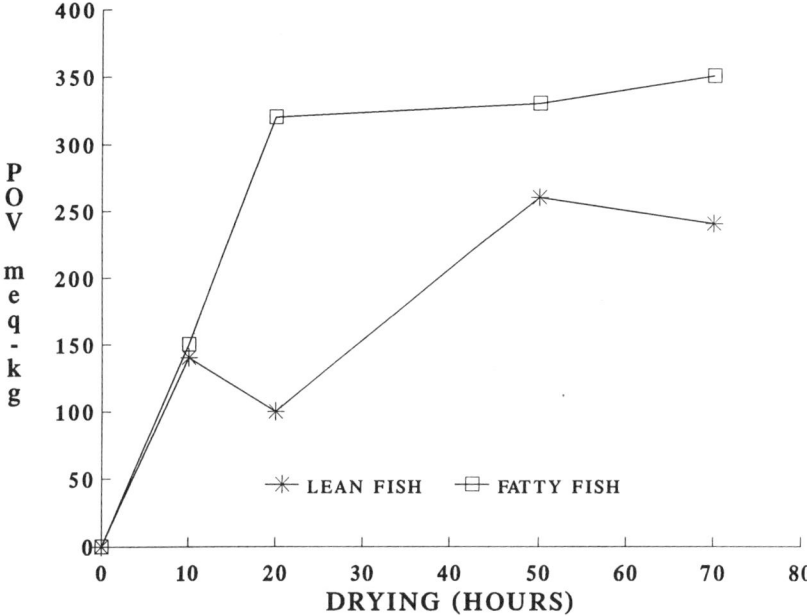

Figure 5. Peroxide values during drying in lean and fatty anchovies.

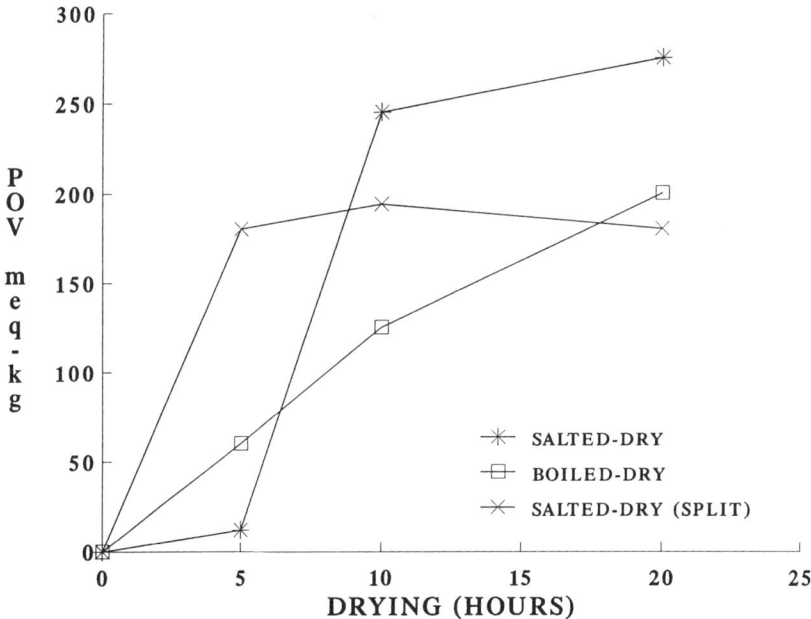

Figure 6. Peroxide values in salted and boiled anchovies during drying.

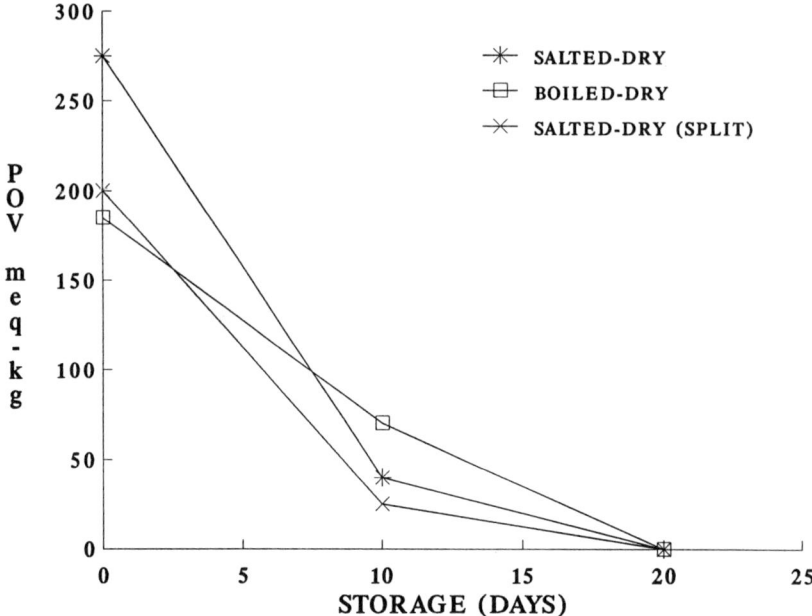

Figure 7. Peroxide values in salt-dried and boiled-dried anchovies during storage at 20°C.

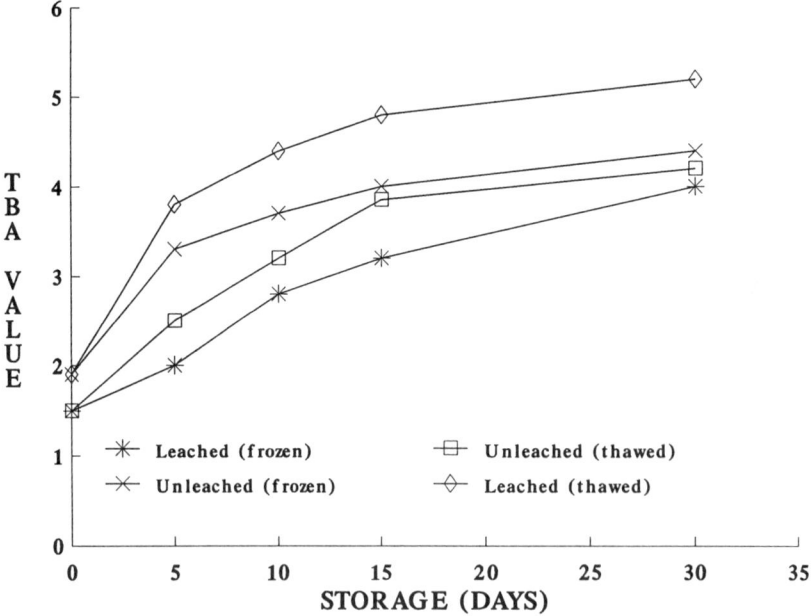

Figure 8. TBA values in washed and unwashed fresh and frozen mullet during refrigerated storage.

thawing time. Ambient oxygen and microbial activity had an insignificant role in lipid oxidation during storage. The lateral tissue (red muscle), along the visceral cavity, and bone marrow exudate appeared to be the most susceptible to the development of oxidative rancidity (Table VII). The rate of TBA change was very rapid when the red muscles had contact with iron surfaces (Figure 9).

Table VI. Effect of Contact with Iron on Lipids in Deboned Fish Muscle

Temperature	2 hour	TBA Value 1 day	3 day
Nonactivated			
leached	1.52	2.43	4.08
unleached	1.85	2.78	4.64
Iron Activated			
leached		7.74	11.45
unleached		9.39	15.81

Table VII. Oxidation of Deboned Fish Flesh According to Muscle Type and Processing Method

	Types of Muscle	TBA Value 3 hour	24 hour
Hand Extracted	light	2.30	2.87
Hand Extracted	dark	4.29	
Deboned	light	2.31	3.15
(nonactivated)	dark	4.89	
Deboned	light	5.89	9.13
(iron activated)	dark	16.04	

The effects of smoking on lipid oxidation of boiled and dried anchovies (*Engraulis japonica*) after drying and storage were investigated by Takiguchi *(51)*. Boiled anchovies were smoked for 30, 60, and 360 min at 40-65°C and air dried at 25°C. Total smoking and air-drying was adjusted to 30 h. During the drying process, POV in samples increased, with decreases in percentages of the remaining highly unsaturated fatty acids depending on smoking times. Increasing rates of oxidized acid contents in the non-smoked control and sample smoked for 30 min were higher than those of samples smoked for 60 or 360 min.

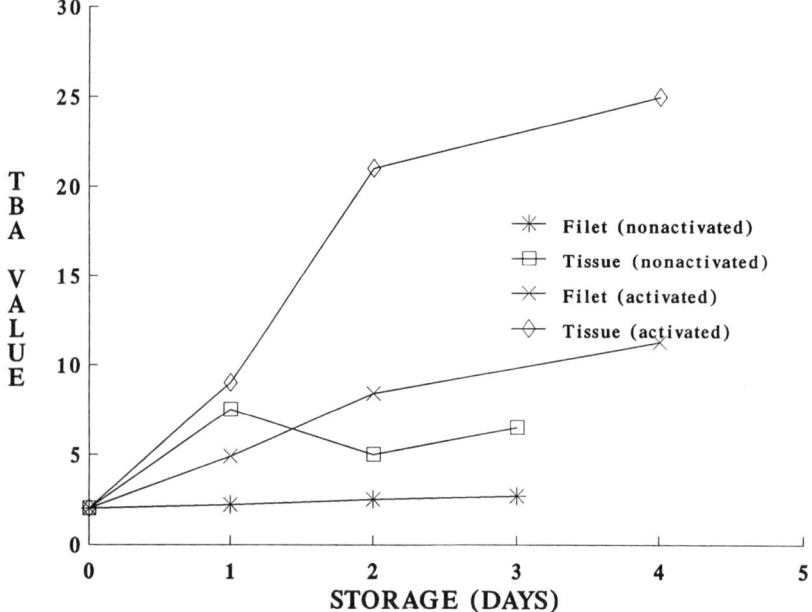

Figure 9. TBA values during storage in lateral and visceral cavity tissue exposed to iron surfaces.

The lipid stability of smoked Great Lakes whitefish (*Coregonis clupeaformis*) was affected by salt level, presence of nitrite, and the type (wood vs liquid) and level of smoking used during processing *(52)*. The addition of salt significantly (p<0.01) increased the levels of oxidation (Table VIII). Of those samples containing salt, the 4% water-phase salt (20% brine) were the least oxidized. Wood smoke exhibited significantly (p<0.01) a greater antioxidant activity than any of the levels of liquid smoke tested. The addition of nitrite to the fish significantly (p<0.01) reduced the level of oxidation, and did not result in the formation of N-nitroso compounds (Table IX).

Table VIII. TBA Values of Smoked Whitefish With and Without Nitrite Stored at 4°C

	Days of Storage			
	0	7	14	22
Control	1.06	2.07	1.94	2.04
Control + N	0.32	0.42	0.34	0.41
10% Brine	1.06	2.30	2.16	2.85
10% Brine + N	0.45	0.67	0.52	0.87
20% Brine	0.97	1.76	1.60	2.54
20% Brine + N	0.54	0.59	0.41	0.83
30% Brine	0.85	1.77	2.27	2.68
30% Brine + N	0.22	0.48	0.44	1.86

Table IX. TBA Values of Smoked Whitefish With and Without Nitrite Stored at 4°C

	Days of Storage			
	0	7	14	22
W Smoke	0.75	0.99	1.17	1.00
W Smoke + N	0.44	0.52	0.54	0.47
0.7% L Smoke	1.80	3.07	5.17	3.25
0.7% L Smoke + N	0.80	0.72	1.45	1.54
1.4% L Smoke	0.99	1.77	3.09	3.07
1.4% L Smoke + N	0.65	0.61	1.39	1.45
2.1% L Smoke	1.29	1.37	2.52	3.22
2.1% L Smoke + N	0.66	0.51	0.59	0.99

W = Wood; L = Liquid; Brine Content = 7.89%

Metal Catalysis in the Lipid Oxidation of Seafoods

It has been generally known that lipid oxidation is affected by metal ions. Ke and Ackman *(53)* and Yong and Karel *(54)* showed that Fe^{2+} and Cu^{2+} accelerated the oxidation of lipids prepared from mackerel skin and meat. Mizushima et al. *(55)* studied the effect of copper, iron, and hemin on lipid oxidation of fish flesh homogenate and found that they accelerate lipid oxidation. The relative activity of these metal ions and hemin determined at 40°C was as follows: Fe^{2+} > hemin > Cu^{2+} > Fe^{3+}. Oxidation of lipid fractions was reported to occur in the following order: total lipids > phospholipids > triglycerides *(56)*. Of the individual phospholipids, oxidation was more pronounced in the phosphatidyl ethanolamine than in the phosphatidyl choline system when compared on an equal weight basis. MacLean and Castell *(57)* reported that the addition of 1-50 ppm of copper to fish muscle reduced the induction period of frozen fish muscle to a few days. Trace amounts of copper and other heavy meatal ions induced the rancidity of both lean and fat fish muscle.

The relative importance of hemeprotein and non-heme iron as catalysts of lipid oxidation in various animal tissues has been extensively studied. Yong and Karel *(54)* found inorganic iron and copper to be strong catalysts of mackerel meat lipid oxidation, whereas Khayat and Schwall *(11)* postulated that heme iron is the major catalyst of lipid oxidation in mullet fish. The increased rate of lipid oxidation in cooked meat has been reported to be due to the release of nonheme iron during cooking *(58)*.

Halliwell and Gutteridge *(59)* suggested that a low molecular weight pool of soluble metal chelates are responsible for catalysis of lipid oxidation *in vivo*. The reduced states of copper and iron are capable of promoting lipid oxidation by catalyzing the decomposition of hydrogen peroxide and lipid peroxides to free radicals such as the hydroxyl radical *(60)*. Transition metals may be reduced *in situ* by an enzyme in the sarcoplasmic reticulum (SR) which uses NADH as a reductant *(61)*. SR-catalyzed lipid oxidation has been reported in thirteen species of Northwest Atlantic fish *(62)*.

The effect of haemoglobin (Hb) concentration and ferritin was studied on lipid oxidation in raw and heated mackerel muscle *(63)*. Lipid oxidation catalyzed by Hb increased with increasing concentration of Hb up to about 3 mg/g in raw muscle systems and thereafter remained constant as the level of Hb was increased to 10 mg/g. In heated systems, the prooxidative effect increased as the concentration of Hb increased to 10 mg/g. In raw ferritin systems, oxidation was not observed. Cooked systems, however, showed significant lipid oxidation. Ascorbic acid exerted a prooxidant effect in the ferritin system while nitrite (100 mg/kg) inhibited both Hb and ferritin-induced oxidation.

The influence of myoglobin, ferrous and ferric irons, nitrite, EDTA and ascorbate on lipid oxidation in cooked water-extracted and non-extracted mackerel (*Scomber japonica*) meat were investigated to elucidate the mechanism involved in oxidative rancidity during refrigerated storage *(64)*. Both myoglobin and ferrous iron accelerated lipid oxidation of the mackerel meat. EDTA inhibited the lipid oxidation accelerated by ferrous iron, but not that accelerated by myoglobin. Also, EDTA inhibited lipid oxidation in cooked non-extracted mackerel meat. Non-heme iron catalysis appeared to be related in part to lipid oxidation in the cooked meat. The addition of nitrite, in combination with ascorbate, resulted in a significant inhibition of lipid oxidation. It was concluded from these results that nitric oxide ferrohemochromogen, formed from added nitrite and myoglobin present in the mackerel meat, in the presence of a reducing agent, possessed an antioxidant activity which can be partially attributed to the function as a metal chelator.

Effect of Antioxidants

Fish can become rancid at sub-freezing temperatures, unless adequate precautions are taken to prevent oxygen from coming in contact with the product. Preventive measures include opaque packaging, vacuum packing, maintaining low temperatures, and prevention of overdrying.

One preventive measure of oxidative rancidity is the use of single or combined antioxidant treatments, such as vitamin E (tocopherol), propyl gallate, BHA, BHT, TBHQ, vitamin C (ascorbic acid), EDTA, phosphates, and citrate *(65)*. Specific uses for these additives are carefully regulated and may not be applicable to certain seafoods.

The storage life of fresh salmon and trout was increased when treated with combinations of either BHA or TBHQ and EDTA or citric acid *(66)*. Ke et al. *(16)* studied the antioxidative potency of TBHQ and other antioxidant compounds on oxidation of mackerel skin lipids. They found the order of effectiveness for inhibiting the oxidation in mackerel skin lipids to be TBHQ > α-tocopherol > tempeh oil > BHA > BHT, at concentrations of 0.02% for all synthetic compounds and 0.1% and 5% for α-tocopherol and tempeh oil, respectively. They also reported that TBHQ not only was the most powerful antioxidant for the unsaturated marine lipids but also retarded the formation of carbonyls from lipid hydrolysis and secondary oxidation reactions. This is in contrast to information reported by Hwang and Regenstein *(67)* who found that ascorbic acid and erythrobic acid had better antioxidant activities than tocopherols, rosemary extracts, and TBHQ. The latter antioxidants were reported to have only limited activity.

Zama et al. *(68)* reported that the rate of oxygen absorption in a mixture of mackerel oil and egg albumin containing added hemin decreased as the concentration of α-tocopherol increased when the water activity (a_w) of the system was 0.75.

The addition of 2% ethylene diaminetetraacetic acid (EDTA) was shown to effectively chelate the nonheme iron and thus significantly reduce lipid oxidation *(58)*. Functioning cooperatively with vitamin E, selenium is a vital factor in protection of lipids from oxidation as part of the enzyme glutathione peroxidase, which detoxifies products of rancid fat *(69)*.

Morrissey and Tichivangana *(70)*, in experiments with pork, chicken, and mackerel meat, reported that the antioxidative effect of nitrite was apparent even at 200 ppm and also that nitrite and nitrosylmyoglobin behaved synergistically toward the inhibition of lipid oxidation.

Smoked and non-smoked fall Atlantic mackerel (*Scomber scombrus L.*) were used by Aminullah Bhuiyan et al. *(71)* to study changes in oxidative rancidity and composition of major lipid classes and fatty acids. After smoking, there was an increase in thiobarbituric acid (TBA) values and peroxide (POV) values, but the values were still acceptable of good quality (Table X).

Table X. Lipid Stability of Atlantic Mackeral After Smoking

	TBA (μmoleMA/kg fish)	POV (meq/kg fat)
Non-Smoked	3.98	1.30
Smoked	6.86	8.96

Effect of Atmosphere and Packaging

The most obvious precaution against oxidative deterioration is preventing oxygen from coming into contact with the product. Ranken *(72)* reported that vacuum packaging or controlled atmosphere packaging (carbon dioxide and nitrogen) of meat and meat products is effective in preventing unacceptable color and rancidity development. Vacuum packaging was able to improve the sensory scores of salmon stored at -18°C *(73)*. In a subsequent study, Ke et al. *(74)* studied lipid oxidation in Atlantic mackerel (*Scomber scombris*) stored under vacuum and with CO_2 in laminated polybags. Vacuum packaging extended the storage life to more than 1 year at -26°C, even for the fatty fall fish, and to at least 6 months at -18°C. Packaging in CO_2 had no more beneficial effect than the simple vacuum treatment. Hwang and Regenstein *(67)* tested several antioxidants as well as vacuum packaging for their effect on the peroxide values and free fatty acids (FFA) of frozen minces from Gulf menhaden (*Brevoortia patronus*) and Atlantic menhaden (*Brevoortia tyrannus*). Vacuum packaging was more effective in retarding oxidative rancidity than the additives. Vacuum packaging at -20°C inhibited hydrolytic rancidity while storage at -7°C was unable to prevent FFA development.

The effect of packaging materials on the oxidation of fish sausage lipids was reported by Satomi et al. *(75)*. During storage for 5 weeks at 37°C, both degradation of nitroso-pigments and development of off-flavor proceeded in proportion to oxygen permeability of the packaging materials. When the sausages were packaged by material of zero oxygen permeability, oxidative deterioration of the flavor did not occur (Figure 10). Red colored films, followed by violet, blue, and green, were the most effective in preventing product degradation under fluorescent and sun light. Orange and yellow colors were found to have no effectiveness.

A subsequent study by the authors *(76)* determined lipid oxidation in sausages composed of 100% sardine meat or from mixed meat of 24% sardine and 76% Alaskan pollock surimi packaged in casings composed of: polyvinylidene chloride (Saran) (PVDC); or polyethylene terephthalate/aluminum foil/polypropylene (PET/Al/PP). The sausages were stored at 30°C in darkness for 30 days and oxidative deterioration was measured by the TBA value and flavor score. The PET/Al/PP film was the most effective in inhibiting lipid oxidation compared with the PVDC film. Also the oxidation deterioration of 100% sardine meat sausage was more remarkable than mixed sardine meat sausage. Addition of $NaNO_2$ or $NaNO_2$ plus sodium ascorbate was effective to inhibit oxidation in the sausage (Table XI). When the sausage was prepared from mixed meat with addition of 100 ppm $NaNO_2$ plus 1,000 ppm sodium ascorbate and with PVDC film, no appreciable oxidative deterioration occurred Figure 11). In sausage prepared from 100% sardine meat, lipid oxidation of the sausage proceeded to an appreciable level even before storage, even though PET/Al/PP film was used.

Table XI. Flavor Score of Fish Sausage from 24% Sardine and 76% Surimi During Storage at 30°C

Additive	Storage Time (Days)				
	0	8	13	19	30
None (Control)	5	5	3	3	2.5
$NaNO_2$	5	5	3.5	3	2.5
$NaNO_2$ and NaAs	5	5	5	4.5	4.5

The rates of lipid oxidation of freeze-dried big-eye tuna (*Thunnus obesus*), a typical red muscle fish, and halibut (*Hippoglossus stenolepis*), a typical white muscle fish, were studied as a function of moisture equilibrium relative humidity at 25°C *(77)*. At relative humidities of 0% and 11%, corresponding to below the monomolecular layer of water, the lipids of both fish underwent oxidation, as estimated by TBA values. The rates of oxidation were faster in the big-eye tuna than in halibut, with the greatest differences occurring after 12 days of storage. On day 12, TBA values were 1.8 and 2.1 (optical density, 532 nm) for halibut and tuna, respectively. By day 19, the values increased to 3.4 and 6.5. At higher relative humidities, such as 52% and 71%, neither fish underwent oxidative deterioration.

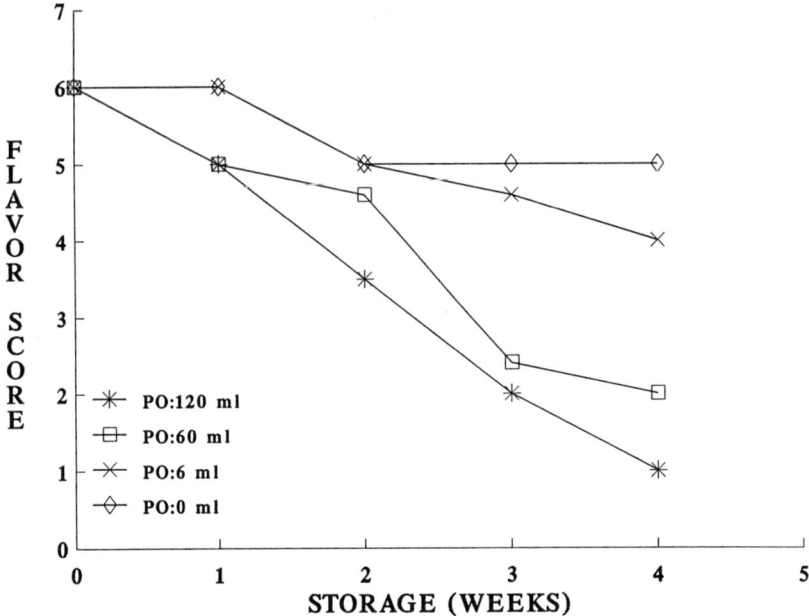

Figure 10. Effect of oxygen permeable packaging on flavor scores of fish sausages during storage at 37°C.

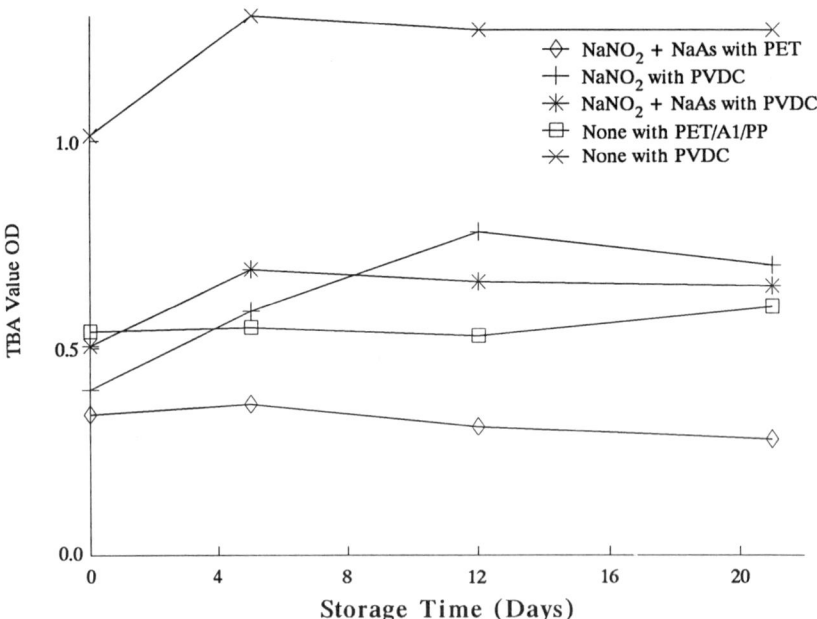

Figure 11. Effect of additives and packaging materials on TBA values of fish sausage during storage at 30°C.

Literature Cited

1. Olcott, H.S. In *Fish in Nutrition*; Heen, E. and Kreuzer, R., Eds.; Fishing News Books Ltd.: London, England, **1962**, pp. 112-116.
2. Banks, A. *J. Sci. Food Agric.* **1952**, 3:250-256.
3. Gray, J.I. *J. Am. Oil Chem. Soc.* **1978**, 55:539-546.
4. Castell, C.H. *J. Am. Oil Chem. Soc.* **1971**, 48:645-649.
5. McDonald, R.E.; Kelleher, S.D.; Hultin, H.O. *J. Food Biochem.* **1979**, 3:125-134.
6. Shewfelt, R.L.; McDonald, R.E.; Hultin, H.O. *J. Food Sci.* **1981**, 46:1297-1301.
7. Slabyj, B.M.; Hultin, H.O. *J. Food Sci.* **1982**, 47:1395-1398.
8. Fujimoto, K. *Nippon Suisan Gakkaishi.* **1970**, 36:850-853.
9. Fujimoto, K.; Abe, I.; Kaneda, T. *Nippon Suisan Gakkaishi.* **1971**, 37:40-43.
10. Dawson, L.E.; Gartner, R. *Food Technol.* **1983**, 37:112-116.
11. Khayat A.; Schwall, D. *Food Technol.* **1983**, 37(7):130-140.
12. Han, T.J.; Liston, J. *J. Food Sci.* **1987**, 52:294-299.
13. Han, T.J.; Liston, J. *J Food Sci.* **1988**, 53(6):1917,1919.
14. Karel, M. In *Autoxidation in Food and Biological Systems*; Simic, M.G. and Karel, M., Eds.; Plenum Press: New York and London, **1980**, pp. 191-206.
15. Quast, D.G.; Karel, M. *J. Food Sci.* **1972**, 37:584-588.
16. Ke, P.J.; Ackman, R.G.; Linke, B.A.; Nash, D.M. *J. Food Tech.* **1977**, 12:37-47.
17. Hardy, R.; Smith, J.G.M. *J. Sci. Food Agric.* **1976**, 27:595-599.
18. Deng, J.C. *J. Food Sci.* **1978**, 43:337-340.
19. Takama, K.; Zama, K.; Igarashi, H. *Bull. Fac. Fish. Hokk. Univ.* **1971**, 24(4):290-300.
20. Shono, T.; Toyomizu, M. *Bull. Jap. Soc. Sci. Fish.* **1973**, 39(4):417-421.
21. Smith, J.G.M.; McGill, A.S.; Thompson, A.B.; Hardy, R. In: *Advances in Fish Science and Technology*; Connell, J.J. Fishing News Books Ltd.: London, England, **1980**, pp. 303-307.
22. Tsukuda, N. *Bull. Tokai Regional Fish. Res. Lab.* **1976**, 84:31-42.
23. Hardy, R.; McGill, A.S.; Gunstone, F.D. *J. Sci. Food Agric.* **1979**, 30:999-1006.
24. Awad, A.; Powrie, W.D.; Fennema, O. *J. Food Sci.* **1969**, 34:1-9.
25. Hiltz, D.F.; Lall, B.S.; Lemon, D.W.; Dyer, W.J. *J. Fish. Res. Bd. Can.* **1976**, 33:2560-2567.
26. Botta, J.R.; Shaw, D.H. *J. Fish. Res. Bd. Can.* **1978**, 35:452-456.
27. Siu, G.M.; Draper, H.H. *J. Food Sci.* **1978**, 43:1147-1149.
28. Jadhav, M.G.; Magar, N.G. *Fish Technol.* **1970**, 7:158-163.
29. Tomas, M.C.; Anon, M.C. *Int. J. Food Sci. Technol.* **1990**, 25:718-721.
30. Reddy, S.K.; Nip, W.K.; Tang, C.S. *J. Food Sci.* **1981**, 46:353-356.
31. Srikar, L.N.; Seshadari, H.S.; Fazal, A.A. *Int. J. Food Sci. Technol.* **1989**, 24:653-658.
32. Srikar, L.N.; Hiremath, G.G. *J. Food Sci. Technol.* **1972**, 9(4):191-193.

33. Fazal, A.A.; Srikar, L.N. *J. Food Sci. Technol.* **1987**, 24:303-305.
34. Quaglia, G.B.; Audisio, M. *Proceedings of IV International Congress of Food Science and Technology;* 1974; Vol 1, 682-688.
35. Bosund, I; Ganrot, B. *J. Food Sci.* **1969**, 34:13-18.
36. Bosund, I; Ganrot, B. *Lebens. Wissen. Technol.* **1970**, 3:71-73.
37. Sen, D.P.; Bhandary, C.S. *Leben. Wiss. u Technol.* **1978**, 2:124.
38. Mai, J.; Kinsella, J.E. *J. Food Sci.* **1979**, 44:1619-1624.
39. Aubourg, S.; Perez-Martin, R.; and Gallardo, J.M. *Int. J. Food Sci. Technol.* **1989**, 24:341-345.
40. Yamaguchi, K.; Toyomizu, M. *Bull. Jap. Soc. Sci. Fish.* **1984a**, 50(12):1905-1908.
41. German, J.B.; Kinsella, J.E. *J. Agri. Food. Chem.* **1985**, 33:680-683.
42. Cho, S-Y.; Endo, Y.; Fujimoto, K.; Kaneda, T. *Nippon Suisan Gakkaishi.* **1989**, 55(3):541-544.
43. Yamaguchi, K.; Nakamura, T.; Toyomizu, M. *Bull. Jap. Soc. Sci. Fish.* **1984**, 50(5):869-874.
44. Nair, P.G.V.; Gopakumar, K.; Mair, M.R. *Fish Technol.* **1976**, 13(2):111-114
45. Yamaguchi, K.; Toyomizu, M. *Bull. Jap. Soc. Sci. Fish.* **1984b**, 50(12):2049-2054.
46. Ohshima, T.; Wada, S.; Koizumi, C. *Bull. Jap. Soc. Sci. Fish.* **1986**, 52(3):511-517.
47. Takiguchi, A. *Nippon Suisan Gakkaishi.* **1989**, 55(9):1649-1654.
48. Takiguchi, A. *Bull. Jap. Soc. Sci. Fish.* **1986**, 52(6):1029-1034.
49. Takiguchi, A. *Nippon Suisan Gakkaishi.* **1987**, 53(8):1463-1469.
50. Lee, C.M.; Toledo, R.T. *J. Food Sci.* 1977, 42(6):1646-1649.
51. Takiguchi, A. *Nippon Suisan Gakkaishi.* **1988**, 54(5):869-874.
52. Cuppett, S.L.; Gray, J.I.; Booren, A.M.; Price, J.F.; Stachiw, M.A. *J. Food Sci.* **1989**, 54(1):52-54.
53. Ke, P.J.; Ackman, R.G. *J. Am. Oil Chem. Soc.* **1976**, 53:636-640.
54. Yong, S.H.; Karel, N. *J. Am. Oil Chem. Soc.* **1978**, 55:352-357.
55. Mizushima, Y.; Takama, K.; Zama, K. *Bull Faculty Hokkaido U.* **1977**, 18(4):207.
56. Tichivanagana, J.Z.; Morrissey, P.A. *Int. J. Food Sci. Technol.* **1984**, 8:47-57.
57. MacLean, J.; Castell, C.H. *J. Fish. Res. Bd. Can.* **1964**, 21:1345-1359.
58. Igene, J.O.; King, J.A.; Pearson, A.M.; Gray, J.I. *J. Agric. Food Chem.* **1979**, 27:838-842.
59. Halliwell, B.; Gutteridge, J.M.C. *TIBS.* **1986**, 11:372-375.
60. Kanner, J.; German, J.B.; Kinsella, J.E. *CRC Crit. Rev. Food Sci. Nutr.* **1987**, 25:317-364.
61. McDonald, R.E.; Hultin, H.O. *J. Food Sci.* **1987**, 52:15-21,27.
62. Decker, E.A.; Erickson, M.C.; and Hultin, H.O. *Comp. Biochem. Physiol.* **1988**, 91(b):7-9.
63. Apte, S; Morrissey, P.A. *Food Chem.* **1987**, 25:127-134.
64. Ohshima, T.; Wada, S.; Koizumi, C. *Nippon Suisan Gakkaishi.* **1988**, 54(12):2165-2171.

65. Ladikos, D.; Lougovois, V. *Food Chem.* **1990**, 35:295-314.
66. Sweet, C.W. *J. Food Sci.* **1973**, 38:1260-1261.
67. Hwang, K.T.; Regenstein, J. *J. Food Sci.* **1988**, 54:1120-1124.
68. Zama, K.; Takama, K.; Mizushima, Y. *J. Food Proc. Preserv.* **1979**, 3:249-264.
69. Pigott, G.M.; Tucker, B.W. *Seafood: Effect of Technology on Nutrition;* Marcel Dekker, Inc.; New York and Basel.
70. Morrissey, P.A.; Tichivangana, J.Z. *Meat Sci.* **1985**, 14:175-190.
71. Aminullah Bhuiyan, A.K.M.; Ratnayake, W.M.N.; and Ackman, R.G. *JAOCS.* **1986**, 63(3):324-328.
72. Ranken, M.D. *Food Sci. Technol. Today.* **1987**, 1:166-168.
73. Yu, T.C.; Sinnhuber, R.O.; Crawford, D.L. *J. Food Sci.* **1973**, 38:1197-1199.
74. Ke, P.J.; Nash, D.M.; Ackman, R.G. *J. Inst. Can. Sci. Technol. Aliment.* **1976**, 9(3):135-138.
75. Satomi, K.; Sasaki, A.; Yokoyama, M. *Nippon Suisan Gakkaishi.* **1988a**, 54(3):517-521.
76. Satomi, K.; Sasaki, A.; Yokoyama, M. *Nippon Suisan Gakkaishi.* **1988b**, 54(12):2211-2215.
77. Koizumi, C.; Iiyama, S.; Wada, S.; Nonaka, J. *Bull. Jap. Soc. Sci. Fish.* **1978**, 44(3):209-216.

RECEIVED February 28, 1992

Chapter 12

Seafoods and Fishery Byproducts

Natural and Unnatural Environments for Longer Chain Omega-3 Fatty Acids

R. G. Ackman and H. Gunnlaugsdottir

Canadian Institute of Fisheries Technology, Technology University of Nova Scotia, P.O. Box 1000, Halifax, Nova Scotia B3J 2X4, Canada

The highly unsaturated fatty acids characteristic of fish lipids include those with five (20:5n-3) and six (22:6n-3) ethylenic bonds. These are part of the total fatty acid intake in food systems in many parts of the world and there is really no other original food source except food fish and shellfish. In most fish the edible muscle is a friendly environment for highly unsaturated fatty acids, especially in frozen storage. However a variety of species-specific problems can be attributed to subdermal fats, or to trimethylamine oxide as a source of oxygen, or to enzyme activity even in frozen storage. Actual oxidation in minces is a man-made situation. One way to transfer these oxidation-susceptible fatty acids to our diet is to feed fish meal as a protein source to the broiler chicken. The oxidation status of the lipids in the original fish meal has been shown to have little effect on flavor when fish meal is fed to fowl at practical levels. The growing aquaculture industry also uses this protein source and salmonids benefit from the long-chain n-3 fatty acids, but fish oil and cheap fats are also involved. Fish meal and fish silage are both unnatural environments for these fatty acids, and oxidation does take place. Nutritional applications of longer-chain fatty acids from these sources require new and detailed studies.

Fish (and shellfish) provide an almost unlimited variety of food resources which are, even in the North American region, an endless source of confusion in the popular and scientific nomenclature (*1-3*). For the purposes of this paper shellfish will be omitted and in the first section the discussion will focus on fish fillets, the most popular and widespread form of fish eaten in our western society. The form of the fish fillet is critical to the access of atmospheric oxygen but this is very important only in the case of fatty fish. The fish or fish product which is physically intact provides a "natural" environment for external and internal oxidation processes, and this environment can be degraded through filleting, skinning of fillets, mincing, and ultimately in the production of fish meal. Fish meal and fish silage, commercial products for non-human use, will be the subject of later sections of this paper.

The Distribution of Fat in Fish: Skin Fat as a Quality Factor

The extremes for fat distribution in fish can be typified by two quite different fish. One is the North Atlantic cod, *Gadus morhua* (*4*), where the muscle lipid is $\leq$ 1%. The fish energy reserve is stored in the liver, which is 30-50% fat and the source of cod liver oil (*5,6*). The other fish is the North Atlantic mackerel, *Scomber scombrus*. Figure 1 shows the distribution of fat in whole mackerel from Norwegian waters (*7*), for spring and fall fish. The important points to note are that the light muscle has a variable level of fat, and that the subdermal fat layer has nearly half of the total fat in the fish.

The mackerel is noted for the rapidity with which oxidation can develop (*8-10*), and the skin lipids are of special interest since skin is regarded as the main source of oxidative problems even if left on the fish. Lindsay and colleagues (*11,12*) showed that in some species there was a potential for *in situ* biochemical oxidation of fish skin lipids, probably by lipoxygenase activity, leading to the species-specific aromas of many species of fish. These aromas are basically alcohols, ketones and aldehydes, especially the latter. There is no record of such an examination being followed up for Atlantic mackerel, but one treatment to reduce biochemically-generated aromas in fish (*11*) was foreshadowed by a pilot study on mackerel fillets with microwave conducted by Ke et al. (*13*). Figure 2 shows the peroxide values developing in skin and meat lipids during subsequent frozen storage. The objective, to discourage enzyme activity in mackerel by the microwave treatment, was not attained, possibly since only a modest temperature rise was detected during that treatment. In support of the theory that the limited energy was absorbed by intra- and intercellular fluids rather than by the subdermal fat layer, where one would expect the enzymes to be least active, the muscle enzymes remained relatively active, leading to greater development of free fatty acids in muscle fat than in skin fat (Figure 3). Overall, enzymes were probably little affected. It is regrettable that the type of product treatments sought by the fishing industry are all too often something of this type - fast and cheap. Research into fundamental biochemical processes is therefore usually not supported by that industry. An in-depth study on mackerel stability conducted in the United Kingdom and reported (*14*) in 1976 agreed with our findings in Halifax (*9*) that the lowest possible frozen storage temperature (-30°C) was mandatory to preserve quality in mackerel for several months.

The observation that subdermal oil removed from mackerel goes rancid faster than oil from muscle implied a lack of some unspecified antioxidant (*8,9*). The later findings by Lindsay and colleagues (*11,12*) that the subdermal fat might be subjected to continuous natural enzymatic "biochemical autoxidation" processes supports this view, since the presence of an antioxidant in the skin fats would likely counteract the benefits of that enzyme process.

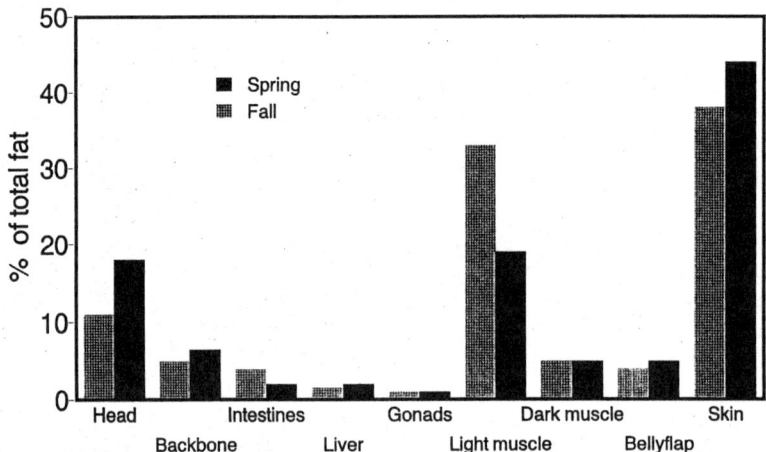

Figure 1. Distribution of total fat in various body parts and organs of mackerel *S. scombrus* of Norwegian origin. Note that the percentage of fat in a given part (e.g., mackerel dark muscle) can be higher than in a larger part (e.g., light muscle), which may contain more of the total fat. (Redrawn from ref. 7. Canadian Fisheries and Marine Service Translation No. 4119, Ottawa).

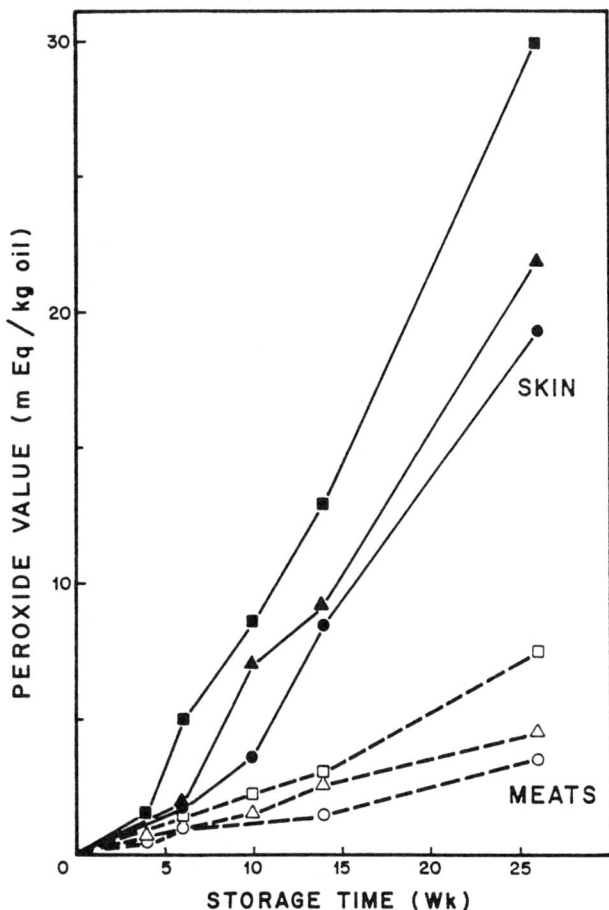

Figure 2. The changes in peroxide values in the skin (solid line) and meat (broken line) samples of mackerel fillets frozen at −15 °C after microwave pretreatment for 0 (○ ●), 10 (△ ▲), and 20 sec (□ ■), respectively. (Reproduced with permission from ref. 13. Copyright 1978 Institute of Food Technologists.)

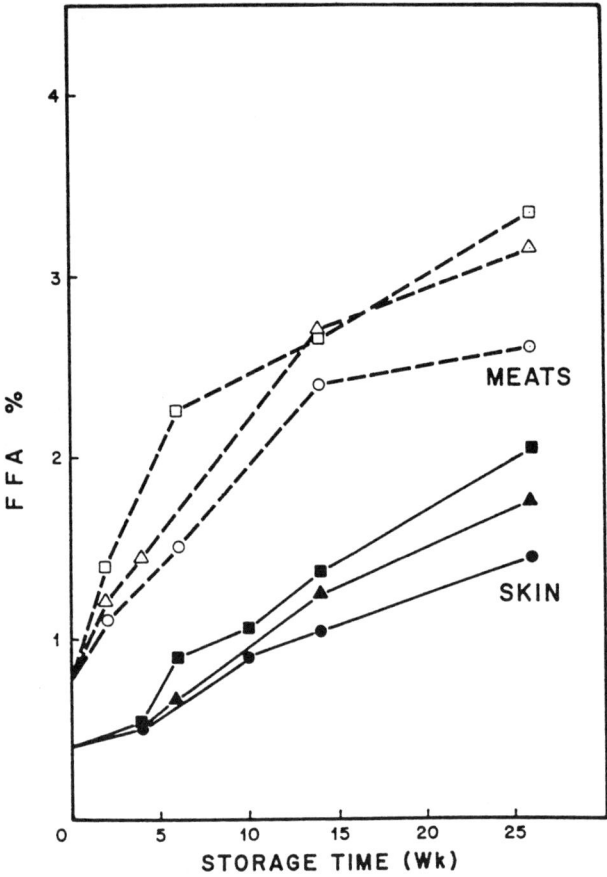

Figure 3. The changes in free fatty acids in the skin (solid line) and meat (broken line) samples of mackerel fillets frozen at −15 °C after microwave pretreatment for 0 (○ ●), 10 (△ ▲), and 20 sec (□ ■), respectively. (Reproduced with permission from ref. 13. Copyright 1978 Institute of Food Technologists.)

Table I. Recommended Frozen Storage Life of Fish and Shellfish (months)[a]

Product Type	Prevailing Storage Temperature		
	-18°C	-25°C	-30°C
Fatty fish; sardines, salmon, ocean perch	4	8	12
Lean fish; cod, haddock	8	18	24
Flat fish; flounder, plaice, sole	9	18	24
Lobster, crabs	6	12	15
Shrimp	6	12	12

[a]Source: International Institute of Refrigeration (IIR); Adapted from Infofish International (No. 2) March/April, 1991.

The popular Atlantic redfish (*Sebastes sp.*), sold in the U.S. market as ocean perch, is attractive because it is packaged as the fillet with the red skin on. This fish has deposits of almost pure triglyceride fat at the nape of the neck and also on each side on the flanks. In larger fish these fat bodies can pass through the fillet and it is important that the filleting operations not smear this fat over the inner surfaces where it can readily oxidize. The oil produced commercially from redfish filleting scrap has a low iodine value (115-125) and only modest levels of highly unsaturated fatty acids (*15*). It is thus likely that the physical spreading of this oil in a thin layer, exposing it to oxygen, is primarily responsible for occasional reports of rancidity in this species and for the short recommended storage life shown in Table I. The phospholipids are unexceptional in fatty acid composition. Carotenoids are a possible factor in fish lipid oxidation (*16*), but these are not obvious in the subdermal fat or fat bodies.

Deep skinning to remove subdermal fat can lead to an economic loss of fillet meat. In the case of the Antarctic fish "marmorbarsch" (*Notothenia rossi mormorata*), the fillets in summer and autumn have about 9.2% fat, but deep skinning reduces the fillet fat to 5.3% with a concomitant loss of approximately 12% of fillet weight (*17*). This is necessary because the subdermal fat is easily oxidized. The fillet lipid apparently should then be normal in composition (i.e., $\leq$ 1% phospholipids and 4-5% triglycerols. Much the same problem exists with wax esters for the deep-water New Zealand orange roughy (*Hoplostethus atlanticus*) (*18*). These have a purgative effect in man. Deep skinning (3-4 mm) removes most of the subdermal fat and the pearly white flesh is a popular export item (*18*). The low level ($\leq$40 mg/g) of wax esters left in or on this meat has no effect on man and there are no problems with fillet oxidation because the wax esters do not contain polyunsaturated fatty acids in any quantity (*18,19*). These two examples are intended to show the potential diversity of subdermal fats which are often not considered as factors in fishery product processing and storage.

Fillets and Fillet Lipids

Fish fillets are seldom homogeneous. There are two types of muscle in fish, white and red (or dark). The latter is normally the outer layer and may be localized under the skin and along the flanks but is usually a greater portion of a transverse section

Table II. Lipid Contents (w/w%) of Various Tissues in Dorsal Portions of Japanese Sardines Caught in Different Seasons

Season	White Muscle Lipid			Red Muscle Lipid		
	Total	Polar	Neutral	Total	Polar	Neutral
April	0.89	0.70	0.19	2.74	1.89	0.85
June	4.30	0.46	3.84	15.5	1.81	13.7
August	8.66	0.76	7.90	14.8	2.10	12.7
October	5.14	0.62	4.52	10.0	1.86	8.1

SOURCE: Reproduced from ref. 23. Copyright 1988 Tokyo University of Fisheries.

of muscle near the tail. In many species the dark muscle is a substantial part of the fish (20,21). It often contains lipid droplets (20). The dark muscle is easily visible in the Atlantic mackerel and some lipid data has been recorded for this species (22), agreeing with Figure 1 in most respects except that dark muscle did show seasonal variations in the fat. One critical point in this muscle is the greater content of phospholipids, 1.6% in the dark muscle versus 0.5% in the white muscle. This reflects the greater metabolic activity of this tissue. Recent data (23) for the dorsal portion of the body of the Japanese sardine (Sardinops melanosticta) is given in Table II. This dark tissue is also richer in heme pigments and this is often considered to be an important catalyst in non-enzymatic oxidation processes (24-26). The tendency for specialists to work on one narrow aspect of fish quality is occasionally overcome. Thus Nishimoto et al. (27) showed that at 0°C skipjack tuna (Katsuwonus pelamis) phospholipids and neutral lipids both yielded free fatty acids which then readily oxidized. The term "surface muscle" used by these authors evidently refers to dark muscle and was more susceptible to these processes than deep muscle in round fish.

There has been much confusion in the origin of free fatty acids between phospholipids and neutral lipids during frozen storage, and it may in fact be another case of species-specific differences (28). Our recent work (29) with the Atlantic salmon (Salmo salar) indicates that free fatty acids primarily came from the polar phospholipids, although the fillets contained much more neutral lipid (Table III). In this work, total lipid was recovered and carefully examined. For this extraction the Bligh and Dyer chloroform-methanol solvent system (4) was used. There is no obvious better solvent system, although the less toxic mixture hexane-isopropanol has considerable potential (30). The fractionation of the lipids by thin-layer chromatography and their determination by flame ionization (Chromarod-Iatroscan TLC-FID) is now an established technology for lipid class analyses (31,32). Titration of recovered fish muscle lipids for determination of free fatty acids requires correction for the phospholipids (33), as shown in Table IV. In our salmon study the source of free fatty acids was identified as the phospholipids. In the case of TLC-FID of some Atlantic cod (G. morhua) fillets the free fatty acids were revealed to be a mixture of saturated (probably 14:0 and 16:0) and polyunsaturated (probably 20:5n-3 and 22:6n-3) fatty acids (34), perhaps mostly from the phospholipid phosphatidylcholine (35). This is as yet not confirmed for cod.

Table III. Lipid Classes (g/100g tissue) of Muscle of Atlantic Salmon (October Sample) Fed Different Diets and Analyzed Before and After Frozen Storage [a]

Diet Fat	Treatment[b]	Lipid Class			
		Phospholipid	Cholesterol	Triglyceride	FFA[c]
Herring oil	F	0.67±0.03	0.07±0.01	5.14±0.03	Tr
	S	0.58±0.05	0.02±0.01	5.09±0.15	0.19±0.1
Canola oil	F	0.66±0.11	0.02±0.01	5.95±0.20	Tr
	S	0.65±0.03	0.02±0.01	5.73±0.12	0.23±0.01
EPA/DHA	F	0.54±0.15	0.05±0.01	3.76±0.13	0.04±0.01
concentrate	S	0.43±0.11	0.04±0.01	3.75±0.15	0.16±0.03
Egg lipid	F	0.56±0.06	0.06±0.01	3.81±0.15	0.01±0.005
	S	0.44±0.01	0.05±0.01	3.72±0.12	0.23±0.05

[a] SOURCE: Reproduced from ref. 29. Copyright 1991 Food and Nutrition Press.
[b] F = frozen briefly at -30°C; S = first frozen at -30°C, then stored at -12°C for 3 months.
[c] FFA = free fatty acids
Note: Generally 8 Chromarods were used. Standard deviations are given for lipid from one fish, but results are averaged for 2 fish.

Table IV. Analysis of Total Lipids from Frozen Hake (*Merluccius capensis*) Fillets, and Hake Mince, Stored at -18°C (g kg⁻¹) [a]

Test Material	Storage Time (Days)	Lipids in Wet Fish	FFA[b] in TL[b]	P[b] in TL	Genuine FFA[c] in TL	Genuine FFA in Wet Fish	Organoleptic Rating (0-5)
Fillets	0	11.8	42.8	13.7	4.0	0.05	4.9
	59	8.5	199.7	15.7	155.2	1.32	4.3
	102	8.7	257.2	15.1	214.6	1.87	3.5
	165	26.1	154.8	6.0	137.9	3.60	2.9
	221	13.0	253.6	7.6	232.1	3.02	2.1
	305	34.9	203.3	4.2	191.3	6.68	1.8
	Mean	17.2	--	--	--	--	--
Mince	0	15.8	47.2	12.8	10.9	0.17	4.7
	59	13.7	154.3	11.2	122.6	1.68	2.0
	102	15.2	180.0	10.2	151.9	2.31	2.0
	165	14.8	221.0	10.1	192.5	2.85	1.0
	221	14.8	241.7	9.9	213.8	3.16	1.0
	305	14.4	350.8	9.6	323.6	4.66	0.5
	Mean	14.8	--	--	--	--	--

[a] SOURCE: Reproduced from ref. 33. Copyright 1991 Elsevier.
[b] FFA = Free fatty acids; TL = Total lipids; P = Phosphorus
[c] Genuine FFA = FFA - 2.83 x P.

Table V. Comparison of Fatty Acids of Non-polar (NP; Primarily Triacylglyceride) and Polar (P; Primarily Phospholipid) Lipid Classes in a Typical North Atlantic Marine Fish with Lipids of One Coldwater and Two Warmwater Freshwater Fish

	Marine		Freshwater			
	Capelin[a]		Rainbow Trout[b]		NA Catfish[c]	Indian Murrel[d]
			Lipid Type			
Fatty Acid	NP	P	NP	P	Total	Total
14:0	8.0	1.8	2.8	1.3	1.4	1.6
16:0	9.6	17.6	18.2	19.6	18.3	19.0
16:1	13.9	4.5	5.7	2.4	4.6	4.0
18:1	8.1	9.0	32.1	13.2	50.0	30.1
18:2n-6	1.2	1.0	19.4	8.8	12.0	4.9
18:3n-3	0.3	0.5	0.5	0.3	0.9	1.2
18:4n-3	0.8	0.4	0.6	0.2	0.1	TR
20:1	14.6	2.8	1.8	0.5	1.3	3.0
20:4n-6	0.2	0.5	0.8	3.7	0.5	10.3
20:5n-3	9.6	14.9	1.4	4.9	0.4	1.0
22:1	13.4	2.2	1.0	-	0.2	-
22:5n-6	TRA	0.1	0.2	1.4	0.3	2.8
22:5n-3	0.8	2.5	0.4	1.2	0.6	1.6
22:6n-3	7.3	33.6	4.5	32.0	1.2	0.6

[a]*Mallotus villosus* (39)
[b]Farmed, *Oncorhynchus mykiss* (40)
[c]Farmed, North American *Ictalurus punctatus* (41)
[d]Indian Sal, *Channa striatus* (42)

Long lists of fatty acids of fish fillets are given in various sources (2,36-38) but relatively few concern this review. Table V shows that, for diverse fish species (39-42), the key to understanding oxidation is the access to the highly unsaturated fatty acid, mostly 20:5n-3 and 22:6n-3. In fish muscle the triacylcerides are mostly found in lipid droplets (20). Small amounts may accompany membrane phospholipids but it is the free fatty acids from phospholipids that contain the 22:6n-3 highly unsaturated acid shown to be a problem during frozen storage (43). Figure 3 and Table IV show how these acids are liberated and it is generally accepted that once liberated they are more freely attacked by oxygen. The most frequently used method for following oxidation in fish tissues is the 2-thiobarbituric acid (TBA) test which has recently been critically reviewed (44,45). Several variations in TBA (or TBRS) technology are used by different fisheries workers and this hinders comparisons between studies. In fact this method may measure the susceptibility to oxidation in a "system", rather than actual oxidation products. The latter are an exceptionally diverse lot of chemicals (45). Among the most offensive are aldehydes, which can in some cases be measured individually reasonably well (9), but *in toto* are summed as the anisidine value, a measure mostly of 2-alkenals and 2,4-dienals. Figure 4 shows the effects of different antioxidants, compared to BHT, on the *p*-anisidine value of minced sardine (*Clupea pilchardus* L.) tissue (*46*). Peroxides may increase up to a point, but then may decay faster than they are being formed (Figure 5). The *p*-anisidine values could however continue to increase under normal circumstances.

These results (*46*) are typical of oxidation in all fish minces (*47*). Minces are often prepared by squeezing the flesh through a "deboner", which also rejects most of the skin, but not necessarily the subdermal fat. In effect conditions for oxidation are exaggerated over the "natural" environment of lipids in the fish muscle. There is also a high probability of disrupting cells and increasing enzyme activity (*48*), even during subsequent frozen storage. The converse view is that stabilizing agents of all types can be blended in, but effectively oxidation of polyunsaturated fatty acids will proceed by normal fish oil reactions, including metal-catalyzed oxidation processes on oil droplets or bulk separated fish oil (*25,26,49-51*). There is an enormous literature on this type of oxidation which is basically similar to the problems regularly encountered with fish oils (*8,49-51*). The situation is not unlike that of the proposed use of unhydrogenated fish oil in margarine (*52*), but an added complication is the presence of the fish phospholipids which could be acting as antioxidants (*53,54*), perhaps by sequestering metal catalysts.

Modern and sensitive analytical methods are gradually clearing up the problems of lipid hydrolysis and oxidation in cold storage (*55,56*), and further advances in technology for examining oils and fats will no doubt apply to fish in due course (*57*). An occasional problem in filleting fish leaves a rough texture or gaps in the fillet. In these cases more access of oxygen can be expected, but such fish often have poor keeping quality (*58*) for other reasons (*59*). There is almost no limit to the species requiring study, and fish from warmer waters may go rancid more rapidly or react differently from fish from colder waters (*60*).

Biochemical Oxidation Processes *in situ*

The autoxidation of fatty acids such as 22:6n-3 can produce a number of hydroxydocosahexaenoates (*61,62*). The various studies on biochemically generated fish aromas and flavors (*11,12,16,51,63*) indicate a high degree of specificity in this type of reaction, and 2,4,7-decatrienal isomers were long identified with "fishy" aromas (*49,51,64*). Since the 1950 era, fisheries food science has, however, often focused on "fishy" odors from volatile bases. Trimethylamine has long been regarded as a "spoilage" indicator in marine fish (*59*), but confusion follows from both a bacterial origin and one associated with TMAO (trimethylamine oxide) present naturally in many marine fish. Moreover it now appears that the presence of TMAO has to be extended to freshwater fish (*65*).

It has come to be accepted that TMAO can be enzymatically degraded (*66*) to produce dimethylamine (DMA) and formaldehyde (CH_2O). Thus the DMA is another index of quality and the formaldehyde helps account for myofibrillar protein binding (i.e., toughening) in fish muscle. Non-enzymatic breakdown of TMAO seems to occur more readily in squid, etc., than in fish (*67*). Some of this oxygen may be available for *in situ* oxidation of the highly unsaturated fatty acids such as 22:6n-3, but the reaction complexity in such systems is daunting (*48,56,68*).

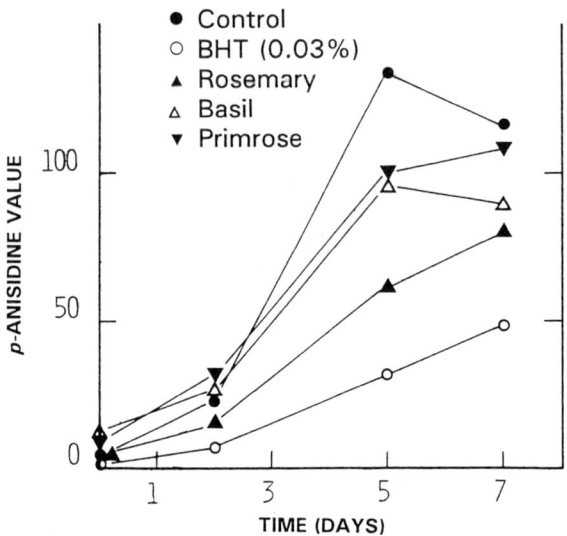

Figure 4. *p*-Anisidine values of muscle (mince) lipids of sardine held at 0 °C with the addition of vegetable antioxidants from rosemary, basil, and primrose, compared to BHT. (Reproduced with permission from ref. 46. Copyright 1985 Rivista Italiana delle Sostanze Grasse.)

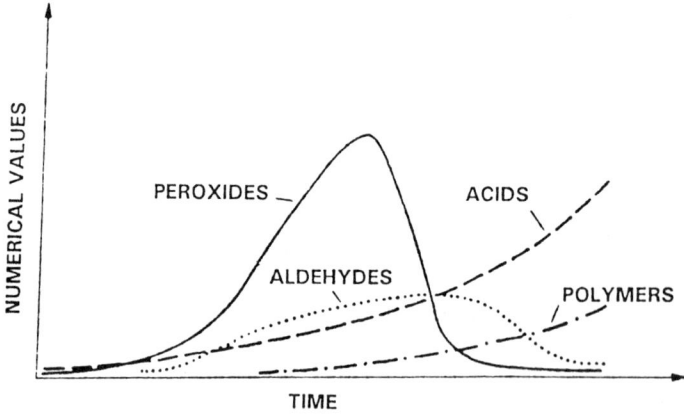

Figure 5. General overview of the complex oxidation processes occurring in fish oils. (Reproduced with permission from Omega-3 News.)

A factor in the development of rancidity during frozen storage of fish is tissue tocopherol. Thus a problem with Atlantic flatfish caught in July and August was that fillets became rancid after only a few months of storage. Our work, never published except in the form of Figure 6 (*69*), shows that in fish caught in these two months, the measurable alpha-tocopherol in the fish fillet disappeared completely during four months of frozen storage. Presumably increased oxidation of polyunsaturated fatty acids *in situ* then followed. The reason for the disappearance of tocopherol is unknown, but the fish were at that time feeding actively and depositing new fat in the muscle. It is therefore uncertain if a dietary factor or some type of increased muscle enzyme activity was involved.

Overall it is imperative that generalizations on *in situ* oxidation be avoided without reference to exact species. As one example, the Japanese red sea bream *Pagrus major* does not show the oxidative changes during storage at 0°C found in other fish from that area (*55*). The North Atlantic capelin *Mallotus villosus* (Table V) also has very superior keeping qualities in the round in frozen storage (*39,70*), compared to herring and mackerel, but no explanation is apparent in either case. There must be unknown protective effects or agents present since as far as is known, the lipid environment is "normal" in all of these fish. The concept of freezing fish round during glut landings or during poor markets, and processing later, has a fatal attraction for the fishing industry. Unfortunately it is impossible to generalize on keeping times and eventual product quality after thawing and refreezing. A comparison was made (*48*) of Japanese sardine (*Sardinops melanosticta*) with a Japanese trachurid fish called horse mackerel (*Trachurus japonica*). Extensive lipid hydrolysis, including that of triglycerides, took place when horse mackerel frozen at -20°C for 35 days was thawed and was then stored at 5°C for 8 days. This hydrolysis process did not take place to the same extent with the sardine, and in those fish all polyunsaturated fatty acids of total lipids hardly changed, whereas some were lost, because of oxidation, in the case of the horse mackerel. The authors postulate membrane disruption from ice crystals formed during the freeze-thaw cycle as being responsible for their observations.

Other Factors

Sexual maturation is, of course, the driving force in natural salmon migrations, yet is scrupulously avoided in farmed salmon for growth reasons. It is unreasonable to expect the frozen products to behave identically in frozen storage, especially if the farmed salmon has nearly twice the muscle fat content (*2*).

The preservative effects of smoking on fish are well-known (*71-75*) but it is curious that, in muscle lipids, there are combined effects from salt (*76*), and heat and water loss, which are at times contradictory. Possibly the strong smoke flavor benefits from combination with lipid oxidation products, but much research remains to be done with this preservation process. The complex interactions possible, involving enzymes, metal ions, etc., are not peculiar to fish and researchers must be aware of other investigations, on fowl in particular (*76*), since this muscle may be closer to fish muscle than are beef and pork.

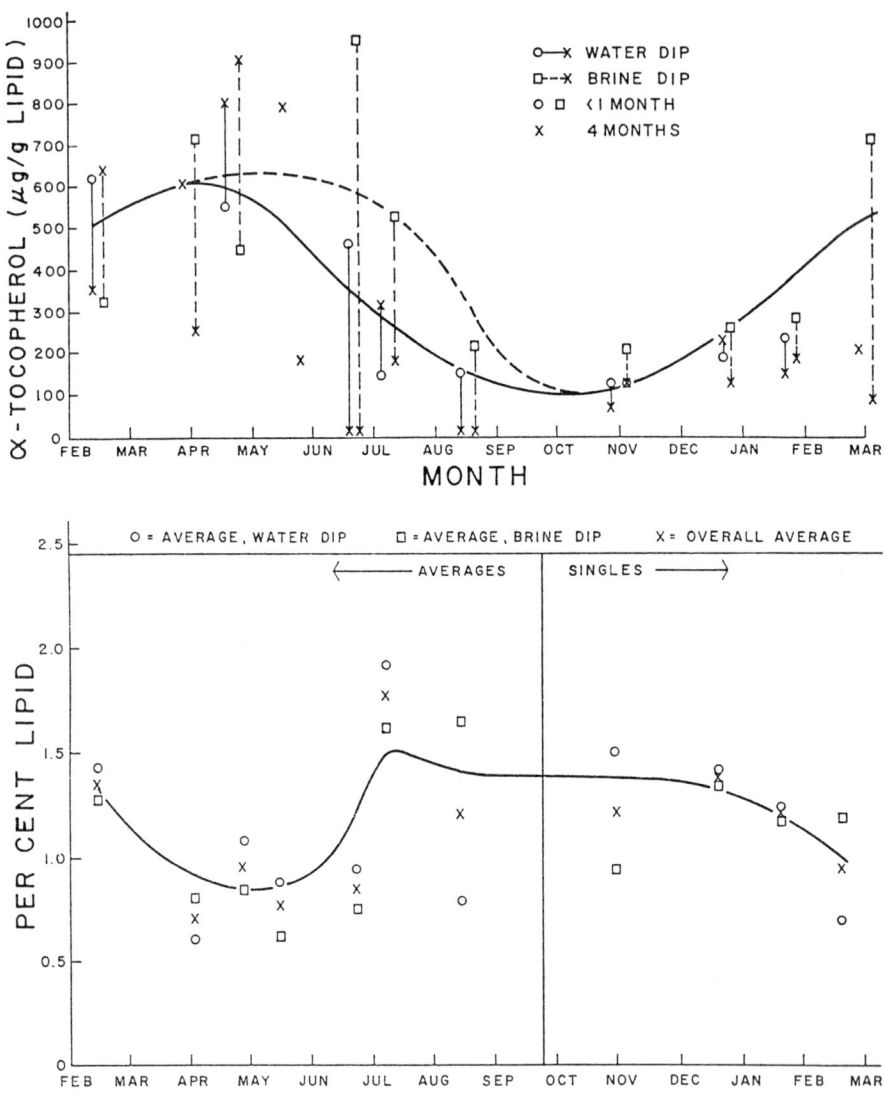

Figure 6. Relationship between seasonal effects on lipid contents of sole fillets and on disappearance of tocopherol during frozen storage. Above: Total lipid extractable by the Bligh and Dyer (4) method. Below: Variations in α-tocopherol contents of commercial sole fillets shortly after being frozen and after four months of frozen storage at −20 °C. "Dip" refers to a prepackaging brine treatment. (Reproduced with permission from ref. 69. Copyright 1974 Food and Agriculture Organization of the United Nations.)

Fish Meal - The Unnatural Environment

The fish meal industry deals with 6-7 millions of tons of product per year. This meal is commonly sold on the basis of protein content, often 65%, and almost invariably contains about 8-10% fat, usually declared on the basis of extraction with diethyl ether (*77*). Some fish meals, produced from scraps in factories processing lean fish such as cod and haddock, have as little as 3-5% fat and are usually referred to as "white" fish meals. For this review, the term fish meal will refer to the product made from whole bodies of fatty fish such as menhaden, anchovy, etc., where there is concurrent production of fish oil, also a commercial commodity produced on a scale of 1.4 million tonnes per year (*78-83*), and currently a material of wide interest for human health reasons (*84-88*).

The basic process for fish meal production (*77,81,83*) is to grind up the specific fish filleting scrap, bycatch, or any other raw material. This is cooked, perhaps for 15 minutes at 90°C, depending on the operator's views on the quality and type of raw material, and "pressed" in a continuous press of single or multiscrew design to remove water and oil. The "presscake" is dried by exposing it to either indirect or direct-heated gases (flame drying) or drying by steam-heated metal discs. Often decanter sludge from oil clarification, and more commonly process water as the evaporator protein concentrate (called stickwater, 40-50% solids), are added back at this stage (*89-91*). After grinding, an antioxidant (commonly ethoxyquin at 150, 400, or 1000 ppm (*92,93*)) is added, and the product is bagged or stored for bulk shipment. Some meals are pelletized before shipment.

The lipids in fish meal are nutritionally important as energy (*94,95*) and yet the lipid quality of this material has received only moderate attention in recent years, and that from one laboratory (*89,92,96,97*). Clearly the residual fish oil and muscle lipids are in a completely unnatural environment. Our interest in enriching the omega-3 acids EPA and DHA in the meat of the common broiler chicken led us to consider that the surface fish meal lipids had already oxidized, but heat would have broken down peroxides, and the resulting adverse volatile flavor products would have already been dissipated by the final drying process. As a result the sensory panelists did not object to eating meat from birds fed diets containing as much as 12% fish meal (*98-100*).

Fish muscle contains about 1% phospholipids. The raw material for fish meal manufacture is often held too long, or under adverse conditions, prior to processing (*90,97,101*). Under these conditions both phospholipids and triglycerides will generate free fatty acids in the raw material (*97,102*). Assuming that the phospholipids are more or less integral with the protein, when the nominal 18% protein in the raw material is converted to the 65% protein of fish meal, this should be accompanied by approximately 3-4% of phospholipids. The balance of the nominal 10% of lipid in fish meal should be mostly triglyceride with up to 10% free fatty acids, although both phospholipids and free fatty acids are interactive variables (*97*). The oil produced will have a fatty acid composition very similar to that of this meal triglyceride, although the dominant 22:6n-3 of the phospholipids (Table V) can be recognized in the meal total fatty acids (*103*). Cholesterol will be ignored in this discussion, although cholesterol oxides (*54*) are undoubtedly formed.

Table VI. Growth of Atlantic Salmon Fed for 18 Weeks on Fish Meals Prepared at Different Temperatures

Meal drying temperature (°C)	60	70	80	90	110
Salmon weight gain (g)	289	290	278	260	217
Relative gain	100	100	96	90	75

SOURCE: Reproduced by permission of the International Association of Fish Meal Manufacturers.

A complication in the oxidation of fish meal lipids is the interaction of oxidation products with amino acids (e.g., formation of Schiff bases). The potential loss of amino acids, especially lysine, tryptophan etc., is important in human nutrition (71,104) and in animal nutrition, including fish fed in aquaculture (105,106). Surprisingly, this effect may be offset readily by tocopherol (107). A less severe reaction system is that of sun-drying of fish, which has recently been examined on a model basis (108). With a very highly refined fish oil these authors were able to obtain browning of the oil above 50°C in the absence of all protein. Evidently temperatures are critical in this process. It was commonly believed for decades that in fish meal manufacture, the surface temperature of the meal did not exceed 90-95°C due to the rapid evaporation of water. This product was the basis for the fish meal used by the broiler chicken industry and was standardized on the basis of extensive digestibility studies in that species. It is now apparent (Table VI) that the Atlantic salmon is a much more sensitive species and fish meal used in their diets should be manufactured by a LT or "low-temperature" process. This definition is usually accepted as $\leq 80°C$ (I. Pike, International Association of Fish Meal Manufacturers, U.K., personal communication, 1991), and applies to cooking as well as drying stages. The latest technology, vacuum drying, ensures even lower surface temperatures.

We concur with de Koning et al. (92,96) that the extraction of lipid from fish meal is best carried out by the Bligh and Dyer (4) procedure. For chloroform-methanol methods, Table VII compares the original Bligh and Dyer monophasic system with the U.S. (109) biphasic modification (Smith-Ambrose-Knobl), and both with the Hara and Radin (30) system based on hexane and isopropanol. The latter system avoids chlorinated solvents, such as $CHCl_3$ (or CH_2Cl_2, sometimes used as an alternative). The test samples were, as indicated, different in respect to ethoxyquin stabilization, and were stored in different atmospheres.

In a separate study (Table VIII) the effect of ethoxyquin stabilization is again clearly confirmed by the superior recovery of lipid from the treated meals. The drop in lipid recovery for the unstabilized meal is almost identical to that observed in the South African study after 90 days (96).

The recoveries of the two most polyunsaturated fatty acids (20:5n-3, 22:6n-3) are very clearly modified by the lack of antioxidant (Tables IX and X). On the other hand, recoveries of saturated and monounsaturated fatty acids are not affected. These results are expressed differently from those of the corresponding studies from South Africa (96) where for 20:5n-3 the w/w% of fatty acids in 100 g methyl esters

Table VII. Amount of lipid extracted by three different procedures, Smith-Ambrose-Knobl (SAK), Bligh and Dyer (BD) and Hara and Radin (HR), from ethoxyquin-treated and untreated menhaden meal (batch 1) stored under nitrogen (N_2), air and oxygen (O_2) for 60 days

Treatment and Storage Conditions	Extraction Methods		
	SAK (%)	BD (%)	HR (%)
With ethoxyquin			
Under N_2	13.26 ±0.15[a]	12.98 ±0.13[a]	11.31 ±0.13[c]
Under air	12.77 ±0.22[a]	12.46 ±0.10[a]	10.94 ±0.17[c]
Under O_2	12.99 ±0.07[a]	12.53 ±0.07[b]	11.05 ±0.04[c]
Untreated			
Under N_2	13.14 ±0.04[a]	12.90 ±0.18[a]	11.32 ±0.08[c]
Under air	12.59 ±0.20[a]	11.95 ±0.14[b]	11.07 ±0.22[c]
Under O_2	12.68 ±0.13[a]	12.25 ±0.21[a]	10.38 ±0.13[c]

SOURCE: Reproduced from ref. 110. Copyright 1992 Helga Gunnlaugsdottir.
Means within a row having different superscripts are significantly different ($p < 0.05$). For each method, data are the mean ±SD of 3 determinations of the lipid content (%) of menhaden meal.

Table VIII. Amount of lipid extracted using the Bligh and Dyer procedure, from ethoxyquin-treated and untreated menhaden meal (batch 2) stored under nitrogen (N_2), air and oxygen (O_2) for 60 days [a]

	Under N_2 (%)	Under air (%)	Under O_2 (%)
With ethoxyquin	11.54 ±0.14	11.64 ±0.06	11.71 ±0.14
Untreated	11.13 ±0.33[a]	9.71 ±0.09[b]	9.80 ±0.25[b]

[a]Gunnlaugsdottir and Ackman, unpublished results. Means within a row having different superscripts are significantly different ($p < 0.05$). Data are the mean ±SD of 3 determinations of the lipid content (%) of menhaden meal.

of the GLC analyses fell from 17.5 to 8.3 in 672 storage days for the fish meal treated with ethoxyquin, and from 13.8 to 1.8 for the untreated meal.

The question of what happens to these fatty acids is one of the objectives of our work (*110*). The extracted lipid already includes the nitrogen in phospholipids, and at least an equal amount of "illegitimate" nitrogen (*92*). Another objective is to clarify the molecular species in the latter materials. If the manufacturing process for fish meal, at least as formerly conducted at higher temperatures (see above), destroys existing peroxides by their interaction with naturally reactive materials such as sulfur-bearing amino acids, then the Table VI results are explained. If the aldehydes present are also polymerized or swept away, it is possible to rationalize the tendency of the final fish meal lipids to rapidly autoxidize as the beginning of a new cycle on the left of Figure 5.

Table IX. Fatty acid composition (mg/g) of total lipids from menhaden meal (batch 2) prepared without ethoxyquin, stored under nitrogen, air and oxygen for 60 days [a]

Fatty acid	under N_2 mg/g	under air mg/g	under O_2 mg/g
14:0	49.60 ± 0.83	47.58 ± 1.81	49.14 ± 1.96
16:0	168.38 ± 2.54	160.26 ± 5.86	163.87 ± 4.86
18:0	32.60 ± 0.25[a]	30.18 ± 1.03[b]	30.71 ± 0.63[b]
16:1n-7	53.71 ± 0.84[a]	50.08 ± 1.82[b]	51.08 ± 1.52[b]
18:1n-9	54.64 ± 0.57[a]	51.46 ± 1.86[b]	52.09 ± 1.32[b]
18:1n-7	17.39 ± 0.57[a]	16.18 ± 0.53[b]	16.62 ± 0.52[b]
Σ 20:1	12.08 ± 0.54	12.13 ± 1.29	12.46 ± 0.5
Σ 22:1	4.28 ± 1.33	4.85 ± 0.01	3.63 ± 0.39
18:2n-6	5.62 ± 0.16[a]	4.05 ± 0.12[b]	4.01 ± 0.21[b]
18:3n-3	4.64 ± 0.05[a]	2.56 ± 0.12[b]	2.40 ± 0.10[c]
18:4n-3	12.34 ± 0.42[a]	5.35 ± 0.15[b]	4.36 ± 0.13[c]
20:4n-6	4.61 ± 0.22[a]	2.08 ± 0.34[b]	1.49 ± 0.07[c]
20:5n-3	41.43 ± 3.71[a]	13.39 ± 0.75[b]	10.63 ± 0.75[b]
22:5n-3	8.40 ± 1.58[a]	3.17 ± 0.08[b]	2.98 ± 0.51[b]
22:6n-3	54.31 ± 1.42[a]	14.62 ± 1.01[b]	10.32 ± 0.28[c]

[a]Gunnlaugsdottir and Ackman, unpublished results. Means within a row having different superscripts are significantly different ($p < 0.05$). Data are the mean ± SD of 3 determinations.

One ultimate objective is to explain why fish oil cannot be added to broiler diets without due care to reduce the intake at least 4 weeks before termination, even if antioxidant is added. A large literature on this concept (111) starts before 1969, and continues to this day (112). Whether "natural" antioxidants (Figure 4) will be effective and practical (113,114) remains to be seen. Our feeding studies with broiler chickens (98-100) included feeding fish meal up to the time of termination. The deposition of obvious fish meal fatty acids (20:5n-3, 22:6n-3) in breast muscle phospholipids was clearly marked, and could easily be measured in other fat and muscle tissues as well. Since no "fishy" flavor was found by sensory panels, it is reasonable to speculate that, in contrast to spreading fish oil in thin films in broiler chicken rations, in the "unnatural" environment of fish meal the lipid residues, including polyunsaturated fatty acids, are partially protected by inclusion in protein and perhaps by binding to protein. The practice of adding an antioxidant is a further benefit. In terms of Figure 5 we feel that fish meal lipids, due to their tendency to oxidize, are mostly on the left side of the figure, whereas fish oils *per se* are mostly on the right by the time the chicken ingests them. The adverse flavor components are thus an integral part of the diet if fish oils are fed.

Fish Silage for Salmonids - Is it a Useful Compromise?

Salmon may, as already mentioned, be especially sensitive to oxidation products in their diets (115-117). They, and trout (107,118), also have a widely recognized

Table X. Fatty acid composition (mg/g) of total lipids from menhaden meal (batch 2) treated with ethoxyquin, stored under nitrogen, air and oxygen for 60 days[a]

Fatty acid	under N_2 mg/g	under air mg/g	under O_2 mg/g
14:0	44.88 ± 2.34	46.74 ± 2.33	48.81 ± 1.9
16:0	150.23 ± 8.27	159.05 ± 7.99	164.66 ± 6.11
18:0	29.08 ± 1.45	30.84 ± 1.66	31.69 ± 1.04
16:1n-7	49.20 ± 2.62	51.95 ± 2.53	53.79 ± 1.96
18:1n-9	48.85 ± 2.83	52.11 ± 2.66	53.68 ± 1.84
18:1n-7	15.78 ± 0.86	16.85 ± 0.83	17.40 ± 0.49
Σ 20:1	10.21 ± 0.87	11.00 ± 0.57	11.24 ± 0.47
Σ 22:1	3.21 ± 1.13	2.85 ± 0.4	3.04 ± 0.17
18:2n-6	5.86 ± 0.23	6.14 ± 0.17	6.30 ± 0.28
18:3n-3	5.69 ± 0.35	5.96 ± 0.3	6.15 ± 0.21
18:4n-3	16.39 ± 1.67	18.65 ± 1.04	19.06 ± 0.69
20:4n-6	6.82 ± 0.60	7.04 ± 0.35	7.13 ± 0.31
20:5n-3	68.40 ± 4.56	73.39 ± 4.40	74.45 ± 2.60
22:5n-3	13.66 ± 2.11	16.10 ± 2.99	14.81 ± 1.90
22:6n-3	100.92 ± 5.51	109.05 ± 6.77	109.16 ± 3.96

[a]Gunnlaugsdottir and Ackman, unpublished results. No significant difference was found between samples (p>0.05). Data are the mean ± SD of 3 determinations.

need for the long-chain polyunsaturated fatty acids found in fish oils and lipids. A compromise source of high quality protein and of these fatty acids is fish silage, especially since the raw material is often available near the farming sites. Basically, the natural enzymes of fish entrails are allowed to digest ground-up fish or fish scrap. Formic or other acids are added to act as preservatives by keeping the pH down to 4 or slightly less (119-121). Fish silage has also been fed to broiler chickens and pigs, but the same caveat of withdrawal well before marketing already mentioned for fish oils, applies.

The silage can be as thin as soup or as thick as porridge, but if the raw material is fatty, some free oil is usually evident. Antioxidants can be mixed in, and in fact have shown favorable results in storage (122). However, this particular study was conducted with dogfish (*Squalus acanthias*) heads and scrap and it is not certain if the urea present in the bodies of this species may have been partially involved in this preservation process. Certainly the protein is not adversely affected by heating and the nutritional values of this type of product have been very good (123). The diets made with fish silage always include several vegetable materials to give it suitable physical properties and to supply other nutrients (124). We have also been studying the fate of tocopherols in this system, since they are an important consideration in commercial salmonid culture (107) and are intimately involved with, and affected by, oxidation processes. The hypothesis is that we are starting again on the left of Figure 6 and in yet another "unnatural" environment for long-chain polyunsaturated fatty acids.

Conclusions

1. Fish is a valuable source of the long-chain highly-unsaturated omega-3 fatty acids desirable in human nutrition.
2. Even in the favorable environment of natural fish muscle, degradative and oxidative processes proceed, albeit slowly at low temperatures.
3. Many natural variables cannot be controlled.
4. It has been proposed that the growth in chicken consumption in the USA offsets the decline in fish and shellfish consumption as far as the supply of 20:5n-3 and 22:6n-3 is concerned.
5. A full understanding of the oxidation processes in fish meal is necessary for efficient utilization of the lipids in this protein resource for food enrichment in fatty acids.
6. Adding additional fish meal to the ration for enrichment of broilers in omega-3 fatty acids differs from merely adding fish oil to the diet of broiler chickens.
7. The particular needs of the farmed salmon industry require that natural antioxidants such as tocopherols be investigated.

Acknowledgments

This work was supported in part by the Natural Sciences and Engineering Research Council of Canada and in part by the International Association of Fish Meal Manufacturers.

Literature Cited

1. Martin, R.E.; Doyle, W.H.; Brooker, J.R. *Mar. Fish. Rev.* **1983**, *45*, 1-20.
2. Ackman, R.G. *Prog. Food Nutr. Sci.* **1989**, *13*, 161-241.
3. Anon. *Fish List: FDA Guide to Acceptable Market Names for Food Fish Sold in Interstate Commerce*, **1988**, Health and Human Services, Food and Drug Administration, Center for Food Safety and Applied Nutrition: Washington D.C. 51 p.
4. Bligh, E.G.; Dyer, W.J. *Can. J. Biochem. Physiol.* **1959**, *37*, 911-917.
5. O'Keefe, S.F.; Ackman R.G. *Proc. Nova Scotia Inst. Sci.* **1987**, *37*, 1-7.
6. Jangaard, P.M.; Brockerhoff, H.; Burgher, R.D.; Hoyle, R.J. *J. Fish Res. Board Canada.* **1967**, *24*, 607-612.
7. Lohne, P. *Meld. fra. S.S.F.* **1976**, No 3, pp 9-14.
8. Ke, P.J.; Ackman, R.G. *J. Amer. Oil Chem. Soc.* **1976**, *53*, 636-640.
9. Ke, P.J.; Nash, D.M.; Ackman, R.G. *Can. Inst. Food Sci. Technol. J.* **1976**, *9*, 135-138.
10. Ke, P.J.; Ackman, R.G.; Linke, B.A.; Nash, D.M. *J. Food. Technol.* **1977**, *12*, 37-47.
11. Josephson, D.B.; Lindsay, R.C.; Stuiber, D.A. *J. Food Sci.* **1987**, *52*, 596-600.
12. Josephson, D.B.; Lindsay, R.C.; Olafsdottir, G. In *Seafood Quality*

Determination; Kramer, D.E.; Liston, J. Eds; Elsevier Science Publishers: Amsterdam, **1987**, pp. 27-47.
13. Ke, P.J.; Linke, B.A.; Ackman, R.G. *J. Food Sci.* **1978**, *43*, 38-40.
14. Hardy, R.; Smith, J.G. *J. Sci. Food. Agric.* **1976**, *27*, 595-599.
15. Ackman, R.G.; Ke, P.J. *J. Fish Res. Board. Can.* **1968**, *25*, 1061-1065.
16. Karahadian, C.; Lindsay R.C. In *Flavor Chemistry - Trends and Developments*, Teranishi, R; Buttery R.G.; Shahidi, F. Eds.; ACS Symposium Series 388, American Chemical Society: Washington D.C. **1989**, pp 60-75.
17. Papajewski, H.; Schrieber, W. *Fette Seifen Anstrichm.* **1979**, *81*, 166-168.
18. Grigor, M.R.; Thomas, C.R.; Jones P.D.; Buisson D.H. *Lipids* **1983**, *18*, 585-588.
19. Buisson, D.H.; Scott, D.N. *Infofish Marketing Digest* **1985**, No. 6, 19-23.
20. Shindo, K.; Tsuchiya, T.; Matsumoto, J.J. *Bull. Japan Soc. Sci. Fish.* **1986**, *52*: 1377-1399.
21. Obataki, A.; Heya, H. *Bull. Japan Soc. Sci. Fish.* **1985**, *51*, 1001-1004.
22. Ackman, R.G.; Eaton, C.A. *Can. Inst. Food Sci. Technol. J.* **1971**, *4*, 169-174.
23. Ohshima, T.; Wada, S.; Koizumi C. *J. Tokyo Univ. Fish.* **1988**, *75*, 169-188.
24. Ohshima, T.; Wada, S.; Koizumi C. *J. Tokyo Univ. Fish.* **1989**, *76*, 19-27.
25. Decker, E.A.; Hultin, H.O. *J. Food Sci.* **1990**, *55*, 947-950, 953.
26. Decker, E.A.; Hultin, H.O. *J. Food Sci.* **1990**, *55*, 951-953.
27. Nishimoto, J.; Takebe, M. *Mem. Fac. Fish. Kagoshima Univ.* **1977**, *26*: 111-118.
28. Hardy, R. In *Advances in Fish Science and Technology*; Connell, J.J., Ed.; Fishing News Books: Farnham, U.K., **1980**: 103-111.
29. Polvi, S.M.; Ackman, R.G.; Lall, S.P.; Saunders, R.L. *J. Food Process. Preserv.* **1991**, *15*, 167-181.
30. Hara, A.; Radin, N.S. *Anal. Biochem.* **1978**, *90*, 420-426.
31. Ackman, R.G.; McLeod, C.A., Banerjee, A.K. *J. Planar Chromatogr.* **1990**, *3*, 450-490.
32. Ohshima, T; Ackman, R.G. *J. Planar Chromatogr.* **1991**, *4*, 27-34.
33. de Koning, A.J.; Mol, T.H. *J. Sci. Food Agric.* **1991**, *54*, 449-458.
34. Ohshima, T.; Ratnayake, W.M.N.; Ackman, R.G. *J. Amer. Oil Chem. Soc.* **1987**, *64*, 219-223.
35. Ohshima, T.; Wada, S.; Koizumi, C. *Bull. Japan. Soc. Sci. Fish.* **1984**, *50*, 2091-2098.
36 Krzynowek, J.; Murphy, J.; Maney, R.S.; Panunzio, L.J. *Proximate Composition and Fatty acid and Cholesterol Content of 22 species of Northwest Atlantic Finfish*; NOAA Technical Report NMFS 74; U.S. Department of Commerce: Springfield, VA, **1989**, 35 p.
37 Gooch, J.A.; Hale, M.B.; Brown, T. Jr.; Bonnet, J.C.; Brand, C.G.; Regier, L.W. *Proximate and Fatty Acid Composition of 40 Southeastern U.S. Finfish Species*; NOAA Technical Report NMFS 54, U.S. Department of Commerce; Springfield VA, **1987**, 23 p.
38. Paul, A.A.; Southgate, D.A.T.; Russell, J. *First Supplement to McCance and Widdowson's "The Composition Foods"*. Elsevier/North-Holland Biomedical Press: London, 1978, 113 p.

39. Ackman, R.G.; Ke, P.J.; MacCallum, W.A.; Adams, D.R. *J. Fish. Res. Board Can.* **1969**, *26*: 2037-2060.
40. Frigg, M.; Prabucki A.L.; Ruhdel, E.U. *Aquaculture* **1990**, *84*, 145-158.
41. Nettleton, J.A.,; Allen, W.H.; Klatt, L.V.; Ratnayake, W.M.N.; Ackman, R.G. (1990) *J. Food Sci.* **1990**, *55*: 954-958.
42. Sen, P.C.; Ghosh, A.; Dutta, J. *J. Sci. Food Agric.* **1976**, *27*, 811-818.
43. Hardy, R.; McGill, A.S.; Gunstone, F.D. *J. Sci. Food Agric.* **1979**, *30*, 999-1006.
44. Hoyland, D.V.; Taylor, A.J. *Food Chem.* **1991**, *40*, 271-291.
45. Frankel, E.N. *J. Sci. Food Agric.* **1991**, *54*, 495-511.
46. Pizzocaro, F.; Caffa, F.; Gasparoli, A.; Fedeli, E. *La Riv. Ital. Sost. Grasse* **1985**, *LXII*, 351-356.
47. Careche, M.; Tejada, M. *Food Chem.* **1990**, *36*, 113-128.
48. Koizumi, C.; Chang, C-M; Ohshima, T.; Wada, S. *Nippon Suisan Gakkaishi* **1988**, *54*, 2203-2210.
49. Ke, P.J.; Ackman, R.G.; Linke, B.A. *J. Amer. Oil Chem. Soc.* **1975**, *52*, 349-353.
50. Hsieh, T.C.Y.; Williams, S.S.; Vejaphan, W.; Meyers, S.P. *J. Amer. Oil Chem. Soc.* **1989**, *66*, 114-117.
51. Karahadian, C.; Lindsay, R.C. *J. Amer. Oil Chem. Soc.* **1989**, *66*, 953-960.
52. Young, F.V.K.; From, V.; Barlow, S.M.; Madsen, J. *INFORM* **1990**, *1*, 731-741.
53. Komatsu, I.; Yasuda, T.; Suzuki, T.; Fukunaga, K.; Suzuki, S.; Takama, K. *Bull. Fac. Fish. Hokkaido Univ.* **1990**, *41*, 232-239.
54. Nawar, W.W.; Kim, S.K.; Ki, Y.J.; Vajdi, M. *J. Amer. Oil Chem. Soc.* **1991**, *68*, 496-498.
55. Miyazawa, T.; Kikuchi, M.; Fujimoto, K.; Endo, Y.; Cho, S-Y.; Usuki, R,; Kaneda, T. *J. Amer. Oil Chem. Soc.* **1991**, *68*, 39-43.
56. Fujimoto, K.; Mohri, S.; Hasegawa, K.; Endo, Y. *Food Rev. Int.* **1990**, *6*, 603-616.
57. Yang, G.C.; Qiang, W.; Morehouse, K.M.; Rosenthal, I.; Ku, Y.; Yurawecz, P. *J. Agric. Food Chem.* **1991**, *39*, 896-898.
58. Ackman, R.G.; Ratnayake W.M.N.; Ohshima, T.; Ke P.J. *Proc. Nova Scotia Inst. Sci.* **1986**, *36*, 107-113.
59. Gill, T.A. *Food Rev. Int.* **1990**, *6*, 681-714.
60. Joseph, J.D.; Seaborn, G.T. *Preliminary Studies in Marine Lipid Oxidation*; NOAA Technical Memorandum NMFS SEFC-95; U.S. Dept. Commerce: Springfield, VA, 1982, 92 p.
61. VanRollins, M.; Murphy, R.C. *J. Lipid Res.* **1984**, *25*, 507-517.
62. Cho, S-Y.; Miyashita, K.; Miyazawa, T.; Fujimoto, K.; Kaneda, T. *J. Amer. Oil Chem. Soc.* **1987**, *64*, 876-879.
63. Josephson, D.B.; Lindsay, R.C. *J. Amer. Oil Chem. Soc.* **1987**, *64*: 132-138.
64. Lindsay, R.C. *Food Rev. Int.* **1990**, *6*, 437-455.
65. Anthoni, U.; Børresen, T.; Christophersen, C.; Gram, L.; Nielsen P.H. *Comp. Biochem. Physiol.* **1990**, *97B*, 569-571.
66. Reece, P. *J. Sci. Food Agric.* **1983**, *34*, 1108-1112

67. Spinelli, J.; Koury, B. *J. Agric. Food Chem.* **1981**, *29*, 327-331
68. Kanner, J.; Salan, M.A.; Harel, S.; Shegalovich, I. *J. Agric. Food Chem.* **1991**, *39*, 242-246.
69. Ackman, R.G. In *Fishery Products*; Kreuzer, R. Ed.; Fishing News Books: West Byfleet, U.K., **1974**, pp. 112-131.
70. Botta, J.R.; Lauder, J.T.; Downey, A.P., Saint, W. *J. Food Sci.* **1983**, *48*, 1512-1515, 1536.
71. Bhuiyan, A.K.M.A.; Ackman, R.G.; Lall, S.P. *J. Food Process. Preserv.* **1986**; *10*, 115-126.
72. Bhuiyan, A.K.M.A.; Ratnayake, W.M.N.; Ackman, R.G. *J. Food Sci.* **1986**, *51*, 327-329.
73. Bhuiyan, A.K.M.A.; Ratnayake, W.M.N.; Ackman, R.G. *J. Amer. Oil Chem. Soc.* **1986**, *63*, 324-328.
74. Allam, M.H.; El-Kalyobi, M.H.; Abou-Arab, A.A. *Ann. Agric. Sci., Fac. Agric., Ain Shams Univ., Cairo, Egypt*, **1988**, *33*, 355-366.
75. Beltrán, A.; Moral, A. *Lebensm.-Wiss. U.-Technol.* **1990**, *23*, 255-259.
76. Takiguchi, A. *Nippon Suisan Gakkaishi* **1989**, *55*, 1649-1654.
77. Windsor, M.; Barlow S. *Introduction to Fishery By-Products*; Fishing News Books: Farnham, U.K., **1981**.
78. Bimbo, A.P. *J. Amer. Oil Chem. Soc.* **1987**, *64*, 706,708-715.
79. Haumann, B.F. *J. Amer. Oil Chem. Soc.* **1989**, *66*, 1531-1543.
80. Knight, P. *Oils & Fats International* **1990**, (4), 50.
81. Bimbo, A.P. In *Marine Biogenic Lipids, Fats, and Oils*; Ackman, R.G. Ed.; CRC Press, Boca Raton FL, **1989**, Vol II, pp 401-433.
82. Hjaltason, B. In *Nutritional Impact of Food Processing*; Somogyi, J.C., Müller, H.R., Eds.; Bibl. Nutr. Dieta.: Basel, Karger, **1989**, No. 43; 96-106.
83. Young, F.V.K. In *Nutritional Evaluation of Long-Chain Fatty Acids in Fish Oil*; Barlow, S.M.; Stansby, M.E. Eds.; Academic Press: London, **1982**, 1-23.
84. *Fish Oils in Nutrition*; Stansby, M.E., Ed.; Van Nostrand Reinhold New York, N.Y., **1990**.
85. Ackman, R.G.; Ratnayake, W.M.N. In *Health Effects of Fish and Fish Oils*; Chandra, R.K., Ed.; ARTS Biomedical Publishers, St. John's, Newfoundland, **1989**, pp. 373-393.
86. Burr, M.L. *Trends in Food Sci. Technol.* **1991**, 2(11): 17-20
87. Leaf, A.; Weber, P.C. *New Engl. J. Med.* **1988**, *318*, 549-57.
88. Knapp, H.R. *J. Amer. Coll. Nutr.* **1990**, *9*, 344-351.
89. de Koning, A.J.; Milkovitch, S.; Fick, M.; Wessels, J.P.H. *Fette Seifen Anstrichm.* **1986**, *88*, 404-406.
90. del Valle, J.M.; Aquilera, J.M. *J. Sci. Food Agric.* **1991**, *54*, 429-441.
91. del Valle, J.M.; Aquilera, J.M. *Process Biochem. International* **1990**, *25*, 122-131.
92. de Koning, A.J.; Evans, A.A.; Heydenrych, C.; de V. Purcell, C.J.; Wessels, J.P.H. *J. Sci. Food Agric.* **1985**, *36*, 177-185.
93. CORPESCA, *Nutrient Analysis of Chilean Fish Meal*; Intern. Assoc. Fish. Meal Manuf. Potters Bar, U.K. 1986.
94. Opstvedt, J. *Acta Agric. Scand.* **1973**, *23*, 200-208.
95. Opstvedt, J. *Acta Agric. Scand.* **1973**, *23*, 217-224.
96. de Koning, A.J.; Mol, T.H. *J. Sci. Food Agric.* **1989**, *46*, 259-266.

97. de Koning, A.J.; Mol, T.; Przybylak, P.F.; Thornton, S.J. *Fat Sci. Technol.* **1990**, *92*, 193-197.
98. Ratnayake, W.M.N., Ackman, R.G., Hulan, H,W. *J. Sci Food Agric.* **1989**, *49*, 59-74.
99. Hulan, H.W.; Ackman, R.G.; Ratnayake, W.M.N.; Proudfoot, F.G. *Poultry Sci.* **1989**, *68*, 153-162.
100. Hulan, H.W.; Ackman, R.G.; Ratnayake, W.M.N.; Proudfoot, F.G. *Can. J. Anim. Sci.* **1988**, *68*, 533-547.
101. Haaland, H.; Njaa, L.R. *J. Sci. Food Agric.* **1991**, *54*, 443-448.
102. Addison, R.F.; Ackman, R.G.; Hingley, J. *J. Fish. Res. Board Can.* **1969**, *26*, 1577-1583.
103. Haque, A.; Pettersen, J.; Larsen, T.; Opstvedt, J. *J. Sci Food Agric.* **1981**, *32*, 61-70.
104. Smith, G.; Hole, M.; Hanson, S.W. *J. Sci. Food Agric.* **1990**, *51*, 193-205.
105. Watanabe, T.; Nanri, H.; Satoh, S.; Takeuchi, M.; Nose, T. *Bull. Japan. Soc. Sci. Fish.* **1983**, *49*, 1083-1087.
106. Watanabe, T.; Takeuchi, M. In *Marine Biogenic Lipids*; Ackman, R.G., Ed.; CRC Press: Baton Rouge, 1989, Vol II, 457-479.
107. Sekiya, T.; Murata, H.; Sakai, T.; Yamauchi, K.; Yamashita, K.; Ugawa, M.; Kanai, M.; Shimada, M. *Nippon Suisan Gakkaishi* **1990**, *57*, 287-292.
108. Smith, G.; Hole, M. *J. Sci. Food Agric.* **1991**, *55*, 291-301.
109. Smith, P., Jr.; Ambrose, M.E.; Knobl, G.N., Jr. *Commercial Fish. Rev.* **1964**, *26* (7), 1-5.
110. Gunnlaugsdottir, H. *Oxidation, quality, and extractability of lipids of menhaden fish meal.* M.Sc. Thesis, Technical University of Nova Scotia, Halifax, **1992**.
111. Miller, D.; Leong, K.C.; Smith, P.,Jr. *J. Food Sci.* **1969**, *34*, 136-141.
112. Huang, Z-B.; Leibovitz, H.; Lee, C.M.; Millar, R. *J. Agric. Food Chem.* **1990**, *38*, 743-747.
113. Haumann, B.F. *INFORM* **1990**, *1*, 1002-1013.
114. Ladikos, D.; Lougovois, V. *Food Chem.* **1990**, *35*, 295-314.
115. Hardy, R.W.; Mugrditchian, D.S.; Iwaoka, W.T. *Aquaculture* **1983**, *34*, 239-246.
116. Ketola, H.G.; Smith, C.E.; Kindschi, G.A. *Aquaculture* **1989**, *79*, 417-423.
117. Desjardins, L.M.; Hicks, B.D.; Hilton, J.W. *Fish Physiol. Biochem.* **1987**, *3*, 173-182.
118. Stone, F.E.; Hardy, R.W.; Shearer, K.D.; Scott, T.M. *Aquaculture* **1989**, *76*, 109-118.
119. Stone, F.E.; Hardy, R.W. *J. Sci. Food Agric.* **1986**, *37*, 797-803.
120. Jangaard, P.M. *Bull. Aquaculture Assoc. Can.* **1991**, *91-1*, 28-53.
121. Gildberg, A.; Raa, J. *J. Sci. Food Agric.* **1977**, *28*, 647-653.
122. Mowbray, J.C.; Rossi, H.A.; Chai, T. *J. Agric. Food Chem.* **1988**; *36*, 1329-1333.
123. Åsgard, T.; Austreng, E. *Aquaculture* **1985**, *49*, 289-305.
124. Lall, S.P. *Bull. Aquaculture Assoc. Can.* **1991**, *91-1*, 63-74.

RECEIVED February 19, 1992

METHODOLOGIES FOR ASSESSING LIPID OXIDATION PRODUCTS

Chapter 13

Gas Chromatographic Analyses of Lipid Oxidation Volatiles in Foods

J. R. Vercellotti, O. E. Mills[1], Karen L. Bett, and D. L. Sullen

Agricultural Research Service, U.S. Department of Agriculture, Southern Regional Research Center, 1100 Robert E. Lee Boulevard, New Orleans, LA 70124

Volatiles analysis, by direct GC and olfactory "sniffer port", was improved by external closed inlet device (ECID) with a wide bore glass capillary column as both trap and separation medium. A study was made of recovery of standards either directly from vegetable oil or from Tenax GC or Carbopack B/Carbosieve SIII. Recovery of volatiles from cartridges was either carried out by thermal desorption or by solvent elution. Roasted peanuts were analyzed for volatiles by direct GC from the ECID and "aromagrams" generated. During lipid oxidation volatiles rose dramatically to the high ppm range while off-flavors intensified in the olfactory portion of the aromagram; simultaneously, positive olfactory attributes became imperceptible as many heterocycles or thio-derivatives (ppb level) disappeared at high peroxide values.

Advances in Lipid Oxidation Analyses by Gas Chromatography as Affecting Food Flavor Measurements

Walter G. Jennings wrote a landmark chapter in 1981 on "Analysis of Food and Flavor" which highlighted the difficulties encountered with analyzing reactive or often unstable food flavor volatiles. He thoroughly considered lipid oxidation products from the point of view of sample preparation, instrumentation, and applications to flavor and aroma volatiles *(1)*. Takeoka et al. *(2)* updated the work of Jennings *(1)* and described new concepts of direct injection of headspace samples which can eliminate errors that accompany conventional methods of purge and trap. Parliment *(3)* detailed in the same volume sample preparation techniques for gas liquid chromatographic analysis of biologically derived aromas. Alternative concepts of enrichment of headspace volatiles from samples were described, and the "Mixor" solvent extraction/centrifugation tube for concentrating flavors was introduced.

[1]Current address: New Zealand Dairy Research Institute, Private Bag, Palmerston North, New Zealand

More recently excellent reviews by G. A. Reineccius have been published which historically and practically put into perspective the analysis of food flavors by gas chromatography with emphasis on lipid oxidation products *(4,5)*. Grob has studied headspace gas analysis and the role and design of concentration traps specifically suitable for capillary gas chromatography *(6)*. He pointed out the importance of coating materials bonded to the walls of the capillary column as well as gas flow conditions. Grob concluded that the favored trap for concentration of volatiles for capillary gas chromatography should be a wall coated column as solid support onto which the volatiles are concentrated, and which maintains continuity with high capillary resolution. Smaller packed cartridges containing highly activated solid phase adsorbent and serving as trap or both a trap and run columns have rather different separation efficiencies. In terms of evaluating packed trap efficiency, Zlatkis et al. *(7)* as early as 1973 evaluated Tenax, Carbosieve (graphitized carbon), and Porapak P. Later, Bishop and Valis *(8)* also evaluated the efficiencies of sorbent materials (Tenax, Carbopack, Carbosieve, etc.), confirming the need to be careful in estimating recoveries with various sorbents and molecular structures.

Porous polymer trapping methods using Tenax GC as support with ethyl ether elution of volatiles were skillfully advanced by Steinke who studied off-flavors in Lake Michigan salmon *(9)*. The completed method was published later by Olafsdottir et al. *(10)*. In general, the sample is placed or suspended in a large round bottom flask equipped with a sparge tube, thermometer, and exit adapter that holds in place a cartridge or Pasteur pipet containing Tenax porous polymer through which the sparge gas carries the organics to be trapped. The volatiles are concentrated for GC after eluting them from the Tenax with ethyl ether. The Lindsay group *(9,10)* preferred to use solvent elution from the Tenax solid support rather than thermal desorption to avoid possible thermal rearrangement or fragmentation.

Michael et al. in 1980 effectively used purging from round bottom flasks onto Tenax traps with headspace purge techniques *(11)* to concentrate volatile environmental samples in biological matrices. Porous polymers were also used for collection and concentration of plant volatiles by Cole *(12)*. In that system Porapak Q and Tenax GC were activated at elevated temperatures and the volatiles desorbed from the support contained in 100 x 1.5 mm tubes by heating the Porapak Q to 150° C or the Tenax GC to 200° C employing an injection heater which desorbed the volatiles directly onto the gas chromatography column *(12)*.

Buttery and Teranishi have explored many ways of concentrating and separating lipid oxidation volatiles. The importance of lipid derived volatiles to vegetable and fruit flavors and their analysis has been reviewed with many pertinent references by Buttery *(13)*. In particular, as cited from the previous paper *(13)*, the relationships of olfactory sensitivity to dilution thresholds have been put on sound quantitative basis by Buttery, Teranishi, and Ling over the years. Buttery et al. *(14)* have used Tenax trapping with solvent elution of volatiles. In their work distillation of a food sample with carrier sweep and mild vacuum onto a large Tenax trap concentrates adequate quantities of lipid oxidation and other pertinent, e.g., sulfur or terpenoid, flavor volatiles to permit further fractionation, olfactory study, and identification by GC/MS. Two landmark papers coming from this work by the group pertain to delineation of key compounds in tomato flavor volatiles *(15,16)*. Seven compounds in varying concentrations were found to contribute collectively to tomato paste aroma

(15,16). Few other examples of analyses and duplication of authentic vegetable aromas composed of complex mixtures or concentration differentials exist in the literature. This work not only pointed to the optimistic future application of combined volatiles analysis and olfactory definition of natural aromas, but also established highly accurate quantitation of the individual stimuli effecting the threshold impression of tomato aroma.

Galt and MacLeod *(17)* concentrated meat flavor volatiles on Tenax using a combination of sparging and vacuum for identification of components by GC-MS after thermal desorption. Suzuki and Bailey demonstrated quantitation of concentrated ovine fat volatiles onto Tenax using 2-methyl-3-octanone as internal standard *(18,19)*.

An important clue that lipid oxidation caused rancidity in foods came from the GC work of Fritsch and Gale *(20)* using hexanal as a measure of oxidative deterioration in corn, oat, and wheat cereals. The authors found by GC that the onset of rancid odors occurred when the hexanal content exceeded about 5 ppm.

Fisher et al. *(21)*, Lovegren et al. *(22)*, Dupuy et al. *(23)*, and St. Angelo et al. *(24)*, developed procedures for determining lipid oxidation volatiles in vegetable oils, raw peanuts, legumes, and a variety of foods or materials which can be used as indicators of inherent quality on any product that can be weighed into the glass tube used for direct gas chromatography. During the course of this work Legendre, Fisher, and Dupuy developed techniques for on-column concentration of volatiles and efficient separation in direct gas chromatography using an external closed inlet device (ECID) *(25)* for Tenax GC-8% polymetaphenyl ether (PMPE) as a packed column; and Dupuy et al. *(26)* extended the apparatus for capillary application. The packed column and capillary ECID methods for food volatiles are covered extensively in Dupuy et al. *(19)*.

Waltking concluded from the correlation of GC data with sensory scores of vegetable oils that GC of lipid oxidation products carried out according to methods as above can be more precise and accurate than panels using AOCS Intensity and Quality Scoring Scales *(27)*. Min *(28)* also compared flavor qualities of vegetable oils by gas chromatography and confirmed the results reported in the collaborative study by Waltking *(27)*. In Min's correlation lipid oxidation products from soybean, corn, and hydrogenated soybean oil were determined by gas chromatography and correlated with sensory data from 94 trained panel members from 8 different laboratories *(28)*.

Using the Dupuy-Legendre-Fisher methods Vercellotti et al. *(29,30)* combined sparging and vacuum with Tenax traps for GC analysis of a variety of foods, such as vegetable oils, peanuts, meat, fruit, packaging materials, residual solvents, etc. The Dupuy-Legendre-Fisher methods of volatiles analysis permitted use of direct gas chromatography methods with thermal desorption from traps as well as concentration on packed or capillary columns for both concentration as well as separations *(29,30)*.

Commercial devices for volatiles concentration and GC measurement have resulted from these pioneering studies and include registered trademarks for automated or manual equipment from, e.g., Tekmar, Envirochem, Supelco, Chrompack, Carlo Erba, and Scientific Instrument Services. Many standard methods for environmental quality or industrial quality control (e.g., lipid oxidation products in vegetable oils and cereal products) have now been established by these purge and trap methods.

Peanut Flavor Volatiles Analysis by Concentration on Solid Supports

Fore et al. *(31-34)* devised methods for quantitating peanut butter flavor volatiles, used this data to correlate GC profiles with sensory scores, and were able to employ the technique to assess shelf life stability of peanut butters. These methods have been extended to many other foods such as vegetable oils, meat, chocolate, or fish and have proven to be applicable to higher resolution gas chromatography on capillary columns *(21,26)*. Buckholz et al. *(35,36)* have considered the effect of roasting time, commercial size, and peanut variety on the quality of flavor volatiles produced and measured these volatiles by GC through purge and trap techniques with Tenax GC. Pattee and Singleton *(37)* reviewed the significance of volatile flavor compounds to peanut quality and included many references from their own gas chromatographic work going back to 1965 as well as results from other laboratories through the 1960's and 1970's. Pattee et al. *(38)* also considered the effects of peanut size and time in storage for varieties of Virginia peanuts with respect to changes in flavor. How *(39)* described effects of variety, availability of precursors, roasting, and modified storage atmosphere on the chemical composition, headspace volatiles, and sensory flavor profiles of peanuts. Sanders et al. *(40,41)* reported the influence of peanut maturity and curing conditions on the flavor character notes of peanuts. Young and Hovis (42) reported a headspace method for rapid analysis of volatiles in raw and roasted peanuts. Of the peaks produced from headspace analysis, eight were related to specific flavor characteristics of peanuts involved in objectionable flavor *(42)*.

A paper by Mason and Waller on the isolation and localization of the precursors of roasted peanut flavor *(43)* is often cited and recommends a maximum volatiles sweep run temperature for peanuts with a maximum of no more than 132° C. We have confirmed Mason and Waller's work *(43)* on minimum temperature required to produce roasted peanut flavor volatiles by Maillard reaction in raw peanuts. However, more recent, sophisticated work by Frankel and Selke on dynamic headspace capillary gas chromatographic analysis of soybean oil volatiles *(44)* has shown that temperatures above 60° C should be avoided when sweeping volatiles from lipid containing samples because inherent peroxides break down and amino acid or other carbonyl condensations occur creating flavor artifacts.

The present paper reports improved gas chromatography of lipid oxidation products and other flavor principles using the ECID and wide bore glass capillary column. An example is given of degree of peanut quality using volatiles analysis from samples widely differing in degree of oxidation, purged at optimum carrier flow rate as well as temperature and separated on a wide bore capillary column with a simple postcolumn splitter that divides the sample to a flame ionization detector and human olfactory sampling port ("sniffer port") *(45)*. These simultaneous measurements of volatiles give two complementary profiles of the peanut samples when a trained sensory panelist perceives, interprets, and describes olfactory attributes corresponding to the region of the chromatogram where compounds of known or unknown identity elute. Targeting reproducible peanut quality characteristics through quantitative measurement of volatiles and olfactory profile should considerably reinforce methods such as color card or reflectance colorimeter values *(46)* used by the industry. Determining roast quality in peanuts by consistently reproducible gas chromatography and sensory characterization has extensive economic ramifications.

Much interest exists in the flavor of homogenized roasted peanuts in peanut butter because some 450,000 metric tons of this product are made in the United States alone each year. The ECID with direct gas chromatography is an ideal method to measure peanut butter flavor potential *(31-34)* because the homogenized paste is more representative of an entire lot of peanuts than individual peanuts themselves. Also, peanut butter is a value-added product intended for the consumer which should represent the final production stage of this commodity. This method is suitable for volatiles analysis on raw, ground roasted peanuts, and pastes made from roasted peanuts with similar conditions.

Experimental Procedures

Preparation of the Sample Tube for ECID Direct Gas Chromatography. Peanuts selected from 1989 crop Florunners (E. J. Williams, Tifton, GA) were roasted (8 kg) as described in Sanders et al. *(40,41)*. Two portions of the roasted peanut sample (2.5 kg each) were stored in polyethylene bags loosely closed to the atmosphere for two and four months, respectively. Peroxide values (PV) were taken on each sample as a measure of relative oxidation. A folded plug of fine glass wool (50 to 55 mg) (8 micron, Pyrex Fiber Glass, Corning #3950) was placed about 5 mm above the base of an 84 x 9 mm o.d. glass cartridge used as a sample tube for direct gas chromatography of roasted peanut or paste volatiles *(19,21-25)*. Approximately 1 gram of ground roasted peanut were weighed into the glass tube. Homogenized peanut paste (500 mg) was applied to the inner wall of the cartridge with a syringe and large diameter needle. A second plug of glass wool (50 mg) was then placed on top of the roasted peanut sample and a final weight taken for comparison of volatiles loss after purging. Vegetable oil samples (100-200 mg) were deposited with weighing on a plug of glass wool (350 mg, *ca.* 40 mm in length), folded over on the end from which volatiles exit and positioned 10 mm in from the lip of the glass cartridge to prevent oil from creeping out into the lines. After depositing the oil, care being made not to smear the oil on the outside of the glass or drip through, a folded piece of glass wool 10 mm in length (*ca.* 100 mg) is inserted to cap the oil sample on the carrier gas entry end of the tube, again keeping this piece of glass wool also 1 cm from the other lip of the cartridge to prevent oil creeping out during heating. A final weight is taken so that loss on heating can be compared by weighing after purging.

Direct Gas Chromatography Procedure Using ECID and Packed Column. The peanut sample tube prepared as above was placed, with the glass wool plug at the bottom, in the heated injection port barrel of the inlet device (ECID, Scientific Instrumentation Service, River Ridge, LA) *(19,21-25)*. The carrier gas was swept over the sample in the glass cartridge and through the glass wool before entering the column at a rate of 20 to 40 ml/min to secure maximum volatiles recovery. With a maximum block inlet temperature of 125° C for roasted peanuts, the sample was purged with nitrogen for 28 minutes at 20 ml/min carrier gas flow onto the top of a Tenax GC-8% PMPE packed column (10 feet x 1/8 inch; Ni 200). The six port VICI valve and transfer lines were heated to 190° C during all operations and also baked overnight with carrier gas flowing when samples were not being run to purge the system and prevent "ghosting" (the process of carrying over condensed volatiles from a previous run which appear as extraneous peaks in a subsequent run). The packed Tenax column was cooled to about 20° C with a wet towel during volatiles concentration to allow at least 28 minutes of stripping time and still keep the first peak (methanol) on the column.

The sample cartridge was removed from the injection port after completion of volatiles stripping. To separate the concentrated flavor compounds, the column was temperature programmed first to 50° C as fast as the oven could heat (about 2 minutes), held for 2 minutes at 50° C, increased at 3° C/minute to 245° C, and then held for a total run time of about 85 minutes. Data collection and analysis was accomplished with a Hewlett-Packard 3359 Laboratory Automation System computer. Base lines on computer reconstruction analysis of curves were used which followed the lower valleys of the profile. Each sample was run gas chromatographically in duplicate and means of peak areas used in tabulating results after checking standard deviations.

Wide Bore Capillary Gas Chromatography with the ECID for Olfactory Analysis of Oils and Roasted Peanut Samples. Volatiles from standards in oil, peanut oil, or ground roasted peanuts (ECID sample prepared as above) were directly injected into a 60 m x 0.75 mm i.d. glass capillary column coated with 1 µm phase thickness of 5% diphenyl, 94% dimethyl, and 1% vinylpolysiloxane (SPB-5, Supelco, Inc., Bellefonte, PA). A Hewlett-Packard 5790 gas chromatograph was used, and the ECID attached through the carrier gas line to the inlet of the capillary splitter. Helium flow rate through the ECID was 20 to 40 ml/min through the sample cartridge, depending on conditions needed. For peanut or other vegetable oils, 40 ml/min is advised to collect maximum amounts of volatiles (6). After the "splitless" mode of the capillary injection port divides the sample flow to the column by venting about two thirds of the carrier stream through the purge vent, flow through the column was 14-15 ml/min. After separation in the wide bore capillary, the flow was split into two equal volumes with an SGE (Austin, TX) uncoated, deactivated fused silica capillary run through a two-hole ferrule to lengths of capillary carrying 7.5 ml/min effluent each, one going to the FID and one to the sniffer port, respectively. The olfactory assay attachment was also from SGE (Austin, TX). A stopwatch was used to time from the beginning of the injection as zero minutes, and retention times were recorded by starting the integrators at injection. Plots reflect integration of the total run time (120 min). Samples were heated at a desirable temperature in the ECID (as described elsewhere in this work, e.g., 125° C/15 min for roasted peanuts) with the sampling valve opened to the column and trapped on the column at +1° C in the splitless mode for the duration of volatiles collection (15 min). The moisture in the sample cannot be permitted to freeze in the capillary, but +1° C both efficiently traps volatiles and permits water vapor to pass through the column. Temperature programming was from +1° C at 4° C/min to 250 ° C (held isothermally for 60 min). The effluent was sniffed continuously with retention times and relative intensities of odors noted on a score sheet. The integration was done on an HP 3390 and data collected and analyzed on a Hewlett-Packard 3359 Laboratory Automation System.

Peanut flavor compounds were trapped on Carbopack B/Carbosieve SIII cartridges activated previously at 250° C/1 hr (see below for preparation of traps).

The concentrated samples were prepared by putting 1.5 g of ground roasted peanuts in a cartridge as above and heating in the ECID at 125° C for 0.5 hr under 40 ml/min carrier sweep and with the purge of volatiles collected on the trapping cartridge from the heated six port valve through a heated transfer line. Volatiles were eluted from the trap with 5 ml methylene chloride. The sample was concentrated with a gentle stream of purified nitrogen (just enough flow to make a small impression on the solvent surface in the vial) and identifications made by GC/MS using a Hewlett-Packard system (Hewlett-Packard Company, Palo Alto, CA). The GC/MS consisted of an HP 5890 Series II GC, an HP 5988A mass spectrometer, and an HP 59872C data system. The GC was fitted with a 50 m x 0.22 mm i.d. Hewlett-Packard Ultra 2 column (cross linked 5% phenyl, 94% methyl, 1% vinylsilicone) with 0.33 µm phase thickness.

Calibration of Flame Ionization Detectors. The flame ionization detectors on the gas chromatographs were calibrated by injecting several concentrations of each of the identified peaks from ground roasted peanut sample that had been spiked with concentrations of each in methanol using a microliter syringe. Response factors were obtained which were linear within the FID response ranges (1 to 50 ppm) found for the peanut butters. Unknowns as well as total volatiles were estimated by using the response factor for pentane (1 to 50 ppm) as external standard by comparison. Concentrations of peaks for the flame ionization detectors were recorded as parts per million (micrograms of volatile per gram of the original sample weight of peanut butter).

Test Solution for Recovery Studies. A test solution was prepared by dissolving 1 µl of each of the following compounds in 5 ml methylene chloride. The compounds used were *trans*, *trans* -2,4-heptadienal, 2-pentylfuran, (1R)-(+)-*alpha*-pinene, *trans*-2-heptenal, (R)-(+)-limonene, pentanal, hexanal, decanal (Aldrich Chemical Co.), 2,3-dichloropyrazine (Pyrazine Specialties), octanal (Alfa Products), methyl 3-hexenoate (K&K Laboratories Inc.), nonanal (Bedoukian Research Inc.) and heptanal (Eastman Kodak Co.).

Adsorbent Traps. Traps consisted of either 250 mg 60/80 mesh Tenax GC (Teklab, Baton Rouge, LA) packed between glass wool in a glass tube (0.7 cm i.d., pathlength 3.5 cm of solid support in an 8.5 cm x 0.7 cm i.d. glass cartridge) or a combination of 290 mg Carbopack B and 106 mg Carbosieve SIII (Supelco Inc, Bellefonte, PA) separated by glass wool in a tube of the same dimensions as used for the Tenax GC trap. The path length of the latter trap was 2 cm through Carbopack B plus 0.5 cm through Carbosieve SIII. Volatiles entered the Carbopack B first.

Purge and Trap Experiments. Two different trapping vessels were evaluated. One consisted of a 500 ml three necked round bottom flask fitted with standard taper 24/40 ground glass joints, and the other was a Tekmar sparging tube, 15 cm long by 2 cm in diameter fitted with appropriate plumbing for efficient transfer of volatiles to the trap (Tekmar, Cincinnati, OH). Vessels were maintained at 65° C during all experiments and the nitrogen flow rate was 50 ml/min of nitrogen filtered through charcoal and molecular sieve traps followed by a 2.5 m packed GC column of activated Tenax GC coated with 8% polymetaphenyl ether.

High quality vegetable oil (10 g, Hunt-Wesson, Inc., Fullerton, CA) was placed in the vessel and the test solution (0.1 ml) was mixed in with a magnetic stirring bar. Unless otherwise stated the vessel used for the trapping experiments was the three-necked round bottom flask; the sample was stirred; sparging was used to obtain volatiles; and vacuum was applied to the trap. Except for kinetic experiments, trapping of volatiles was carried out for 16 h.

Traps were eluted with about 6 ml distilled diethyl ether, 10 µl of 2-chloropyrazine (b.p. 153-154° C) (Pyrazine Specialties) internal standard solution was added and solvent was evaporated to about 100 µl with a gentle stream of nitrogen. The internal standard solution consisted of 12.8 mg 2-chloropyrazine dissolved in 5 ml methylene chloride. A blank sample was prepared in which all of the above steps were carried out except for the addition of the test solution.

Gas Chromatography of Recovery Study Samples. Samples were analysed on an HP 5790 gas chromatograph (Hewlett-Packard Company, Palo Alto, California) fitted with 60 m x 0.75 mm i.d., SPB-5 glass capillary column bonded with 1 µm phase thickness as described above. When samples were injected directly the oven was temperature programmed from 35°C to 155°C at 4°C/min. Helium carrier gas was used at a flow rate of 7.5 ml/min through the column.

Oil samples (100 mg) were applied to glass wool in a glass tube (8.5 cm x 0.7 cm i.d.). A second plug of glass wool was placed on top, as described above for oil sample preparation. The tube was then inserted into the ECID port at 100°C. Carrier gas was swept through the tube for 15 min transferring volatiles to the top of the capillary column. The GC oven was held at 1°C during the 15 min injection period then increased at 4°C/min.

Recovery Calculations. At least three and as many as six repetitions of each recovery sample were run and the means and standard deviations calculated. Since the treatment interactions were not the purpose of the experiment, more sophisticated statistical calculations were not needed and data were plotted as shown in the Figures and Tables for first order comparison. Reproducibility was within a few percent at the concentrations of sample applied, and more than adequately precise for the semiquantitative applications used in comparison of commodity samples such as roasted peanuts.

In purge and trap experiments the factor which adjusted the area of the internal standard to 400,000 counts was determined. All peak areas of test volatiles were then multiplied by that factor. Peak areas of the volatiles equivalent to 100% recovery were determined by adding 0.1 ml test solution plus internal standard direct to 6 ml diethyl ether and evaporating this solution to approximately 0.1 ml for GC analysis. The above calculation based on 400,000 counts for the internal standard was made and percentage recovery was calculated. No adjustment was necessary for blank values as no significant contribution was made by the blank to any of the test compound peaks.

Results and Discussion

Recovery Studies and Variables in Purge and Trap GC. The following parameters will affect the performance of a purge and trap system for gas chromatography of volatiles from foods either with direct gas chromatography (ECID with sample in a cartridge for desorption) or trapping from headspace of a large vessel onto a solid support.

1) Temperature of sample.
2) Time for which volatiles are collected.
3) Nitrogen flow rate for purging or sparging.
4) Geometry of the sample vessel.
5) Application of vacuum to the adsorbent tube.
6) Type of adsorbent material or chromatography column used as the solid phase for concentration.
7) Amount of adsorbent (path length).
8) Sparging sample or purging of headspace.

Table I lists the compounds used in all the recovery experiments and gives their retention times and identification number used in all subsequent figures. It was considered that these substances reflected a typical cross section of lipid oxidation products generated in foods. The terpenes and synthetic chloropyrazine represent polarities and interactions widely found in aromatics.

Table I. Compounds Used for Purge and Trap Evaluation

Number	Boiling Point (Atm, °C)	Compound Name	Retention Time (min)
1	102-103°	pentanal	8.53
2	130-131°	hexanal	12.10
3	153°	heptanal	16.45
4	169°	methyl 3-hexenoate	17.94
5	156°	α-pinene	18.17
6	178°	2-heptenal	18.91
7	163°	2-pentylfuran	20.56
8	171°	octanal	20.98
9	198°	2,4-heptadienal	21.34
10	175-176°	d-limonene	22.37
11	208°	2,3-dichloropyrazine	23.58
12	191°	nonanal	25.39
13	207-210°	decanal	29.60

The Effect of Time. The effect of trapping the standard volatiles in Table I onto Tenax for 4 or 16 hours from vegetable oil contained in a three-necked flask with nitrogen sparging (50 ml/min) at 65° C. (no vacuum) is shown in Figure 1. Since the Tenax trap was eluted with ethyl ether and the volatiles concentrated in a nitrogen stream before injection by syringe into the GC, there was concern that recoveries of standards could be affected by possible evaporation losses during concentration in a nitrogen stream. Repeated checks of the system with simple concentration of the standards in ethyl ether (without trapping) showed, however, that little or no loss occurred during removal of eluting solvent. At the lower retention time end, hexanal was recovered to a greater extent in 4 hours while at the other end decanal had a greater recovery at 16 hours. Because of poorer affinity for Tenax, hexanal may be competitively replaced on the trap by other compounds as time increases. Pentanal was absent at both 4 and 16 hours and *alpha*-pinene had a low recovery at 4 hours, indicating breakthrough from the solid support, Tenax GC. A very low affinity of these compounds for Tenax GC would explain their low recovery. There was some difference in the recovery of most compounds by extending trapping time from 4 hours to 16 hours, but the higher boiling compounds were more resistant to recovery from the oil, creating a bell-shaped profile across retention times with respect to recovery.

A trend towards greater recovery of the higher boiling point compounds with longer sparge time was accentuated when volatiles were trapped from oil in the Tekmar sparging tube under vacuum for 16 hours and 64 hours rather than from the flask without vacuum (data not tabulated). It was found that more 2,3-dichloropyrazine, nonanal and decanal were recovered after the longer time, while much more hexanal, heptanal, methyl 3-hexenoate and 2-heptenal were recovered in the shorter time. Breakthrough using the configuration of the Tekmar sparging tube is probably more understandable since no condenser was placed between the oil and the trap, and displacement of the lower boiling compounds from the solid support by warm sparging vapor much more likely.

Trapping the same sample from the three-necked flask under sparging (cf., same conditions as in Figure 1) but still in the absence of vacuum for an additional 16 hours (total of 20 hours trapping) resulted in additional recovery of all compounds except pentanal (Figure 2), which in Figure 1 broke through after only 4 hours. Recovery in excess of 90% was achieved for 6 of the 13 compounds simply by sparging for 20 hours, but still did not result in complete vaporization of the three highest boiling substances.

When a second trap was placed in series while trapping from the Tekmar sparging tube with sparging and vacuum, only five compounds were found to be adsorbed onto the second trap. Four of them were present at a level of 5% or less while 55% original hexanal content of the oil was trapped on the second trap. No pentanal was found in either trap. The zero recovery of pentanal and significant breakthrough of hexanal into the second trap indicate poor affinity for Tenax GC. In this experiment the problem of low affinity for the solid support was magnified by the use of vacuum which was an additional driving force, no doubt, for displacement from surface active sites of adsorption.

Effect of Geometry of the Sample Vessel. There was little difference in recovery of most volatiles from either vessel, the three necked flask or sparging tube (Figure 3), when sparged for 16 hours under vacuum at 65 ° C. Nonanal and decanal, however, were significantly better recovered from the flask than the tube. The efficiency of recovery may be directly related to the greater surface area of the

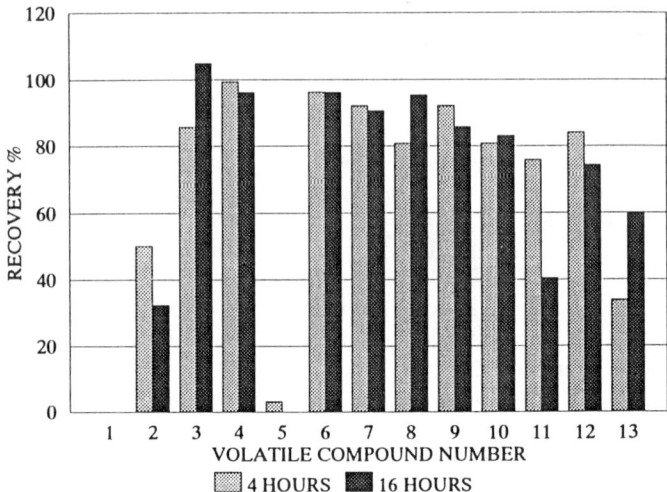

Figure 1. Recovery (%) of standard volatiles in Table I from vegetable oil contained in a three-necked flask with nitrogen sparging (50 ml/min) at 65° C onto a Tenax GC trap without vacuum for 4 hours or 16 hours. Standards eluted with ethyl ether, concentrated, and injected by syringe onto wide bore capillary column.

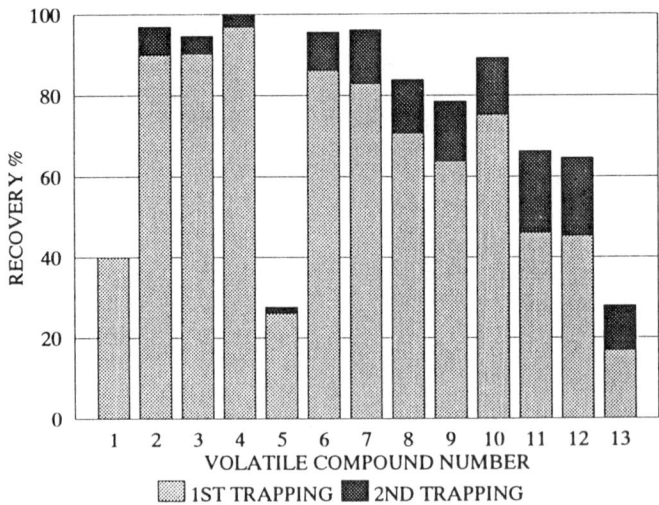

Figure 2. Effect of repeated trapping on % recovery of standards from the same sample in the three-necked flask with conditions as in Figure 1 with first sampling for 4 hours and second, an additional 16 hours for a total of 20 hours sweep.

sample in the sparged flask under vacuum, although one would consider sparging in the tube also to expose greater surface area.

Application of Vacuum to the Adsorbent Tube. In comparing sparging from the three-necked flask with and without vacuum (Figure 4), (Figure 4 is a comparison of the flask results from Figures 1 and 3), sparging alone onto Tenax GC was more efficient at recovering pentanal, hexanal and *alpha*-pinene. There was a general trend towards increasingly greater recovery with increasing retention time or boiling point when vacuum was applied. Vacuum inhibits adsorption of certain lower boiling compounds to the porous polymer, possibly because surface contact is decreased.

Type of Adsorbent Material Used for Concentration. To compare efficiency of solid supports the three-necked flask was used with sparging at 65° C., but without application of vacuum (Figure 5). Except for pentanal and *alpha*-pinene, all compounds were recovered slightly more efficiently when Tenax GC was used than when Carbopack B/Carbosieve SIII was used. The latter graphitized carbon trap is clearly the better choice for certain lower boiling aliphatic compounds as illustrated with pentanal, or terpene hydrocarbons like *alpha*-pinene, which always showed a very low affinity for Tenax GC. A second elution of the Carbopack B/Carbosieve SIII trap did not result in any further recovery of volatiles.

Effect of Sparging versus Purging of Headspace. A fundamental question concerning representative sampling of flavor volatiles involves dynamic headspace *versus* continuous sparging of the sample matrix. In Figure 6, results are shown for recovery of sample in oil from the three-necked round bottom flask with nitrogen flow at 50 ml/min in either the headspace or sparging mode with trapping on Tenax GC. Sparging of volatiles from the oil sample was consistently more efficient for all test compounds than purging the headspace above the oil. Vacuum was not applied to avoid losses as described above.

Recovery of Volatiles from Oil Using the ECID. An assessment of recovery was made by reinserting into the ECID a sample of test solution in oil on glass wool for a further 15 min injection after the initial recovery. The additional recovery is shown in Figure 7. The amount of volatiles recovered from the second injection was on average 19% of the total recovered from the two injections. This experiment was repeated with a 30 min injection time. Figure 8 shows that many of the compounds were completely injected in the first 30 min and although the remainder were incompletely injected the greatest additional amount injected was 13.5% (decanal) of its total for two injections. An alternative interpretation of these recovery results would be that Figures 7 and 8 illustrate a test for completeness of elution of the volatiles.

Reproducibility of the ECID injection of volatiles from oil was determined by injecting four separate samples of the same oil-test compound mixture. Table II shows the greatest percentage standard deviation to be 18.7% for decanal. There is possibly a trend towards an increasing percentage standard deviation as the retention time increases above 20-25 mins.

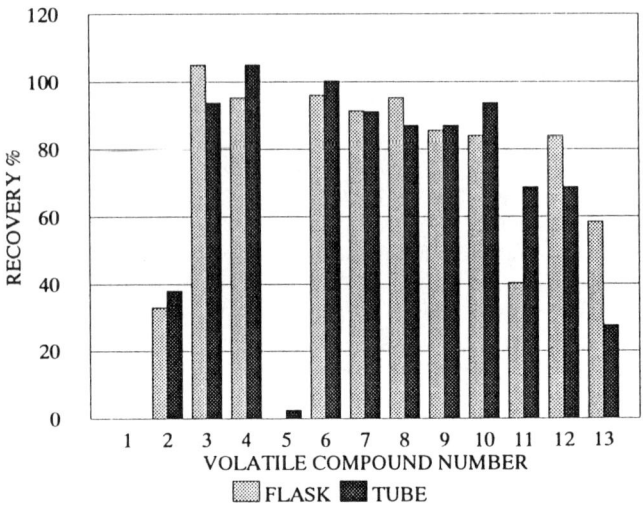

Figure 3. Comparison of % recovery of standards in oil from the three-necked flask *versus* Tekmar sparging tube under vacuum for 16 hours at 65° C with 50 ml/min nitrogen flow to Tenax GC traps with ethyl ether elution.

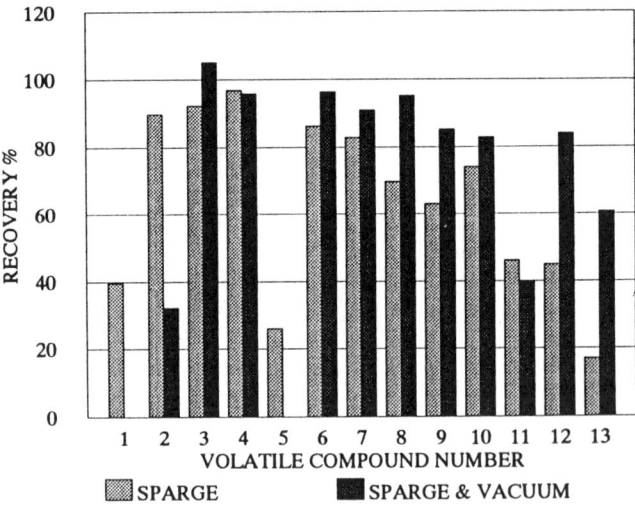

Figure 4. Effect of sparging *versus* sparging-plus-vacuum on % recovery of standards from a three-necked flask for 16 hours at 65° C with 50 ml/min nitrogen flow to Tenax GC traps with ethyl ether elution.

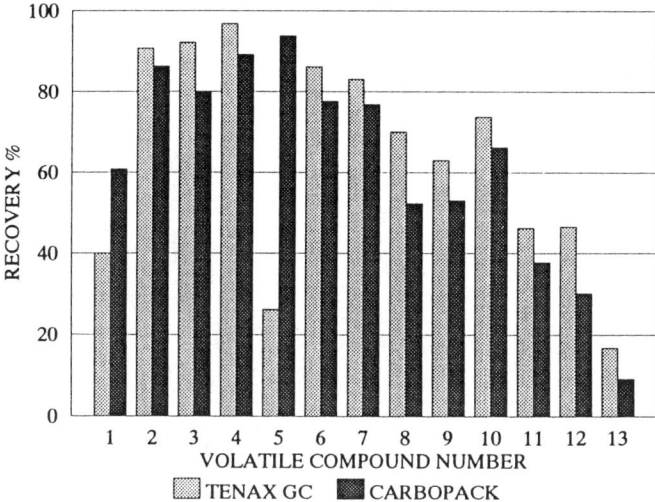

Figure 5. Comparison of the solid supports, Tenax GC *versus* Carbopack B/Carbosieve SIII for efficiency of % recovery of standards from a three-necked flask sparged without vacuum. Ethyl ether elution of traps with syringe injection after concentration.

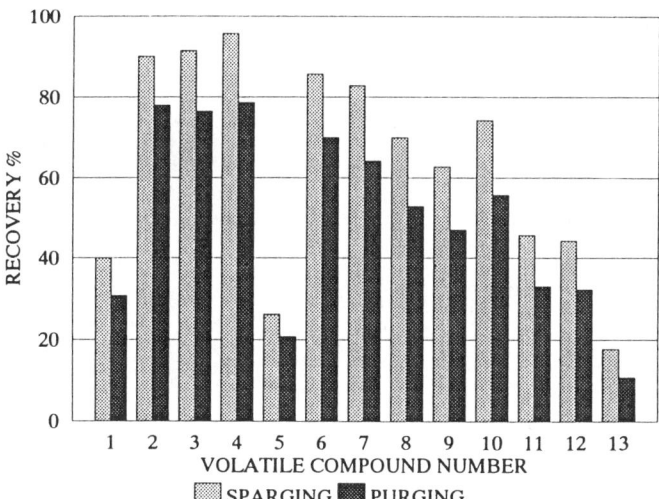

Figure 6. Recovery % of standards from matrix of sample in oil as affected by sparging (50 ml/min nitrogen) through the oil compared with dynamic headspace purging above the oil at the same nitrogen flow rate, trapping on Tenax GC for 16 hours at 65° C from the three-necked round bottom flask followed by ethyl ether elution of traps.

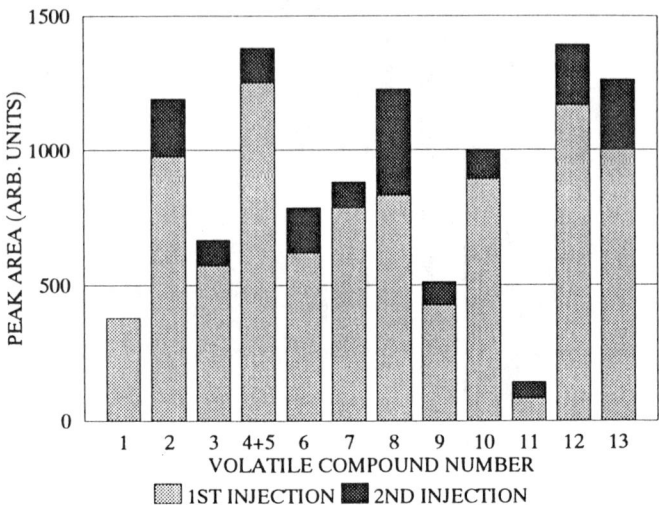

Figure 7. Additional recovery of volatiles from the same standards sample in oil on glass wool after repeating first 15 minute injection at 100° C with column at +1° C for a second 15 minutes with ECID (total injection time 30 minutes). Peak areas are in relative integrator units.

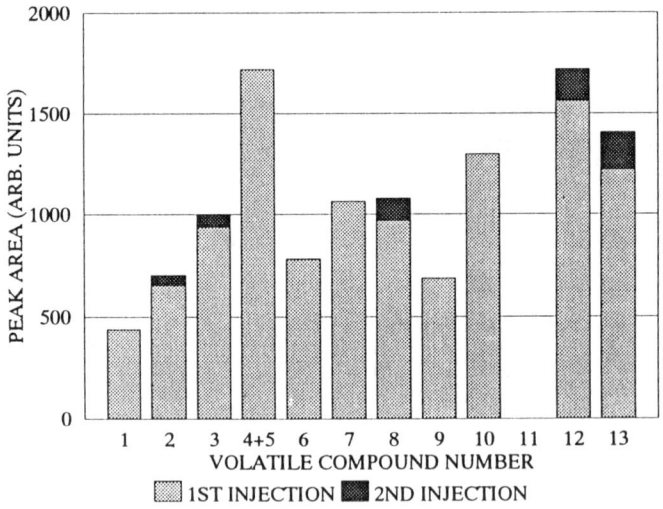

Figure 8. Recovery of standards sample as in Figure 7 after repeating first 30 minute injection at 100° C with column at +1° C for a second 30 minutes with ECID (total injection time 60 minutes). Peak areas are in relative integrator units.

Table II. ECID Evaluation: Reproducibility of Compounds

Number	Mean Area	% SD
1	406	7.9
2	598	10.1
3	738	7.1
4+5	1345	2.4
6	632	6.8
7	775	2.4
8	771	10.4
9	467	5.2
10	925	2.3
12	1067	16.9
13	951	18.7

Problems Arising in Quantitation of Dynamic Purge and Trap Headspace Chromatography of Flavor Volatiles

A very important question arising in sampling of volatiles, e.g., from 50 g. of peanut butter in a 500 ml three necked flask, is that relatively little of the total volatiles are swept out of the peanut butter even in one four hour run. At best, even a long purge and trap sampling run leads only to a "snapshot" of the molecular cross section of how the total volatiles are distributed both quantitatively and qualitatively. The peanut butter is melted at 90° C, leaving a thick glob on the bottom of the flask. What is sampled and concentrated on the Tenax or Carbopack supports as representative of the volatile peanut butter flavor principles is probably no more than a small percentage of the total compounds present possessing sufficient vapor pressure to reach the traps. This was verified when repeated runs on the same sample (6 or 8 times at 90° C) continued to yield significant amounts of the volatiles originally present.

Over a period of time researchers have become well aware that the differences in vapor pressure of food flavor volatiles as well as their relative affinity for the solid supports used (including the packed Tenax column) result in either breakthrough losses on trapping or poor recoveries. The key markers for roasted peanut volatiles ranked here (Table III) for response in the flame ionization detector have a wide range of boiling points and their recoveries, from highest vapor pressure to lowest, probably represent a bell shaped progression of recoveries. The original sample of a product such as vegetable oil or peanut butter probably is, therefore, of a rather different composition for these compounds than is measurable by purge-and-trap, and its cumulative sensory characteristics also differ from this ranking order.

Experiments performed here demonstrate kinds of yield problems that purge-and-trap concentration have on effective sampling of sensory stimuli from flavor volatiles depending on partial vapor pressures and affinities of the compounds for the supports. Examples of recovery effects in answer to questions posed in the

Table III. FID Response Factors for Roasted Peanut[a] Marker

Chemical Peaks	Calibration Factor[b]
MeOH	10000
Unk1	30000
EtOH	15000
Pentane/Acetone	30000
Unk2	30000
Methylpropanal	20000
Methylbutanal	25000
Pentanal	25000
Methylbutanol	20000
N-Methylpyrrole	15000
Hexanal	25000
Unk3	30000
Hexanol/Methylpyrazine	17000
Dimepyz/2-pentylfuran	15000
Unk4	20000
Methylethylpyrazine	15000
4-Carbon Sub't Pyrazines	15000
Unk5	30000
Vinylphenol	20000
Decadienal	20000
Unk6	30000
Total Volatiles One	30000
Total Volatiles Two	30000

[a] (0.5g samples)
[b] counts/ppm

beginning are given in each of the graphs for the experiments. It cannot be emphasized enough that when sweeping a sample in the ECID with carrier gas, high yields of volatiles are achieved when 20 to 40 or more ml/min of carrier flow is applied. This point was made by Grob (6) a number of years ago and is key to making the wide bore capillary as effective as it is to concentrate maximum load of volatiles.

An effective method of creating comparison factors for response standards in volatiles research by purge and trap or direct injection gas chromatography is to dissolve the standard in a very highly refined vegetable oil. In Figure 9 is shown the flame detector response for various concentrations of hexane in a fine vegetable oil that is free of lipid oxidation products. The dilutions of hexane are representative of ranges of volatiles found in foods. The relative quantity of hexane can be used, e.g., as an internal standard trapped from the food sample, or it can be used as an external standard with a standard response slope after regression analysis and fit to a linear slope ($R^2 = 0.939$).

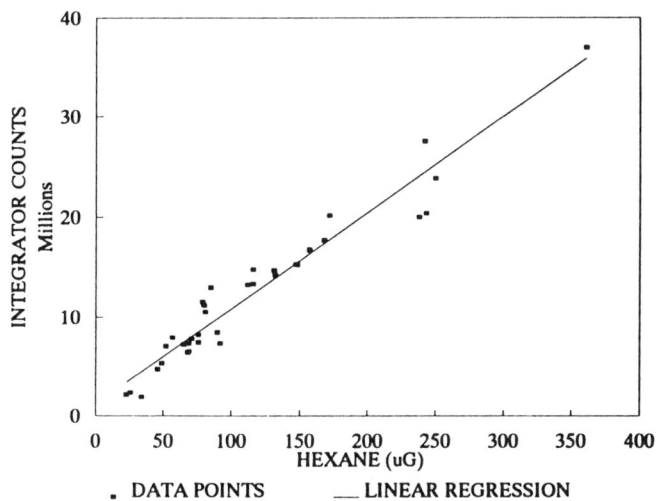

Figure 9. Response curve for ECID injection of dilutions of hexane in vegetable oil (500 mg samples of oil on glass wool heated for 30 min at 100° C onto the wide bore capillary column at +1° C). Response factor derived from regression slope of FID signals ($R^2 = 0.939$).

Other internal standards used in this laboratory in oil solution are 2-undecanone, benzothiophene, and benzothiazole. Alternatively, dilutions of peaks of interest (e.g., nonanal) can be made and injected by direct gas chromatography from an inert solid support or by syringe as serial dilutions. The flame ionization response of a peak by syringe injection as opposed to direct purge onto the column is at least proportional in peak area produced. For semiquantitative comparison the FID responses for roasted peanut marker compounds can be set up as in Table III to represent differences in integrator signal the various organic structures make in the flame (cf., Figures 10-14). Broadening of the peaks on direct injection affects these responses somewhat but the ratios are quite similar. Unknowns and total volatiles in Table III are assigned the external standard value for pentane (30,000 counts/ppm) as an estimation of relative concentration of volatiles present during comparison of samples by GC.

It can be concluded, therefore, that dynamic headspace purge onto an intermediary solid support or by direct gas chromatography onto a packed or capillary column serving both as a concentration support as well as separation column results in representative but nonquantitative sampling of the whole of the volatiles. A good analogy from another technique of analytical chemistry is sampling much like one obtains from a flow injection analysis. It is thus possible to quantitate how much of each compound was concentrated per unit of sampling time (not the whole), assuming that the total might or might not represent new species produced during the heating or trapping time. Even using internal standards or surrogates to ascertain matrix effects, the measured volatiles then may or may not be a true measure of the effective sensory stimulation from the real burden of flavor principles in the original sample.

Adsorption followed by thermal desorption from Tenax GC is the method of choice for many priority pollutants as well as food flavor components *(47)*. Problems with drying the cartridge after trapping from a moist sample are not a real difficulty with the ECID and 0.75 mm wide bore capillary column described in the present paper. In fact, unless the cartridge is saturated with water, no previous drying is necessary since on thermal desorption, the wide bore column acts as a solid support adsorbent, concentrating the organics and letting the water pass through.

Analyses of Roasted Peanut Volatiles by Instrumental and Olfactory Response

What is being measured in dynamic headspace analysis of peanut butter after heating under inert gas sweep and/or vacuum is a combination of natural flavor volatiles and compounds formed through precursors during the heating period used for trapping. The kinetics of the Maillard reaction are such that when precursors are present, the products form slowly even at ambient temperature. In the above recovery studies, four hours of trapping were used as a typical efficient sampling of volatile markers. In peanut butter, four hours at 90°C certainly would also result in a lot of browning reaction products forming in peanut butter and should be taken into account as part of the assessment of a roasted peanut sample. Therefore, in the present work a relatively short inert gas sweep of the sample is performed (15 min) through the cartridge. In this paper, relative quantitation of the roasted peanut components would be possible using external and internal response factors as in Figure 9 or Table III. However, since the purpose of the paper is to present an updated method, data will be limited to qualitative identification of the peanut flavor compounds separated, and the kinds of reproducible olfactory perceptions made at various retention times without attempting quantitation. Another paper in this Symposium (Chapter 18, Bett and Boylston) does an excellent job of quantitation of roasted peanut flavor volatiles and their dynamic changes during shelf life storage. The reader is referred to that Chapter for more definitive statements on the relative quantities of key volatiles in roasted peanut flavor blends.

Delineating olfactory stimuli in aromatics of roasted peanut is a specialized task requiring at least three crop years of samples with results incorporated into principal component, analysis of variance, or other statistically designed experiments. Correlation of identified important volatiles through high resolution capillary GC with sensory attributes from controlled human olfactory bioassay through the sniffer port leads to very valuable information about a product.

In the present report, no attempt is made to create a benchmark listing of attributes and quantities of high impact character stimuli of the roasted peanut samples analyzed by this system applicable to all peanuts. However, as stated above the objective of this research is to set up and test a reproducible gas chromatography system generally applicable to volatiles from many kinds of samples (even those with high moisture such as cooked meat or plant material).

A roasted Florunner peanut (Hunter L = 50, PV = 1.4 meq/kg), used as a control of good quality for the peanut flavor descriptive sensory panel at this Center and well defined through the collective judgements of many panelists, was analyzed in Figure 10 for both olfactory and GC composition identified by mass spectrometry. In chromatograms run under the same conditions as in Figure 10, a moderately rancid (PV = 33 meq/kg) (Figure 11) and very rancid peanut (PV = 111 meq/kg) (Figure 12) were also compared for olfactory composition and GC profiles. The rancid peanuts were also roasted Florunners which would normally have fresh roast flavor similar to the control, but which under industrial storage conditions deteriorated through autooxidation. The positive attributes of roasted peanuts literally disappeared in the rancid samples (cf., Figures 10-12) as shown both by GC retention times of positive components as well as olfactory identification with the sniffer port.

As a test of the reproducibility of the method, the ground roasted peanut with PV = 1.4 meq/kg was run fourteen times with good standard deviation (+/- 2%) for FID responses and olfactory perceptions in sequence. The +/-2% standard deviation is an average of all runs with the lowest being +/-1% and the highest, +/-6%. This was an improvement over recovery experiments (above) for trapping the standards from oil on glass wool where a standard deviation of +/- 18% was observed for decanal. The odors perceived by two trained panelists were more difficult to estimate for intensity reproducibility each run, but the retention times of each listed peanut aroma attribute were reproducible over 14 runs. This method is, therefore, efficient and useful for defining peanut quality. Considering that there is a dynamic range of more than 8 to 10 orders of

252　　　　　　　　　　　　　　　　　　　　　　LIPID OXIDATION IN FOOD

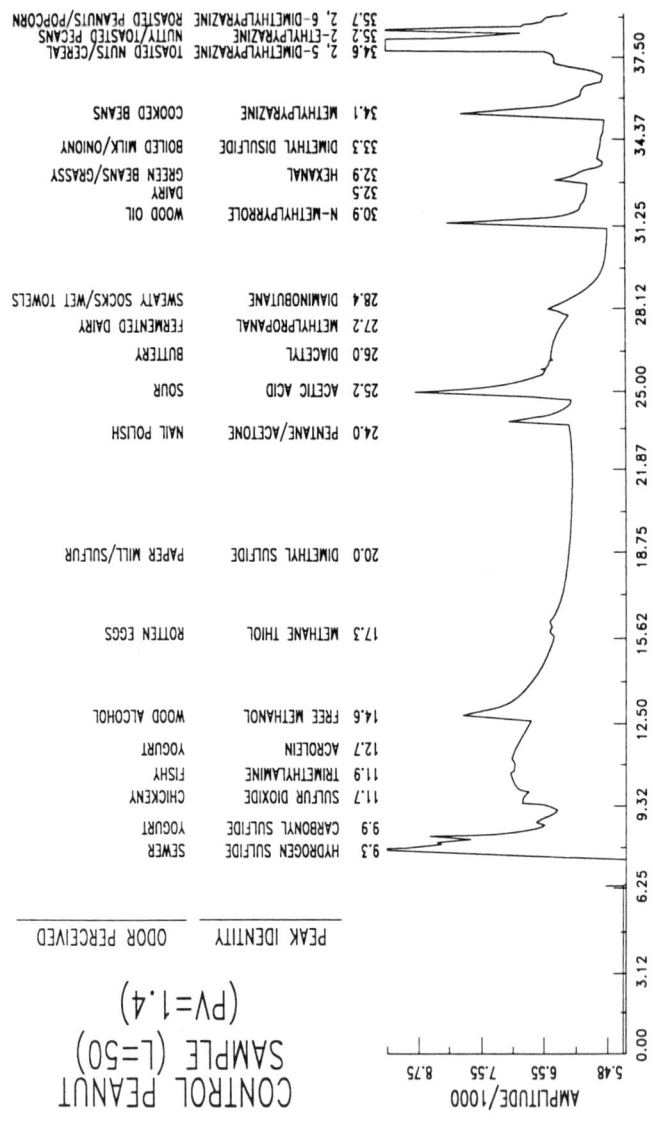

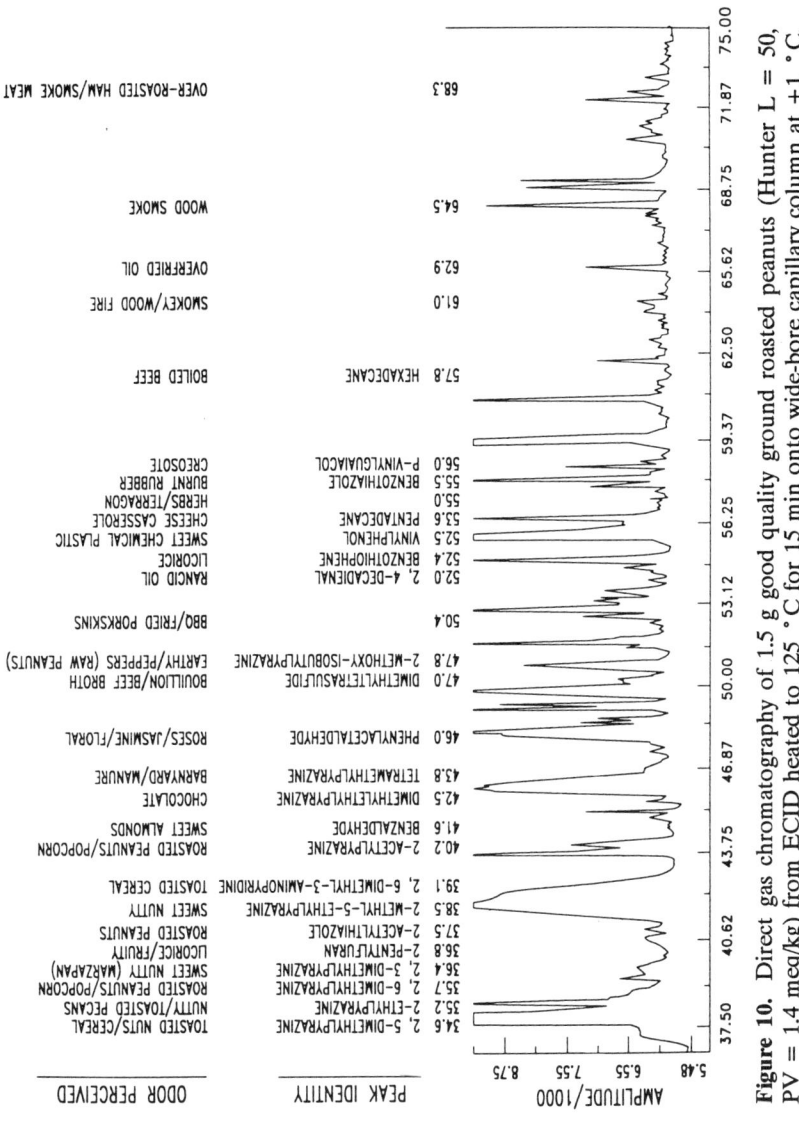

Figure 10. Direct gas chromatography of 1.5 g good quality ground roasted peanuts (Hunter L = 50, PV = 1.4 meq/kg) from ECID heated to 125 °C for 15 min onto wide-bore capillary column at +1 °C. Key peaks identified by GC/MS of volatiles as concentrate eluted from Carbopack B/Carbosieve SIII in separate experiment. Olfactory attributes described by sniffer port (comparison of 14 runs by two trained panelists).

254 LIPID OXIDATION IN FOOD

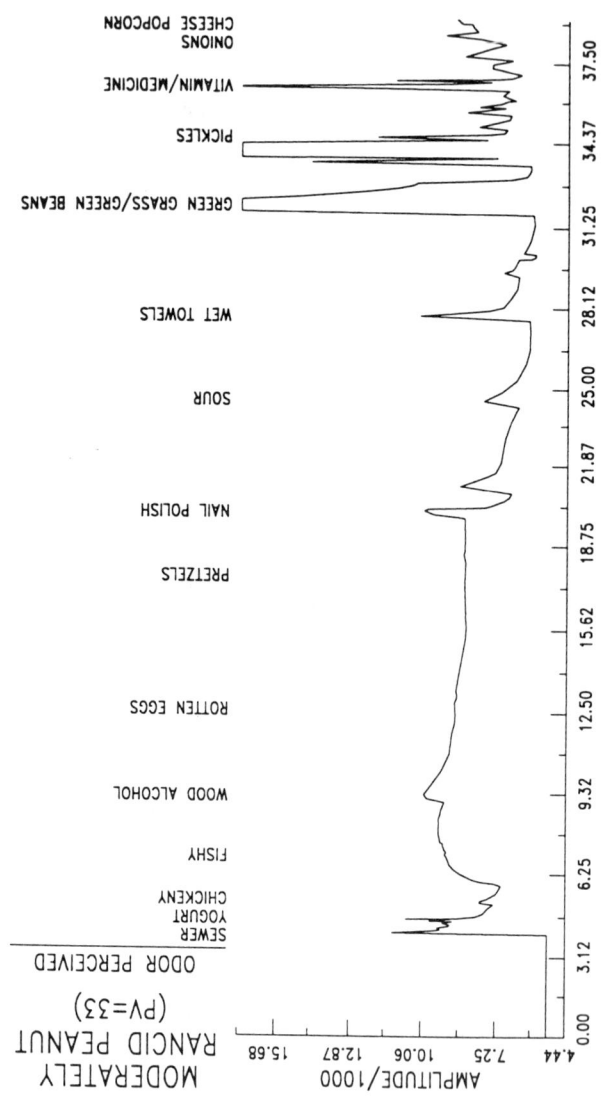

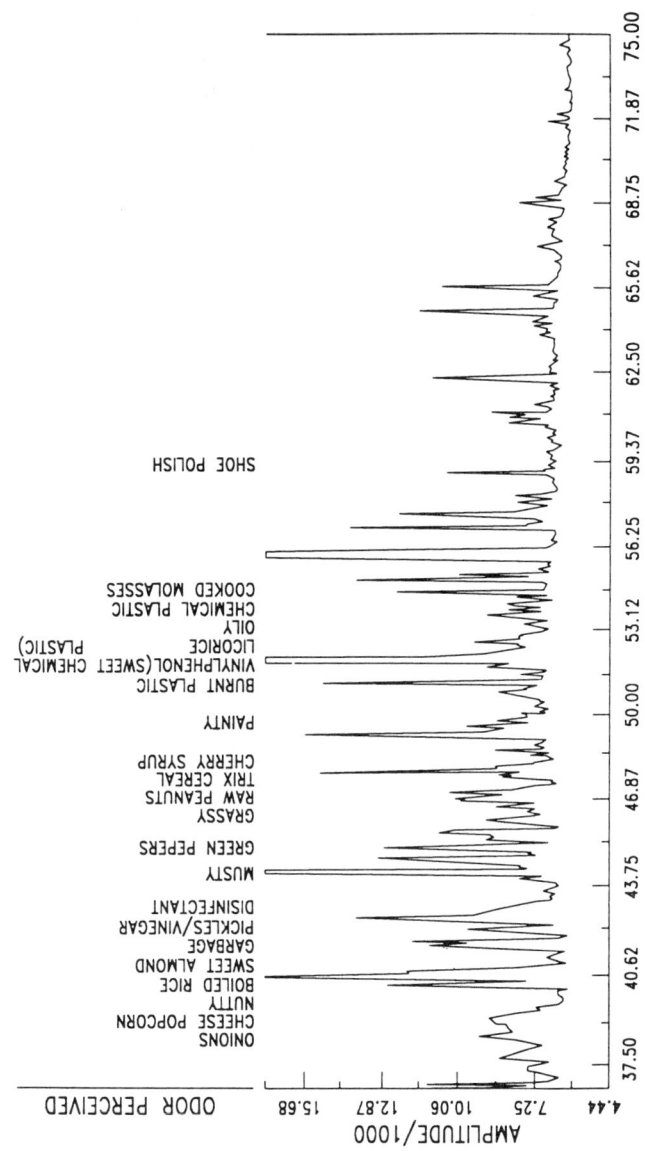

Figure 11. Comparative chromatogram and olfactory attributes by sniffer port for a moderately rancid roasted peanut with PV = 33 meq/kg, run according to conditions in Figure 10.

256   LIPID OXIDATION IN FOOD

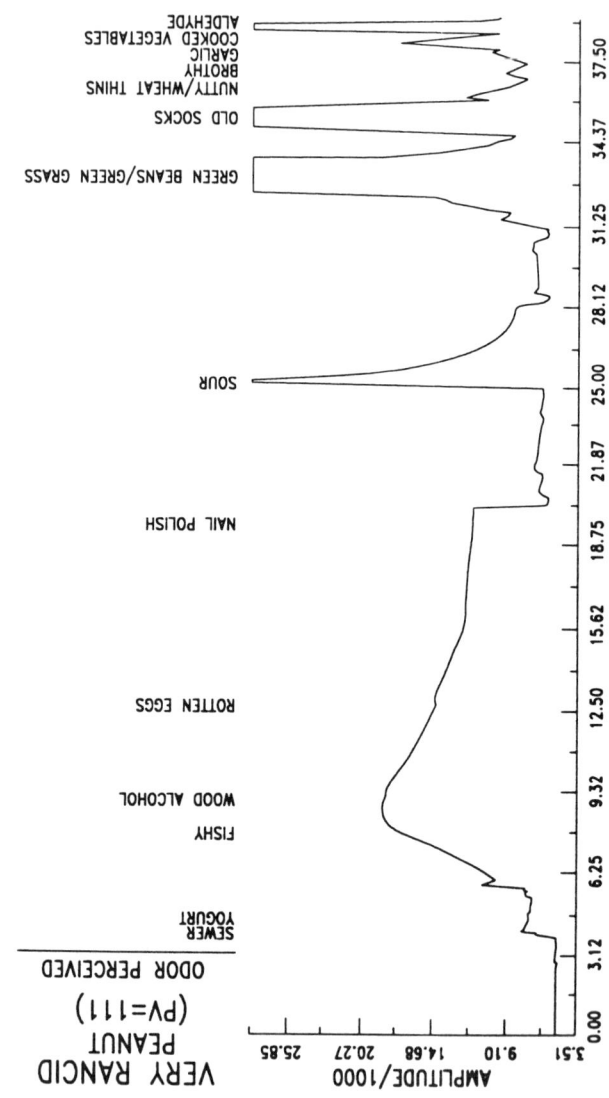

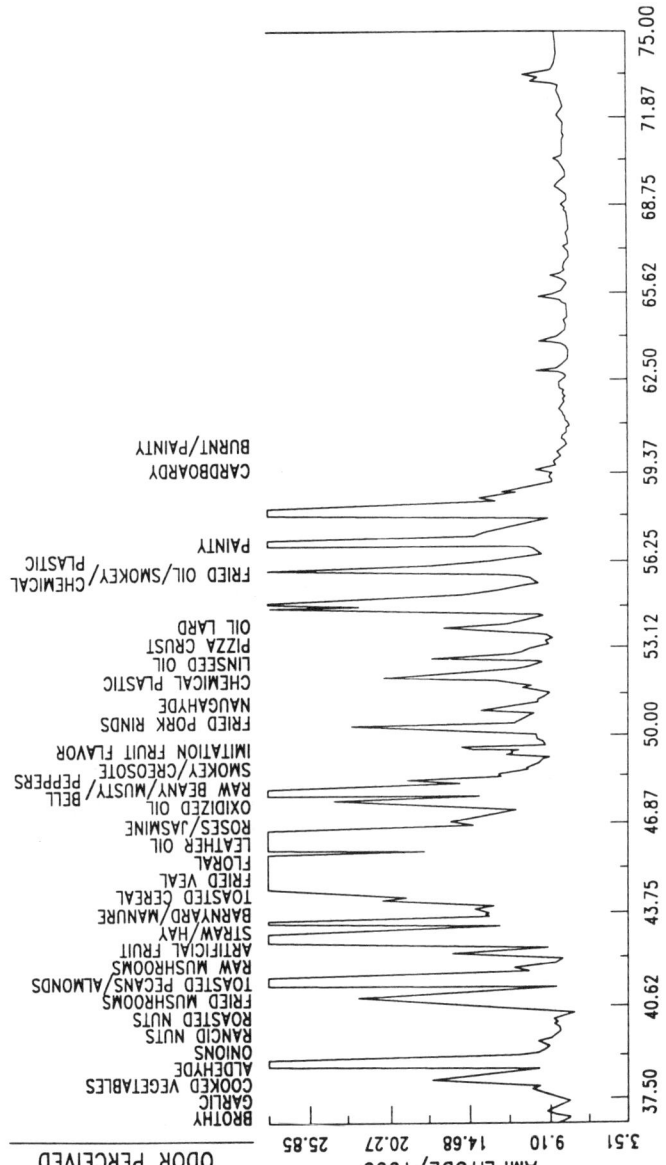

Figure 12. Chromatogram and olfactory attributes for a very rancid roasted peanut with PV = 111 meq/kg, run according to conditions in Figure 10.

magnitude for olfactory identification of low threshold compounds, having an efficient system to concentrate and separate volatiles directly is a useful tool to study food quality.

The mass spectral identification of the components in the control peanut was carried out by purge and trap concentration of the volatiles on Carbopack B/Carbosieve SIII cartridges and injected into the GC/MS after elution with methylene chloride and concentration to small volume. Solid support (Tenax GC or Carbopack B/Carbosieve SIII) concentrates of the same peanut volatiles as from direct gas chromatography were compared either by thermal desorption in the ECID, or after elution with solvents such as ether, hexane, etc., using syringe injection. The compounds identified and listed in Figure 10 have all been reported earlier *(34-36)*, and many more compounds are present than identified in this work. The GC/MS identification was done in this study to confirm the accuracy of trapping, concentration, and identification of these peanut volatiles using the method reported. Although much more work must be done to relate the volatile organic compounds with perceived flavor descriptors, the present profile presents a useful comparison of identified peaks with olfactory responses in that retention time region. For instance, Fischer and Grosch demonstrated clearly by GC and sniffer port that the green/ legume-like aroma of raw peanuts was actually not a single compound but rather a mixture of two separate zones, with quite individual contributions of each component *(48)*. The one zone with raw beany contribution is *n*-hexanal and the other is a complex made up of *gamma*-butyrolactone, *n*-nonanal, benzaldehyde, benzyl alcohol, and 2-methoxy-3-isopropylpyrazine. In the descriptive sensory lexicon for peanuts, "raw beany" is actually a single descriptor *(49)*. It will be interesting to unravel the complexities of many other flavor attributes in addition to peanut descriptors using more sophisticated GC/sniffer port/computer techniques.

It should be noted that the two rancid roasted peanuts in Figure 11 (PV = 33 meq/kg) and Figure 12 (PV = 111 meq/kg) not only have increases in other volatiles over those of the control with its low PV of 1.4 meq/kg (e.g., hexanal), but also many of the key markers for acceptable flavor in the control are much diminished or lacking in the rancid peanuts. Similarly, expected roasted peanut aroma attributes decrease in the rancid peanuts and negative odors are perceived as predominant olfactory characteristics. Typical lipid oxidation volatiles from oil pressed out of the sample of intermediate rancidity (Figure 11) were chromatographed using this method. The retention times and names of the lipid degradation compounds identified by GC/MS are listed in Table IV. It is interesting to note that heterocycles with positive aroma attributes (listed in Figure 10) are more concentrated in a good quality roasted peanut (Figure 10) than in an oxidized sample (Figure 11), where some have disappeared in the baseline or are actually completely absent.

Comparison of the ECID with Packed Column and Wide Bore Capillary

The ECID introduced by Dupuy, Legendre, and Fisher *(19,23,25,26)* is a workable tool for introduction of food flavor volatiles onto a packed or capillary column with high enough capacity to present olfactory stimuli for accurate sensory

Table IV. Lipid Oxidation Products from Oil of a Rancid Roasted Peanut[a] (PV=33 meq/kg)

Compound	Direct Capillary G.C./M.S.[b]
acetaldehyde	5.21
acetone/pentane	11.96
2-butanal	20.33
benzene	20.97
3-penten-2-one	22.49
acetic acid	22.96
pentanal/heptane	23.86
2-pentanol	30.34
1-pentanol	30.45
1-octene	32.03
hexanal	32.86
2-octene	33.45
3-octene	34.23
2-hexenal	38.35
1-hexanol	40.13
octatriene	40.34
2-heptanone	42.40
nonane	42.98
heptanal	43.56
2-butylfuran	48.46
t-2-heptenal	49.41
1-hepten-3-ol	50.76
2,3-octanedione	52.46
2-pentylfuran/ 2-octanone	53.25
t,c-2,4-heptadienal	53.99
octanol	54.51
t,t-2,4-heptadienal	55.30
limonene	57.31
2-Octenal	60.40
3,5-octadien-2-one	62.57
nonanal	65.21
2-nonenal	70.99
benzothiazole	78.68
2-decenal	80.90
t,c-2,4-decadienal	84.06
t,t-2,4-decadienal	86.11
tetradecane	91.96

[a] 1 gram oil heated at 125° for 15 min. With 10 ml/min carrier in splitless injection mode. Cryofocused at -30°C during volatiles collection. Program from -30° (hold 15 min) with 4°/min program to 250°C (hold 30 min).
[b] Retention time from injection (min).

analysis of the GC peaks as well as to afford adequate separation and detection. The ECID also permits direct thermal desorption of concentrated volatiles either in foods, vegetable oils, packaging materials or from solid supports used to concentrate volatiles.

Much larger quantities of compounds of interest can be concentrated for olfactory analysis using direct gas chromatography with the wide bore (0.75 mm i.d.) capillary system described. A direct comparison of the chromatograms achieved by both packed column (Tenax GC-8% PMPE) and wide bore capillary is given in Figure 13. Compared to the packed columns previously used with the ECID, however, the wide bore capillary has much better resolution of peaks and permits much greater distinction of perceived odors in the sniffer port (Figure 10). However, Figure 14 compares the packed column chromatograms of the freshly roasted peanut (cf., Figure 10 for the capillary column) with the two more oxidized samples (cf., Figures 11 and 12 for the capillary column). Figure 14 illustrates that the ECID with packed column provides a clear comparison for making quality judgements about a set of roasted peanut samples.

A definite plus for the packed column with the ECID is the fact that, while less well resolved peaks result, profiles make comparison of samples in quality control much easier, especially with computer replotting and baseline establishment. The packed columns do not need cryofocusing below ambient for concentration of volatiles nor is water content of the sample a particular problem with resolution on the column. The packed columns also admit response factor calibration (Table III), and for the period of volatiles collection ("snapshot" of flavor cross section), the relative quantity of compounds in peaks eluted can be estimated for sake of close sample comparison. Mass spectral analysis has, in fact, repeatedly shown that the principal compound in each of the peaks of the packed column is the one assigned on Figure 13 *(34)*.

Summary and Conclusions

As another perspective of this *Symposium on Lipid Oxidation in Foods*, the present paper highlighted great strides made by researchers in the past decade to advance analytical methodology in food flavor quality evaluation. A significant problem in food flavor stability involves generation of free radicals capable of oxidation of polyunsaturated lipids with simultaneous deterioration of positive contributors to valuable flavor principles.

After presenting an historical overview and major empirical hypotheses from many authors demonstrating the feasibility of approaching quantitation of key food flavor volatiles, new experimental work was presented. The examples demonstrated the usefulness and efficiency of concentrating volatiles at +1° C on wide bore (0.75 mm) glass capillary columns by direct gas chromatography and ECID with the column as both solid concentration support as well as chromatographic separation medium.

Using representative lipid oxidation standard mixtures as well as roasted peanuts or their oil in various states of oxidation as a commodity example, a study to demonstrate the comparative efficiency of direct gas chromatographic sampling of volatiles with purge-and-trap of the same compounds to solid

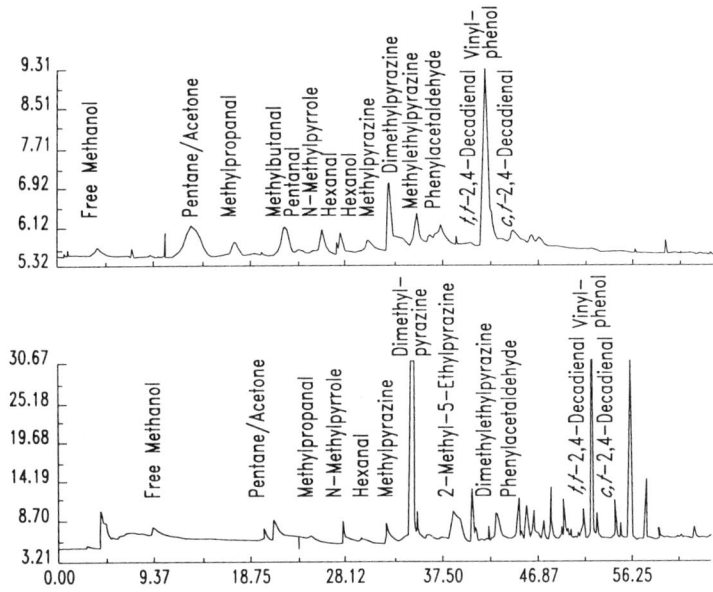

Figure 13. Comparison of chromatograms made of good quality roasted peanut (PV = 1.4 meq/kg) with ECID on packed Tenax GC-8% PMPE and 0.75 mm wide bore DBS-5 glass capillary with identification of important roasted peanut markers.

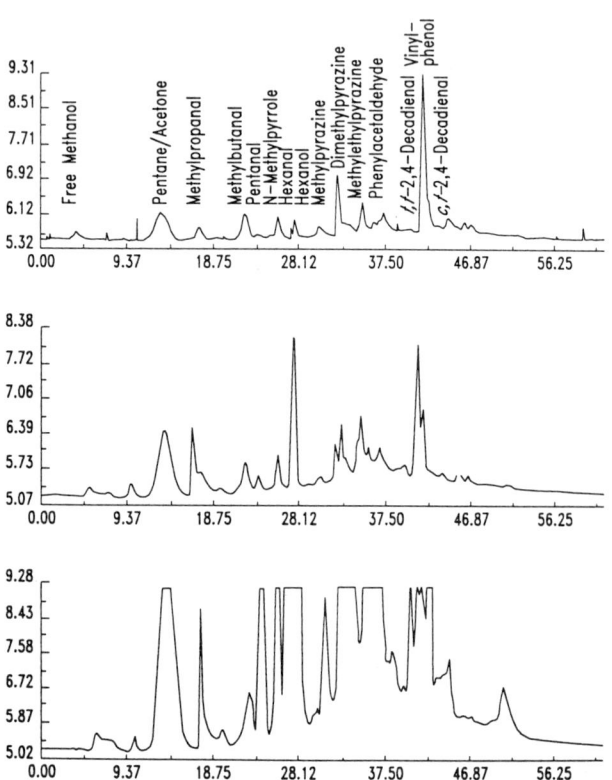

Figure 14. Direct gas chromatography with ECID on packed column of Tenax GC-8%PMPE of the control roasted peanut (PV = 1.4 meq/kg) (cf., Figure 10); moderately oxidized sample (PV = 33 meq/kg) (cf., Figure 11); and very oxidized sample (PV = 111 meq/kg) (cf., Figure 12) illustrating the diagnostic effectiveness of the packed column for quality assessment of foods.

supports such as Tenax GC or Carbopack/Carbosieve with thermal or solvent elution was presented. After demonstrating the efficiency of this direct capillary system, aromagrams of roasted peanuts in various states of oxidation were compared. The continuing usefulness of packed column chromatography coupled directly with the ECID was demonstrated and comparison made with a wide bore capillary column. Simultaneous instrumental detection and olfactory judgement about an individual peak's odor and intensity are possible. The role of lipid oxidation volatiles in food flavor quality was emphasized throughout this work.

Acknowledgments

The editor and authors dedicate this chapter to Dr. Harold P. Dupuy, whose lasting contributions to direct gas chromatographic analyses of food flavor principles have inspired much research that has resulted in making the method an objective assessment of food quality. Continuing discussions with Terri D. Boylston, Timothy H. Sanders, Norman V. Lovegren, and Gordon S. Fisher helped frame major questions of this manuscript, for which the authors express gratitude. Technical assistance by Alex V. Pisciotta and Maurice R. Brett in testing the chromatographic system is also appreciated as are the Florunner peanuts supplied by E. Jay Williams, USDA-ARS, Tifton, GA.

Literature Cited

1. Jennings, W. G. *Applications of Glass Capillary Gas Chromatography*, Marcel Dekker, Inc.: New York, NY, 1981; pp. 511-533.
2. Takeoka, G.; Guentart, M.; Macku, C.; Jennings, W. G. In *Biogeneration of Aromas;* Parliment, T.H.; Croteau, R., Eds.; ACS Symposium Series No. 317; American Chemical Society: Washington, D.C., 1986; pp. 53-64.
3. Parliment, T. H. In *ibid.*; pp 31 - 52.
4. Reineccius, G. In *Flavor Chemistry of Lipid Foods*; Min, D. B.; Smouse, T. H., Eds.; American Oil Chemists Society: Champaign, IL, 1989; pp 26-34.
5. Heath, H. B.; Reineccius, G. A. *Flavor Chemistry and Technology*; AVI Publishing Co., Inc.: Westport, CT, 1986; pp 3-42.
6. Grob, K.; Habich, A. *J. Chromatogr.* **1985**, *321*, 45-58.
7. Zlatkis, A.; Lichtenstein, H. A.; Tishbee, A. *Chromatographia* **1973**, *6*, 67-70.
8. Bishop, R. W.; Valis, R. J. *J. Chromatographic Science* **1990**, *28*, 589-593.
9. Steinke, J. A. *Isolation and Identification of Volatile Constituents from Off-flavored Lake Michigan Salmon* **1978** Ph.D. Dissertation, Univ. of Wisconsin; *Dissertation Abstracts*, **1979**, *39*, 4798; *CA* **91**: 37655), and *Univ. Microfilms*, No. 7823767)
10. Olafsdottir, G.; Steinke, J. A.; Lindsay, R. C. *J. Food Sci.* **1985**, *50*, 1431-1450.
11. Michael, L. C.; Erickson, M. D.; Parks, S. P.; Pellizzari, E. D. *Analytical Chemistry* **1980**, *52*, 1836-1841.
12. Cole, R. A. *J. Sci. Food Agric.* **1980**, *31*, 1242-1249.
13. Buttery, R. G. In *Flavor Chemistry of Lipid Foods*; Min, D.B.; Smouse, T.H., Eds.; American Oil Chemists Society: Champaign, IL, 1988; pp 156-164.

14. Buttery, R. B.; Teranishi, R.; Ling, L. C. *J. Agric. Food Chem.* **1987**, *35*, 540-544.
15. Buttery, R. G.; Teranishi, R.; Ling, L. C.; Turnbaugh, J. G. *J. Agric. Food Chem.*, **1990**, *38*, 336-340.
16. Buttery, R. G.; Teranishi, R.; Flath, R. A.; Ling, L. C. *J. Agric. Food Chem.*, **1990**, 38, 792-795.
17. Galt, A. M.; MacLeod, G. *J. Agric. Food Chem.* **1984**, *32*, 59-64.
18. Suzuki, J.; Bailey, M. E. *J. Agric. Food Chem.* **1985**, *33*, 343-347.
19. Dupuy, H. P.; Bailey, M. E.; St. Angelo, A. J.; Vercellotti, J. R.; Legendre, M. G. In *Warmed-Over Flavor of Meat*; St. Angelo, A. J.; Bailey, M. E., Eds.; Academic Press, Inc.: Orlando, FL, 1987; pp 165-191.
20. Fritsch, C. W.; Gale, J. A. *J. Amer. Oil Chem. Soc.* **1976**, *54*, 225-228.
21. Fisher, G.S.; Legendre, M.G.; Lovegren, N.V.; Schuller, W.H.; Wells, J.A., *J. Agric. Food Chem.*, **1979**, *27*, 7-11.
22. Lovegren, N. V.; Vinnett, C. H.; St. Angelo, A. J. *Peanut Sci.* **1982**, *9*, 93-96.
23. Dupuy, H. P.; Brown, M. L.; Fisher, G. S.; Lovegren, N. V.; St. Angelo, A. J. *APRES Peanut Quality Methods Manual*; Amer. Peanut Res. and Educ. Soc.: Stillwater, OK, 1983. Method QM 1.
24. St. Angelo, A. J.; Lovegren, N. V.; Vinnett, C. H. *Peanut Sci.* **1984**, *11*, 36-40.
25. Legendre, M. G.; Fisher, G. S.; Schuller, W. H.; Dupuy, H. P.; Rayner, E. T. *J. Amer. Oil Chem. Soc.* **1979**, *56*, 552-555.
26. Dupuy, H. P.; Flick, G. J.; Bailey, M. E.; St. Angelo, A. J.; Legendre, M. G.; Sumrell, G. *J. Amer. Oil Chem. Soc.* **1985**, *62*, 1690-1693.
27. Waltking, A. E. *J. Amer. Oil Chem. Soc.* **1982**, *59*, 116A-120A.
28. Min, D. B. *J. Food Sci.* **1981**, *52*, 1453-1456.
29. Vercellotti, J. R.; Kuan, J. W.; Liu, R. H.; Legendre, M. G.; St. Angelo, A. J.; Dupuy, H. P. *J. Agric. Food Chem.* **1987**, *35*, 1030-1035.
30. Vercellotti, J. R.; St. Angelo, A. J.; Legendre, M. G.; Sumrell, G.; Dupuy, H. P.; Flick, Jr., G. J. *J. Food Comp. Anal.* **1988**, *1*, 239-249.
31. Fore, S. P.; Goldblatt, L. A.; Dupuy, H. P. *J. Am. Peanut Res. Educ. Assoc.* **1972**, *4*, 177-185.
32. Fore, S. P.; Dupuy, H. P.; Wadsworth, J. I.; Goldblatt, L. A. *J. Am. Peanut Res. Educ. Assoc.* **1973**, *5*, 59-65.
33. Fore, S. P.; Dupuy, H. P.; Wadsworth, J. I. *Peanut Sci.* **1976**, *3*, 86-89.
34. Fore, S. P.; Fisher, G. S.; Legendre, M. G.; Wadsworth, J. I. *Peanut Sci.* **1979**, *6*, 58-61.
35. Buckholz, Jr., L. L.; Daun, H; Stier, E; Trout, R. *J. Food Sci.* **1980**, *45*, 547-554.
36. Buckholz, Jr., L. L.; Daun, H. In *Quality of Selected Fruits and Vegetables of North America*; Teranishi, R; and H. Barrera-Benitez, H., Eds.; ACS Symposium Series No. 170; American Chemical Society: Washington, D.C., 1981; pp 163-181.
37. Pattee, H. E.; Singleton, J. A. *ibid.*, pp. 147-161.
38. Pattee, H. E.; Pearson, J. L.; Young, C. T.; Giesbrecht, F. G. *J. Food Sci.* **1982**, *47*, 455-460.
39. How, J. S. L. *Effects of variety, roasting, modified atmosphere packaging, and storage on the chemical composition, headspace volatiles, and flavor profiles*

of peanuts. **1984**. Ph.D. Dissertation, North Carolina State University, Raleigh, N.C.
40. Sanders, T. H.; Vercellotti, J. R.; Crippen, K. L.; Civille, G. V. *J. Food Sci.* **54**, 475-477.
41. Sanders, T. H.; Vercellotti, J. R.; Blankenship, P. D.; Crippen, K. L.; Civille, G. V. *J. Food Sci.* **1989**, *54*, 1066-1069.
42. Young, C. T.; Hovis, A. R. *J. Food Sci.* **1990**, *55*, 279-280.
43. Mason, M. E.; Waller, G. R. *J. Agric. Food Chem.* **1964**, *12*, 274-
44. Frankel, E. N.; Selke, E. *J. Amer. Oil Chem. Soc.* **1987**, *64*, 749-753.
45. Acree, T. E.; Barnard, J; Cunningham, D. G. *Food Chemistry* **1984**, *14*, 273-286.
46. Woodruff, J. G. *Peanuts. Production, Processing, Products.* Third edition. AVI Publishing Co.: Westport, CT, 1983; pp 394-395.
47. Pankow, J. F.; Isabelle, L. M. *J. Chromatogr.* **1982**, *237*, 25-39.
48. Fischer, K.-H.; Grosch, W. *Lebensm.-Wiss. Technol.* **1982**, *15*, 173-176.
49. Johnsen, P. B.; Civille, G. V.; Vercellotti, J. R.; Sanders, T. H.; Dus, C. A. *J. Sensory Studies* **1988**, *3*, 9-18.

RECEIVED February 19, 1992

Chapter 14

Characterization of Off-Flavors by Aroma Extract Dilution Analysis

Werner Grosch, Ute Christine Konopka, and Helmut Guth

Deutsche Forschungsanstalt für Lebensmittelchemie, Lichtenbergstrasse 4, D-8046 Garching, Germany

Aroma extract dilution analysis was used to identify those odorants which contribute significantly to the warmed-over flavor (WOF) of boiled beef, as well as to the light-induced off-flavors of soybean oil and milkfat. WOF was due to an increase of the odorants formed by lipid peroxidation, in particular hexanal and trans-4,5-epoxy-(E)-2-decenal. The light-induced off-flavors of soybean oil and milkfat were mainly caused by 3-methylnonane-2,4-dione, which was formed by photosensitized oxidation of furanoid fatty acids, whose concentration levels in these fats were in the range of 0.01 to 0.05 per cent.

The aroma compounds, which contribute significantly to the flavor of a food, can be localized in the capillary gas chromatogram of the volatile fraction by gas chromatography/olfactometry (GC/O) *(1, 2)*. The odor-activity of the eluting compounds can be determined by GC/O of serial dilutions of the extract and expressed as flavor dilution (FD)-factor *(2, 3)*. The FD-factor for a compound is the ratio of its concentration in the initial extract to its concentration in the most dilute extract in which the odor was still detected by the GC/O. One such procedure called Aroma Extract Dilution Analysis (AEDA) was used to identify those odorants which caused off-flavors during the storage of boiled beef in a refrigerator, as well as during the storage of soybean oil and butter oil at room temperature in the presence of light.

This paper summarizes the results of these studies. In the cases of soybean and butter oils the identification of the fatty acids, which act as precursors of the odorants causing an off-flavor, is additionally reported.

Warmed-Over Flavor (WOF)

The flavor of meat is optimal when it is freshly cooked. However, when cooked uncured meat is stored, flavor changes occur rapidly. Tims and Watts *(4)* have reported that meat flavor can deteriorate after only a few hours of refrigerated storage. They characterized the off-flavor of reheated meats with terms such as "warmed-over", "stale" or "rancid".

It has been reported *(5)* that WOF is the result of both, a degradation of important odorants of the fresh boiled meat and a peroxidation of meat lipids. In particular, the phospholipids with a higher concentration of polyunsaturated fatty acids are quite susceptible to autoxidation and, thus, are considered to play a major role as precursors of WOF *(6, 7)*.

More than 150 compounds have been identified in the volatile fraction of stored (4°C, 16 h) and reheated roast beef showing WOF *(8)*. Of these compounds hexanal, 2,3-octanedione and the total volatile content have been recommended as markers to indicate WOF. In addition to hexanal and 2,3-octanedione, a strong increase of further 10 volatiles known as autoxidation products of oleic, linoleic and linolenic acids has been found in a comparative study *(9)* of freshly cooked beef and a four days old sample exhibiting WOF.

In our study *(10)* beef (bull, round, 500 g) was boiled for 60 min in a water-bath. During this period the internal temperature reached 91.5°C. Immediately after boiling the hot meat was ground in a meat grinder. After storage at 4°C for 48 h and subsequent heating at 50°C, the meaty boiled flavor of the beef had changed into an off-flavor characterized as fade, tallowy, green and metallic.

To evaluate the potent odorants that were formed by lipid peroxidation during the storage, the meat samples were extracted with diethyl ether and the volatile fraction was distilled off <u>in vacuo</u> from the non-volatile material.

The volatile fractions obtained from the fresh and from the stored boiled meat were subjected to AEDA. According to the results summarized in Table I, the FD-factors of hexanal, 1-octen-3-one, (Z)-2-octenal, (Z)-2-nonenal, (E,E)-2,4-nonadienal and trans-4,5-epoxy-(E)-2-decenal increased during the storage period, indicating that those which contributed significantly to the WOF were among these carbonyl compounds.

It should be emphasized that AEDA is only a screening-procedure for the important odorants of a

Table I. Storage of Boiled Beef for 48 h at 4°C - Changes in the FD-Factors of Odorants Formed by Lipid Peroxidation

Compound	FD-Factor fresh	FD-Factor stored
Hexanal	4	64
Heptanal	<4	8
(E)-2-Heptenal	8	16
1-Octen-3-one	<4	256
Octanal	4	16
(Z)-2-Octenal	<4	256
(E)-2-Octenal	4	64
Nonanal	8	32
(Z)-2-Nonenal	<4	128
(E)-2-Nonenal	512	256
(E,E)-2,4-Nonadienal	32	256
(E)-2-Decenal	256	128
(E,E)-2,4-Decadienal	512	1024
trans-4,5-Epoxy-(E)-2-decenal	<4	4096

food. The FD-factors are not corrected for differences in the yields obtained after the isolation procedure. Furthermore, the flavor of a food is partially affected by the food composition, since the volatility of the aroma compounds is influenced by their binding to the non-volatile components. By contrast, all the extracted aroma compounds are completely volatilized and then judged by sniffing in the GC/O.

Therefore, to get more exact data, flavor unit (FU)-values *(12)* were calculated at first for the five odorants listed in Table II on the basis of quantitative and flavor threshold data *(13)*. The quantitative analysis was performed using the corresponding odorants labelled with deuterium as internal standards *(13)* and the FU-values were calculated under the assumption that the odorants were mainly dissolved in the lipids of the meat.

Table II shows a dramatic increase of the hexanal content of the boiled beef during storage. trans-4,5-Epoxy-(E)-2-decenal and (E,E)-2,4-nonadienal were also significantly higher in the stored beef. By contrast, the level of (E,E)-2,4-decadienal increased only a little. The highest increase of the FU-values during the storage of the boiled beef was found for hexanal, indicating that this odorant might be mainly responsible for the green, tallowy flavor notes of the WOF. In addition trans-4,5-epoxy-(E)-2-decenal, causing a metallic flavor impression, contributed significantly to the WOF. The results shown in Table II underline the conclusion *(8)*, that hexanal is useful as indicator substance for the assessment of WOF.

Table II. Changes in the Levels of Some Important Odorants during the Storage of Boiled Beef

Compound	Amount[a] (FU-value)[b] in boiled meat			
	fresh		stored[c]	
Hexanal	375	(5)	45000	(616)
1-Octen-3-one	<0.05	(<1)	<0.05	(<1)
(E,E)-2,4-Nonadienal	14	(<1)	223	(2)
(E,E)-2,4-Decadienal	224	(5)	283	(7)
trans-4,5-Epoxy-(E)-2-decenal	0.1	(<1)	16	(5)

[a] Given in µg/kg.
[b] FU-values were calculated by dividing the concentration of the flavor compound by its taste threshold in an oil (13).
[c] The sample was stored for 48 h at 4°C.

Light-Induced Off-Flavors

Soybean oil and milkfat are especially susceptible to photo-oxidation (14, 15). AEDA was the first step to clarify the cause for this instability.

Soybean oil (SBO). SBO was stored at room temperature at a window facing north. An off-flavor which was described as "strawy, lard-like, beany" developed rapidly, and after 30 days its intensity was scored as "very strong". Darkness inhibited the deterioration of the SBO sample.
  A comparative AEDA (16) of the SBO sample stored for 30 days in the presence and absence of light led to the identification of the carbonyl compounds and of the monohydroperoxide listed in Table III. As their FD-factors increased dramatically, when the SBO was stored in daylight, these compounds were considered as the primary odorants of this sample. Among them, 3-methylnonane-2,4-dione having a strawy, beany odor was by far the most important odorant on the basis of its high FD-factor (Table III).
  Quantification of the potent odorants confirmed the results of the AEDA. In each of the 3 SBO samples analysed (Table IV), the highest FU-value was found for 3-methylnonane-2,4-dione. A sensorial study, in which definite amounts of the odorants were added to a deodorized sunflower oil, underlined the importance of the dione for the light-induced off-flavor of SBO. A concentration level of 90 µg/kg of the dione, which corresponded to 60 FU, caused the strawy, beany and green off-flavor notes reminiscent to the flavor of SBO exposed to light. The other potent odorants (hexanal, (Z)-3-hexenal, 1-octen-3-one, (Z)-2-nonenal) enhanced only somewhat the intensity of the odor caused by the dione.

Table III. Potent Odorants of SBO Stored in the Dark and in Daylight

Compound	FD-factor after storage for 30 days in	
	Dark	Daylight
3-Methylnonane-2,4-dione	16	16384
(Z)-3-Hexenal	8	2048
(Z)-2-Nonenal	64	1024
(Z)-1,5-Octadien-3-one	<1	512
1-Octen-3-hydroperoxide	32	512

SOURCE: Data are from ref. 16.

Table IV. FU-Values of Odorants in three SBO Samples Stored for 30 Days in Daylight

Compound	FU-Value[a]		
	Sample A	Sample B	Sample C
3-Methylnonane-2,4-dione	1002	481	137
(Z)-3-Hexenal	60	34	19
Hexanal	46	32	17
1-Octen-3-one	20	36	21
(Z)-1,5-Octadien-3-one	17	20	17
(Z)-2-Nonenal	5.7	7.3	7.7
(E)-2-Heptenal	9.5	4.4	5

[a] The FU-values were calculated by dividing the concentration by the flavor threshold in an oil (13). Quantitative analysis was performed as stable isotope dilution assay (13).
SOURCE: Data are from ref. 13.

To clarify the precursor of the dione SBO was transesterified into methyl esters which were then fractionated (17). Each fraction obtained was photooxidized after addition of meso-tetraphenylporphine as sensitizer and the 3-methylnonane-2,4-dione formed was assayed.
At first, these experiments indicated that the dione did not originate from a component of the unsaponifiable fraction. The precursor was found among the fatty acids and was enriched by a two-fold urea adduct formation. Figure 1 shows the capillary gas chromatogram of the fraction of the fatty acid methyl esters isolated in this way from an unprocessed SBO.
Mass spectrometry of the peaks 1, 2 and 3 indicated three fatty acids having a furan ring (F-I, F-V and F-VI in Figure 2). Such branched furanoid fatty acids (F-acids) with 19 carbon atoms (F-I), 20 carbon atoms (F-V) and 22 carbon atoms (F-VI) are known as constituents of fish liver oils (18). Recently, Hannemann et al. (19) have detected them also in some plants and in mushrooms.

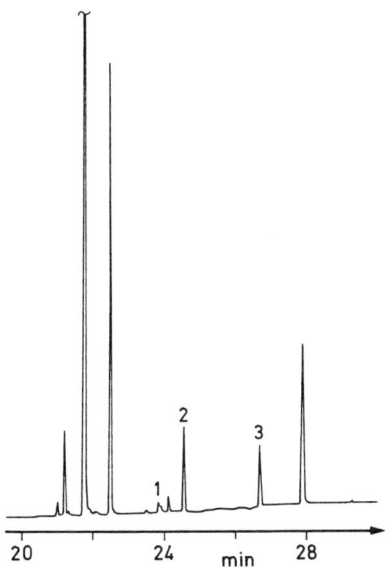

Figure 1. Capillary gas chromatogram of a fraction obtained from unprocessed soybean oil by transesterification and a two-step urea crystallization. Peaks 1, 2 and 3 were identified as furanoid fatty acids F-I, F-V and F-VI; (the numbering refers to Fig. 2)

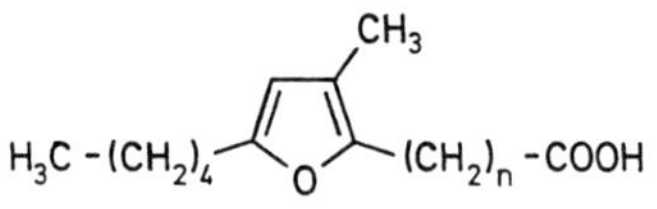

$H_3C-(CH_2)_4$—[furan ring with CH_3]—$(CH_2)_n-COOH$

F-I : n = 8
F-II : n = 10

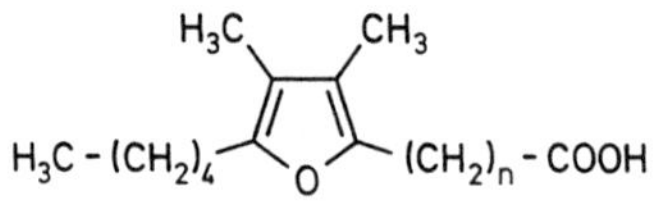

$H_3C-(CH_2)_4$—[furan ring with H_3C and CH_3]—$(CH_2)_n-COOH$

F-III : n = 4
F-IV : n = 6
F-V : n = 8
F-VI : n = 10

Figure 2. Numerical key and chemical structures of furanoid fatty acids

Table V. Formation of 3-Methylnonane-2,4-dione (MND) during Photooxidation of the Furanoid Fatty Acid F-V

Reaction system	MND[a] (μmol)
Methyl ester of F-V (11.9 μmol) and meso-tetraphenylporphine (1 μmol) dissolved in pentane (5 ml)	0.15

[a] After irradiation with a mercury lamp for 15 min at 15°C.
SOURCE: Data are from ref. 17.

The fatty acids F-V and F-VI, containing the same methyl end, were suggested as precursors of the 3-methylnonane-2,4-dione. To prove this suggestion a model experiment was performed. As shown in Table V, the methyl ester of F-V and the sensitizer meso-tetraphenylporphine were dissolved in pentane. The mixture was irradiated at 15°C using a mercury lamp. Capillary gas chromatography indicated that most of the methyl ester F-V was consumed in 15 min and that the dione was formed as a minor product.
Quantitative analysis showed that 1.3 % of the methyl ester was converted into the dione.
According to Schenck (20) the formation of endoperoxides is the predominant reaction of furans with singlet oxygen. However, this reaction cannot explain the formation of the dione. As an alternative, we suggest that F-V reacts with singlet oxygen not only to an endoperoxide, but also directly to a hydroperoxide. Figure 3 indicates that four allylic groups occur in the tetrasubstituted furan ring of the fatty acid. To explain the formation of the dione, we suggest that an "ene" reaction of singlet oxygen with the allylic group located at the carbons-9, -10, and -11 takes place yielding a HOO-group at carbon-11 (Figure 3). A ß-scission of this HOO-group leads to a carbonyl function at carbon-11 and a hydroxyl radical, which, in turn, induces the cleavage of the bond between carbon-13 and the oxygen of the furan ring and combines with carbon-13. The breakdown of the furan ring yields the ketoenol form of the 3-methylnonane-2,4-dione, which is in equilibrium with the diketo form.
A method was developed (17) to quantify F-acids in vegetable oils. It was applied to the samples summarized in Table VI. The SBO A and B, extracted in the laboratory from two samples of soybeans, and the processed SBO (C-E) agreed in the relatively higher amounts of F-V and F-VI compared to that of F-I. With exception of the SBO A, these oils contained 250-300 mg/kg of F-V and F-VI which, in contrast to F-I, were

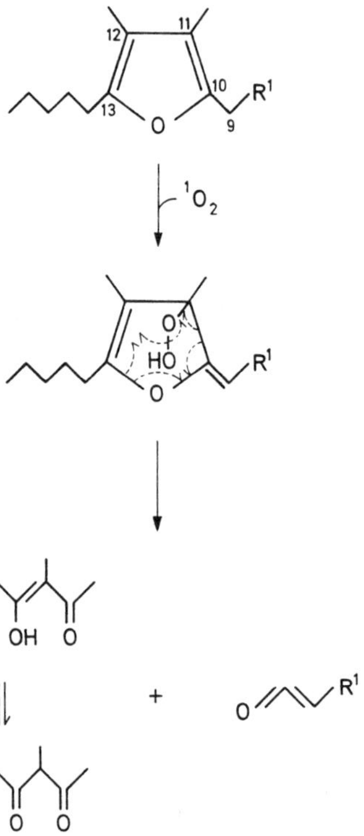

Figure 3. Mechanism proposed to explain the photo-sensitized oxidation of the furanoid fatty acid F-V [R: $(CH_2)_7COOH$] to 3-methylnonane-2,4-dione

Table VI. Amounts of Furanoid Fatty Acids (F-I, F-V and F-VI) in Vegetable Oils

Sample	Concentration (mg/kg)			
	F-I	F-V	F-VI	F-V + F-VI
Unprocessed soybean oil A	23	170	229	399
Unprocessed soybean oil B	10	121	132	253
Refined soybean oil C	13	143	152	295
Refined soybean oil D	13	148	172	320
Refined soybean oil E	7	131	148	279
Wheat germ oil A	9	103	105	208
Wheat germ oil B	3	34	37	71
Corn oil A	30	11	13	24
Corn oil B	13	8	9	17
Rapeseed oil A	4	16	20	36
Rapeseed oil B	n.d.	6	7	13

SOURCE: Reproduced from ref. 17. Copyright 1991.

the precursors of the dione. F-acids were also detected in samples of wheat germ oil, corn oil and rapeseed oil, but they were absent in sunflower and olive oils. One wheat germ oil (sample A) was almost as high in F-V and F-VI as the SBO, the other sample (B) contained lower levels. The rapeseed oil and the corn oil contained much smaller amounts of the dione precursors F-V and F-VI than the SBO.
We suggest that the much lower concentration level of F-acids in the rapeseed oil is one reason for the higher stability of this oil (15) in the presence of light in contrast to SBO.

Butter and butter oil. The buttery, sweet and acidulous flavor of fresh butter oil is mainly caused by diacetyl, acetic acid, butyric acid, guaiacol, vanillin, skatole, (Z)-6-dodecen-γ-lactone, δ-octalactone, δ-decalactone and a number of carbonyl compounds formed by lipid peroxidation (21). Exposure of butter oil for 48 h to fluorescent light changed the flavor; the sample tasted green, strawy and fatty.
A comparative AEDA of the odorants formed by lipid peroxidation indicated (Table VII) that this change was mainly due to the production of 3-methylnonane-2,4-dione, trans-4,5-epoxy-(E)-2-decenal and an increase in the concentrations of (E)-2-nonenal and (E,E)-2,4-decadienal. The concentration levels of the other carbonyl compounds listed in Table VI were not significantly affected by the photooxidation process, as their FD-factors in the fresh and irradiated samples were identical, differing, at the most, within the limit of error of the AEDA.
The amount of 3-methylnonane-2,4-dione was determined in milkfat after photooxidation for 48 h. The data

Table VII. Photooxidation of Butter Oil (BO) - Changes in the FD-Factors of Potent Flavor Compounds Formed by Lipid Peroxidation

No.	Compound	FD-factor in fresh BO	photooxidized BO[a]
1	1-Hexen-3-one	128	64
2	Hexanal	<32	32
3	(Z)-3-Hexenal	64	64
4	(Z)-4-Heptenal	64	64
5	1-Octen-3-one	128	256
6	Unknown	n.d.	32
7	(Z)-2-Nonenal	64	64
8	(E)-2-Nonenal	32	128
9	3-Methylnonane-2,4-dione	n.d.	256
10	(E,E)-2,4-Decadienal	32	128
11	trans-4,5-Epoxy-(E)-2-decenal	n.d.	128

[a] Butter oil (100 g) in a glass culture flask according to Fernbach (16) was exposed to fluorescent light (cool-white fluorescent tube, 18 Watt). The distance between the surface of the sample and the light source was 16 cm. After 48 h of irradiation the volatile flavor compounds were analysed (13).

listed in Table VIII showed the production of 21 µg/kg and 30 µg/kg of the dione in samples of butter and butter oil, respectively.

The formation of 3-methylnonane-2,4-dione in butter exposed to light suggested that F-acids, which act as precursors of the dione, occur also in this animal fat.

To confirm this suggestion, the methyl esters of fatty acids obtained from butter oil were separated by urea crystallization, column chromatography and HPLC. Six F-acids (F-I to F-VI in Figure 2) were identified by capillary gas chromatography/mass spectrometry in the HPLC fractions.

The amounts of the F-acids in 4 samples of butter and in a sample of butter oil were determined by an isotope dilution assay using F-V, which was deuterated at carbons-8 and -9, as internal standard (17).

Table IX indicates different concentration levels of the six F-acids in the butter samples. Sample A was 4-fold higher in F-acids than sample B which contained the lowest concentration. F-V and F-VI, the major F-acids of most of the vegetable oils (Table VI), predominated also in the butter samples with exception of sample B, which contained also F-I and F-II in relatively high proportions.

The model experiments reported in Table V have revealed that the photosensitized oxidation of F-V yielded

Table VIII. Formation of 3-Methylnonane-2,4-dione during Photooxidation of Butter and Butter Oil for 48 h

Sample[a]	3-Methylnonane-2,4-dione[b] (µg/kg)
Butter A (sweet cream)	21
Butter oil	30

[a] The conditions of photooxidation are reported in footnote "a" of Table VI.
[b] 3-Methylnonane-2,4-dione was assayed as described *(13)*. The fresh samples contained less than 0.5 µg/kg of the dione.

Table IX. Amounts of Furanoid Fatty Acids (F-I–F-VI) in Butter and Butter Oil[a]

Sample	F-I	F-II	F-III	F-IV	F-V	F-VI	Total
Butter A (sweet cream)	19	14	24	72	139	208	476
Butter B (sweet cream)	17	30	3	9	23	34	116
Butter C (sour cream)	7	8	6	21	38	93	173
Butter D (sour cream)	6	9	5	15	42	76	153
Butter Oil[b]	13	15	18	52	105	192	395

[a] Results are given in mg/kg.
[b] The butter oil was made from acidified sweet cream butter (NIZO process).

3-methylnonane-2,4-dione as minor product. As the methyl end and the furan ring system of F-III, F-IV and F-VI agree with structural feature of F-V, these F-acids might also act as precursors of the dione.

Conclusions

The examples discussed here demonstrate that AEDA is suitable to evaluate the positions of potent odorants causing an off-flavor in a capillary gas chromatogram. The identification experiments are then focussed on the odorants showing high FD-factors. As the AEDA is only a screening procedure, the results have to be confirmed by quantitative data and by the calculation of flavor unit values. Studies on the light-induced deterioration of soybean oil and milkfat have been completed with the identification of the furan fatty acids which act as precursors of the 3-methylnonane-2,4-dione causing mainly the off-flavor.

Literature cited

1. Acree, T.E.; Barnard, J.; Cunningham, D.G. *Food Chem.* **1984**, *14*, 273-286
2. Ullrich, F.; Grosch, W. *Z.Lebensm.Unters.Forsch.* **1987**, *184*, 277-282
3. Blank, I.; Fischer, K.-H., Grosch, W. *Z.Lebensm. Unters.Forsch.* 1989, *189*, 426-433
4. Tims, M.J.; Watts, B.M. *Food Technol.* **1958**, *12*, 240-243
5. Vercellotti, J.R.; Kuan, J.W.; Spanier, A.M.; St. Angelo, A.J. In *Thermal Generation of Aromas*; Parliment, T.H.; McGorrin, R.J.; Ho, C.-T., Eds.; ACS Symposium Series No. 409; American Chemical Society: Washington, DC, 1989; pp. 452-459
6. Pearson, A.M.; Love, J.D.; Shorland, F.D. *Adv.Food Res.* **1977**, *23*, 1-74
7. Pearson, A.M.; Gray, J.I. In *The Maillard Reaction in Foods and Nutrition*; Waller, G.R.; Feather, M.S., Eds.; ACS Symposium Series No. 215; American Chemical Society: Washington, DC, 1983; pp. 287-300
8. St. Angelo, A.J.; Vercellotti, J.R.; Legendre, M.G.; Vinnett, C.H.; Kuan, J.W.; James, C.; Dupuy, H.P. *J. Food Sci.* **1987**, *52*, 1163-1168
9. Vercellotti, J.R.; St. Angelo, A.J.; Legendre, M.G.; Sumrell, G.; Dupuy, H.P.; Flick, G.J. *J.Compos. Anal.* **1988**, *1*, 239-249
10. Konopka, U.C.; Grosch, W. *Z.Lebensm.Unters.Forsch.* **1991**, *193*, 123-125
11. Gasser, U.; Grosch, W. *Z.Lebensm.Unters.Forsch.* **1988**, *186*, 489-494
12. Guadagni, D.G.; Buttery, R.G.; Harris, J. *J.Sci.Food Agric.* **1966**, *17*, 142
13. Guth, H.; Grosch, W. *Lebensm.Wiss.Technol.* **1990**, *23*, 513-522
14. Radtke, R.; Smits, P.; Heiss, R. *Fette Seifen Anstrichm.* **1970**, *72*, 497-504
15. Sattar, A.; De Man, J.M.; Alexander, J.C. *Lebensm.Wiss.Technol.* **1976**, *9*, 149-152
16. Guth, H.; Grosch, W. *Lebensm.Wiss.Technol.* **1990**, *23*, 59-65
17. Guth, H.; Grosch, W. *Fat Sci.Technol.* **1991**, *93*, 249-255
18. Glass, R.L.; Krick, T.P.; Olson, D.L.; Thorson, R.L. *Lipids* **1977**, *12*, 828-836
19. Hannemann, K.; Puchta, V.; Simon, E.; Ziegler, H.; Ziegler, G.; Spiteller, G. *Lipids* **1989**, *24*, 296-298
20. Schenck, G.O. *Angew.Chem.* **1957**, *69*, 579-599
21. Widder, S.; Sen, A.; Grosch, W. *Z.Lebensm.Unters. Forsch.* **1991**, *193*, 32-35

RECEIVED February 19, 1992

Chapter 15

Sensory Evaluation of Lipid Oxidation in Foods

G. V. Civille and C. A. Dus

Sensory Spectrum, Inc., 24 Washington Avenue, Chatham, NJ 07928

Traditionally, some sensory terms describing lipid oxidation have been common across food products. In many cases, however, researchers also have developed terminology specific to a given food type. Many of these descriptive terms evolved from the sources or reasons for the oxidation rather than a description of the result in the product. This paper examines a common approach to describing flavor notes generated by lipid oxidation across several food products in terms of both the typical characteristics for both fresh samples and aged samples. Meat, peanuts and vegetable oils are used as examples to illustrate the use of common lexicons and common sensory methods to measure responses to lipid oxidation in fresh, aging and aged products.

Food scientists, lipid chemists and sensory analysts have long been interested in finding appropriate words to describe the sensory characteristics of lipid based foodstuffs. The quality of the terminology selected and used may be viewed differently by food scientists than sensory scientists. Food scientists might be content with words which describe the apparent cause of the flavor. Sensory analysts, however, are often more interested in descriptions of the perceived flavor note or characteristics. To the product developer or research chemist, terms such as "rancid," "oxidized," "warmed-over" and "light-struck" describe apparent reasons for the "off-notes." For the sensory analyst these terms suggest the occurrence of specific chemical reactions or probable causes, they do not provide an accurate description of the flavor perception itself. The development of lexicons and methodologies which detect off-flavors caused by oxidation has resulted in terminology that describes sensory characteristics which describe perceptions rather than causes.

Sensory Evaluation of Lipid-Oxidation Off-Flavors: The Past

The terminology and methodology developed for specific product types, such as meats, peanuts and vegetable oils, have been lexicons and methodologies which predominantly focus on the occurrence of off-flavors in each category of products.

Meat and Poultry. Warmed-Over Flavor or WOF is a general term used to describe off-flavor in reheated meat which has been precooked and refrigerated. In food service situations, such as institutional cafeterias or airlines, where the foods, including meats, are cooked ahead and refrigerated and reheated before serving, the presence of "warmed-over" flavor in the meat is regarded as a defect. Meat and poultry research at university and government laboratories has sought to identify the causes, chemical and physical conditions and potential means to reduce or eliminate "warmed-over" flavor. WOF describes the cause of a combination of changes in flavor attributes within a given meat or poultry product. Investigators, aware of this deficiency have tried to develop expanded terminology to describe the individual sensations that are perceived. Table I illustrates some researchers' choices of terms for description of meat flavor.

Table I. Descriptors Used in Some Investigations of Warmed-Over Flavor For Beef

Jacobson & Koehler (1970)	Harris & Lindsay (1972)	Joseph et. al. (1980)
Bland	Greasy	Sour
Sweet	Sulfur	Bitter
Rich	Liver-Giblet	Metallic
Meaty	Warmed-Over[a]	Sweet
Sulfury	Stale[a]	Putrid
Gizzard-Like	Rancid[a]	Salty
Musty[a]		Rancid[a]
Stale "old"[a]		Other off-flavor
Rancid[a]		

[a]These quality descriptors were highest in intensity in meats having WOF or were used most frequently by panelists describing meat with this flavor.
SOURCE: Reproduced with permission from ref. 1. Copyright 1987 Academic.

These researchers used different flavor descriptors (musty, stale, rancid and warmed-over) to describe the effects of reheating precooked meat. Of these off-notes related to reheating, only musty describes a sensory characteristic, whereas stale, rancid and warmed-over describe the process suspected of producing off-notes or flavors.

It should be noted that these earlier meat lists did include a few terms to describe the full flavor of fresh or "on" meat and poultry flavor.

Peanuts. As with meats, the peanut industry has used several terms to describe peanut flavor which included descriptors of flavor sensations, such as roasted peanut, nutty and green, as well as descriptors for the sources suspected to contribute to off-flavors, such as rancid, stale and oxidized, shown in Table II.

The CLER (Critical Laboratory Evaluation Roast) Method uses a quality measurement which originally combined intensity and hedonic responses on a single continuum (2). The updated method includes specific flavor changes in a supplemental comments section only (3).

Table II Descriptors Used in Some Investigations of Peanut Flavor

Oupadissakoon & Young (1984)	Syarief et.al. (1985)	Holaday (1971)
Aroma	Roasted Peanut	Off-Flavor
Flavor	Fruity	Low Level Off-Flavor
Aftertaste	Mold	Low Peanut Flavor
Astringent	Bitter	Good Peanut Flavor
Bite	Sour	
Bitter	Over/Under Roast	
Burnt	Earthy	
Chemical	Oxidized	
Earthy	Petroleum	
Green		
Nutty		
Oil		
Rancid		
Roasted Peanut		
Sour		
Stale		
Sweet		

SOURCE: Adapted from ref. 2, 4-5.

The Edible Oil Industry. The flavor of a high quality oil is generally described as bland. It is when the quality of the oil is suspect and off-flavors are prevalent that chemists and food scientists use more descriptive terminology. As the oil deteriorates, flavor volatile compounds are created at each stage which impart different flavor notes that encompass nutty and beany to rancid and fishy. The terms are divided into terms that refer to a specific characteristic (corny, nutty, fishy) and terms that describe the suspected degrading process (reverted, rancid, oxidized). The AOCS (American Oil Chemists Society) Flavor Quality Scale, shown in Table III, combined intensity and quality into one scale having each grade associated with a specific flavor description.

Table III. AOCS Flavor Quality Scale

Flavor Grade	Description of Flavor[a]
10 (Excellent)	Completely bland
9 (Good)	Trace of Flavor but not recognizable
8	Nutty, sweet, buttery, corny
7 (Fair)	Beany, hydrogenated, popcorn, bacony
6	Oxidized, musty, weedy, burnt, grassy
5 (Poor)	Raw, reverted, rubbery, watermelon, bitter
4	Rancid, painty
3	Fishy, buggy
2	Intensive objectionable flavors
1 (Repulsive)	

[a]Flavor intensity at presented concentration rated slight.
SOURCE: Reproduced with permission from ref. 6. Copyright 1985 American Oil Chemists' Society.

The original AOCS scale is difficult to use since it requires the integration of quality and intensity of both appropriate and off-notes. The original AOCS Flavor Quality Scale has since been revised to incorporate separate ballots for grading and flavor intensity. The terms used to describe oxidized oil not only include terms that refer to a specific flavor sensation, but continue to include terms that are process orientated. The descriptors are so divided in Table IV.

Table IV. A Partial List of Terms Used to Describe Oxidized Oil

Flavor Related Terms	Process Oriented Terms
Buttery	Hydrogenated
Nutty	Oxidized
Beany	Reverted
Grassy	Light-Struck
Watermelon	Rancid
Painty	
Fishy	

SOURCE: Adapted from ref. 7.

Across these three food categories, (meats, peanuts and oils) certain flavor descriptors are common for describing lipid oxidation: stale, rancid, oxidized, fresh and off-flavor. Within each category, other terms have been used specifically to describe some forms of lipid oxidation: meats - warmed-over and musty, peanuts - cardboard and soapy, and oils - fishy, reverted and light-struck.

Sensory Evaluation of Lipid Oxidation Off-Flavors: Recent Development

The recent focus on improved descriptive terminology has resulted in complete lexicons for product categories. The American Society of Testing Materials Committee E18 has a task group dedicated to the development of precise descriptive terminology. The use of descriptive analysis methodology has allowed for more complete lexicons and more precise intensity measurements. The prescribed characteristics of a complete lexicon are as follows: terminology must be orthogonal, based on underlying structure, based on a broad reference set, precisely defined and primary, rather than integrated (7).

Descriptive Analysis. Descriptive analysis methods involve the detection and the description of both the qualitative and quantitative sensory aspects of a product by a trained panel (8). In the terminology development stage, one descriptive analysis method, the Spectrum Method, provides a wide array of flavor references for training panelists. Processing variables, ingredient variables and examples of off-flavors are presented to the panelists in order to develop a list of terms that completely characterize the product's sensory attributes. Intensity references (both general and product specific) are also presented to the panelists, thus insuring quantitative and qualitative analytical sensory data which is both valid and reliable. The development of standard lexicons, standard rating scales, standard methods for preparation and presentation of products and standard panel training all contribute to a uniform analytical sensory technique.

The focus in the past investigations of lipid oxidation has been to describe the "off-notes" and/or quantify them in some way to indicate the quality of the food in question. Only some research in lipid oxidation attempted to describe and quantify the flavor of the fresh product. Current sensory and food science research in these food categories and others involving lipid oxidation have focused on the full descriptions of the "on-flavors" as well as the off-notes, which include those resulting from oxidation.

On-flavor incorporates those flavor characteristics which describe the fresh meat, peanuts or oil. The most current terms for lipid oxidation flavors describe the characteristics (cardboard, painty) rather than the suspected cause or process (stale, rancid or oxidized).

The Meat Industry: Warmed-Over Flavor. The phenomenon of WOF is complex and, despite recent developments (9), its formation is not yet completely understood. Analytical chemists, engaged in solving this mystery, need terminology to describe the flavor notes that arise as WOF develops.

Table V shows a beef lexicon developed by a panel of meat flavor experts convened by the Monell Chemical Senses Center. To create a complete list, the panelists were initially presented with a variety of meats (beef, pork, turkey and chicken), cooking procedures, storage times and reheating procedures. The beef flavor descriptors which were the focus of the research were defined clearly and reference samples were developed for each descriptor (*10*).

Table V. Beef Flavor Descriptions

Attribute	Definition
Cooked Beef Lean	The aromatic associated with cooked beef muscle meat.
Cooked Beef Fat	The aromatic associated with cooked beef fat.
Browned or broiled beef.	The aromatic associated with grilled
Serum/Bloody	The aromatic associated with raw beef lean.
Grainy/Cowy	The aromatic associated with cow meat and/or beef in which grain feed character is detectable.
Cardboard	The aromatic associated with slightly stale beef (refrigerated for a few days only) and associated with wet cardboard and stale oils and fats.
Oxidized/Rancid/Painty	The aromatic associated with rancid oil and fat (distinctly like linseed oil).
Fishy	The aromatic associated with some rancid fats and oils (similar to old fish).
Sweet	Tastes on the tongue associated with sugars.
Sour	Tastes on the tongue associated with acids.
Salty	Tastes on the tongue associated with sodium ions.
Bitter	Tastes on the tongue associated with bitter agents such as caffeine, quinine, etc.

SOURCE: Reproduced with permission from ref. 10. Copyright 1986 Dial Technical Center.

These terms, in conjunction with intensity ratings, allow the researcher to track not only the development of off-flavors, but also the decrease of on-flavors. Table VI shows the differences between a fresh, frozen beef patty and a beef patty which was stored for 5 days then reheated. Note that the intensity of the off-notes cardboard and oxid/rancid/painty increased and the intensity of the cooked beef lean and cooked beef fat notes decreased.

Table VI. Comparison of Fresh Control Baked Beef Patty with Stored Steamed And Baked Patty

	Control	Reheated
Cooked Beef Lean	6	4
Cooked Beef Fat (fresh)	4	1
Browned	4	1
Serum/Bloody	3	2
Grainy/Cowy	2	1
Cardboard	1	4
Oxid/Rancid/Painty	0	3
Fishy	0	0
Sweet	3	2
Sour	1	3
Salty	2	2
Bitter	0	0

SOURCE: Reproduced with permission from ref. 10. Copyright 1986 Dial Technical Center.

A similar approach was developed for chicken. Table VII indicates the differences between fresh cooked chicken patties and patties stored for 3 days.

Table VII. A Comparison Between Fresh Cooked Chicken Patties And Patties Stored For Three Days

	Fresh Cooked	Stored 3 Days
Chickeny	53.3	28.6
Meaty	54.2	32.9
Brothy	35.6	9.3
Liver/Organy	27.2	14.1
Browned	16.7	33.8
Burned	7.3	31.7
Cardboard	7.7	35.6
Warmed-Over	9.5	54.3
Rancid/Painty	5.5	46.1
Sweet	22.3	5.6
Bitter	8.3	23.6
Metallic	10.5	19.5

SOURCE: Adapted from ref 11.

Table VIII. Lexicon of Peanut Flavor Descriptors

Attribute	Definition
Roasted Peanutty	The aromatic associated with medium-roast peanuts (about 3-4 on USDA color chips) and having fragrant character such as methyl pyrazine.
Raw Bean/Peanutty	The aromatic associated with dark-roasted peanuts (about 1-2 on USDA color chips) and having legume-like character.
Dark Roasted Peanut	The aromatic associated with dark roasted peanuts (4+ on USDA color chips) and having very browned toasted character.
Sweet Aromatic	The aromatics associated with sweet material such as caramel, vanilla, molasses, fruit.
Woody/Hulls/Skins	The aromatics associated with base peanut character (absence of fragrant top notes) and related to dry wood, peanut hulls and skins.
Cardboard	The aromatics associated with somewhat oxidized fats and oils and reminiscent of cardboard.
Painty	The aromatic associated with linseed oil and oil based paint.
Burnt	The aromatic associated with very dark roast, burnt starches and carbohydrates, (burnt toast or espresso coffee).
Green	The aromatic associated with uncooked vegetables/grass/twigs, cis-2-hexanal.
Earthy	The aromatic associated with wet dirt and mulch.
Grainy	The aromatic associated with raw grain (bran, starch, corn, sorghum).
Fishy	The aromatic associated with trimethylamine, cod liver oil and old fish.
Chemical/Plastic	The aromatics associated with plastic and burnt plastics.
Skunky/Mercaptan	The aromatic associated with sulfur compounds, such as mercaptan, which exhibit skunk-like character.
Sweet	The taste on the tongue associated with sugars.
Sour	The taste on the tongue associated with acids.
Salty	The taste on the tongue associated with sodium ions.
Bitter	The taste on the tongue associated with bitter agents such as caffeine or quinine.
Astringent	The chemical feeling factor on the tongue, described as puckering/dry and associated with tannins or alum.
Metallic	The chemical feeling factor on the tongue described as flat, metallic and associated with iron and copper.

SOURCE: Reproduced with permission from ref. 12. Copyright 1988 Dial Technical Center.

The Peanut Industry. The peanut industry, like the meat industry, sought to develop a comprehensive flavor lexicon. The USDA-ARS-SRRC and the Monell Chemical Senses Center assembled a panel of peanut experts to develop a complete peanut lexicon (*12*). The panel, presented with peanut paste samples that represented different roast levels as well as a variety of flavors resulting from different growing and handling conditions, developed terms to encompass the treatment variables. Terminology was also developed to describe the flavor notes generated when the oils oxidize.

The complex lexicon is presented in Table VIII and is currently being used in research studies to investigate the development of flavor changes in peanuts. In Table IX, a comparison of fresh and oxidized peanut pastes made from the same peanut source, demonstrates the decrease in the "fresh/on" peanut characteristics (roasted peanutty, sweet aromatic and sweet) as well as the increase in the cardboard and painty terms. This system allows for tracking the relative decrease in fresh notes and the increases in the two flavor characteristics which describe the effects of oxidation (cardboard and painty).

Table IX. Comparison of Fresh Peanut Paste and Oxidized Peanut Paste

Attribute	Fresh	Oxidized
Roasted Peanutty	6.1	2.7
Raw Bean/Peanut	1.7	1.4
Dark Roasted Peanut	2.1	1.3
Sweet Aromatic	3.5	1.3
Woody/Hulls/Skin	1.2	2.1
Cardboard	0	3.4
Painty	0	4.7
Sweet	2.9	1.4
Bitter	1.2	2.0
Astringency	1.6	2.0

SOURCE: Adapted from ref. *12*.

The Vegetable Oil Industry. Through the efforts of the sensory analysts who work with vegetable oils, new studies are being conducted to develop effective sensory methodology. The ASTM E18 task group on edible oils is responding to the needs of the oil industry, which has used the AOCS quality method for so long, while providing better qualification and quantitative description of the oil description of the oil flavors. Table X includes both an overall quality rating and the intensity ratings for individual descriptors. Unlike the descriptors for meat and peanuts, the rancid and painty notes for oil description have not yet been combined. The use, however, of the on-flavor descriptors with intensity allows for characterization of the unique properties of oils from different seed sources.

Table X. Flavor Quality and Descriptive Analysis of Fresh Aged Low Erucic Acid Rapeseed Oil (Canola)

	0-Time	Aged 8 Days, 60 C
Overall Quality	8.2	4.5
Buttery	3.1	0
Nutty	2.3	0
Toasted	1.2	0
Waxy	1.3	3.6
Painty	0	7.4
Rancid	0	5.4
Metallic	0	3.2
Fishy	0	2.1
Sulfur (vegetable)	0	1.7

SOURCE: Warner, K., USDA-ARS-NORTH. REG. RES. CNTR., unpublished data

CONCLUSION

The contribution of sensory analysis to the development of descriptors, definitions, and reference for the use in food analysis have special application to lipid oxidation in foods.

Other research with coffee, nuts, fish, milk products, cereals and grains has been undertaken and continues to develop new descriptors for describing the on-flavor of fresh product and the flavors of lipid oxidation. The move toward a more unified approach to documenting flavor gives researchers high quality analytical sensory tools that can be used to study treatments which cause or retard lipid oxidation in foods.

Literature Cited

1. Melton, S.L., Davidson, PM. and Mount, J.R. In *Warmed-Over Flavor of Meat*; A.J. St. Angelo and M.E. Bailey., Eds.; Academic Press: Orlando, FL, 1987; pp 141-164.
2. Holaday, C.E. *J. Amer. Peanut Res. Educ. Soc.* **1971**, *3 (II)*, 239-241.
3 Fletcher, M.M. In *Peanut Quality - Its Assurance and Maintenance from the Farm to End Product*; H. Ahmed and H.E. Pattee., Eds.; Tech. Bull. 874; Agric. Ex.Stat., Inst. Food and Agric. Sci., Univ. of Florida, Gainsville. **1987**; pp 60-72.
4. Oupadissakoon, C. and Young, C.T. *J. Food Sci.*, **1984**, *49*, 52-58.
5. Syarief, H., Hamann, D.D., Giesbrecht, F.G., Young, C.T. and Monroe, R.J. *J. Food Sci.* **1985**, *50*, 631-638.

6. Warner, K. In *Flavor Chemistry of Fats and Oils*; D.B. Minn and T.H. Smouse, Eds.; American Oil Chemist's Society: Champaign, IL, **1985**, pp 207-222.
7. Civille, G.V. and Lawless, H.L. *J. Sensory Studies*, **1986**, 1 (3/4), 203-206.
8. Meilgaard, M., Civille, G.V. and Carr, B.T. *Sensory Evaluation Techniques*. CRC Press: Boca Raton, FL, **1991**, pp 187-199.
9. Spanier, A.M., Vercellotti, J.R. and James, C. *J. Food Sci.*, in press.
10. Johnsen, P.B. and Civille, G.V. *J. Sensory Studies*, **1986**, *1*, 99-104.
11. Lyon, B.G., Lyon, C.E., Ang, C.Y.W. and Young, L.L. *Poultry Sci.*, **1988**, *67*, 736-742.
12. Johnsen, P.B., Civille, G.V., Vercellotti, J., Sanders, T.H. and Dus, C.A. *J. Sensory Studies*, **1988**, *3*, 9-18.

RECEIVED February 19, 1992

PROCESSING EFFECTS ON LIPID OXIDATION

Chapter 16

Influence of Food Processing on Lipid Oxidation and Flavor Stability

Hans Lingnert

SIK—The Swedish Institute for Food Research, P.O. Box 5401, S–402 29 Göteborg, Sweden

Lipid oxidation may occur during processing and influence subsequent oxidation during storage. Early oxidation during processing is in most cases very difficult to detect or measure with the methods commonly used for lipid oxidation. Still, this limited oxidation is of significant importance to the further oxidation and flavor stability during storage. This is illustrated by examples from experiments performed during manufacturing of potato granules and sausages, processing of cereals, and baking of cookies. The influence of processing on factors that affect lipid oxidation, e.g. prooxidants, antioxidants, oxygen content, and the protection against such oxidation are discussed.

Lipid oxidation during food storage is influenced by several factors: storage temperature, oxygen availability, water activity, exposure to light, antioxidants, prooxidants, etc. However, what is often overlooked is that oxidative stability is to a great extent dependent on what has happened to the food product prior to storage. The development of oxidation in food products during storage is schematically illustrated in Figure 1a. The oxidation proceeds slowly in the beginning, but accelerates later on during the storage period. The acceptability limit (the broken line) for one of the products is reached in half the time required for the other product. One reason for this difference could be that there is a difference in the level of oxidation at the start of the storage period. From harvest or slaughter, some oxidation could take place during storage of raw materials, at various steps of the food process, during packaging, etc., resulting in important, but not easily detectable, differences in oxidation levels at the start of the storage period. This oxidation, as illustrated in Figure 1b, may have important consequences on subsequent oxidation during storage.

This paper will focus on what happens during food processing that is relevant to lipid oxidation. Examples will be taken from research at SIK, demonstrating where food processing conditions influence the oxidative stability. The various ways in which processes may affect lipid oxidation will be discussed.

Methods

In the examples presented lipid oxidation was measured either by sensory evaluation or headspace analysis of the formation of volatiles, or both.

Sensory Evaluation. Descriptive sensory analysis was used to follow the development of rancidity. In the sausage experiment, a ten-member panel was used. The intensity of the rancid flavor was judged on a scale of 0-10. All analyses were performed in three replicates. In the cookie experiment, five judges judged the samples, using a scale of 0-4 for the intensity of rancid flavor. These evaluations were not replicated.

Headspace Analysis of Volatiles. The formation of volatiles during oxidation was analyzed by dynamic headspace gas chromatography as described by Hall and co-workers (*1*). In the cookie experiment, a static equilibrium method, as described in the same paper, was used. Of the volatiles analyzed and identified, hexanal was chosen as an indicator of lipid oxidation.

Potato Granule Production

There are a number of different processes for manufacturing potato granules. We have studied the so called "add-back process" which is rather complicated. In short, the raw potato tubers are washed, peeled and sliced into approx. 2 cm thick slices. These slices are blanched in water at 76°C for 12-15 minutes. The slices are then chilled in water at a temperature of 10°C for approx. 10 minutes, before they are steam-cooked at 102°C for 20 minutes. The steam-cooked slices are then mixed with already produced potato granules (hence the name "the add-back process") and are thereby disintegrated in a gentle way. After conditioning for about 20 minutes the potato granules are air-dried and sifted. Following this step, 85-90% of the potato granules are returned to the mixing stage and the remaining 10-15% are further dried in a fluid bed to give the final product.

Our objective was to study lipid degradation in the potato during this process and the effects of this degradation on subsequent oxidation of the final potato granules during storage (*2*). We could demonstrate a considerable release of free fatty acids in the initial stages of the manufacturing process. This is of importance since free fatty acids constitute the substrate for further enzymatic oxidation by lipoxygenase. Free fatty acids are also considered to be more susceptible to oxidation during storage than esterified fatty acids. In our project we also specifically studied the lipid oxidation in the blanching step.

In model experiments simulating the blanching step, two cm thick potato slices were heated in 76°C water for 5 or 15 minutes and then chilled in a 10°C water bath. Thin slices of the two cm potato slices were removed from the surface and the center, respectively, and hexanal content was analyzed (*3*).

Figure 2 shows that hexanal was formed during this blanching treatment. After heating for 5 minutes the hexanal concentration had increased at the surface as well as in the middle of the potato slices. After 15 minutes the hexanal content in the middle of the slices had increased further, while the hexanal concentration at the surface was lower after 15 minutes than after 5 minutes.

This is a demonstration of lipid oxidation occuring during processing. The probable mechanism is enzymatic oxidation during heating, before potato lipoxygenases were inactivated. Lipoxygenase at the surface was inactivated after 15 minutes of heating while that in the center of the potato slices was probably not totally inactivated. Consequently, oxidation could continue in the center of the potato slices during the chilling period which explains the further increase of hexanal.

Our main conclusion is that lipid oxidation occurs during the potato granule manufacturing process. This early oxidation is in most cases very difficult to measure and has, therefore, rarely been demonstrated. As discussed above, onset of oxidation during the manufacturing process will most probably continue during storage of the final product and influence storage stability.

Cereal Processing

In a large cooperative research project in Sweden various thermal processes for cereals were compared regarding their influence on the physical, chemical and nutritional properties of the processed wheat products. At SIK we studied, among other things, the influence of these processes on lipid oxidation.

Whole wheat grain from one batch was processed in five different ways: roller-drying, steam-cooking, autoclaving, puffing and extrusion-cooking. The process parameters were reported by Siljeström and co-workers (4). All samples were milled. Figure 3 shows the hexanal concentration of the processed wheat products immediately after production and after storage of the milled samples for 6 months at +30°C in darkness (5). The samples were stored in 500ml glass bottles, with 50 g samples in each bottle.

As seen in Figure 3, there were differences in the hexanal concentration after processing. For example, the hexanal level was considerably higher in the extrusion-cooked sample than in the steam-cooked sample. However, the most obvious result was that the extrusion-cooked and the roller-dried samples were less stable by far than the three other samples. The hexanal concentration in the extrusion-cooked and the roller-dried samples increased 10-100 times during the storage period. The most plausible explanation for the extensive lipid oxidation in these two samples is enzymatic oxidation during processing. Roller-drying and extrusion-cooking were performed on water suspensions of the whole wheat grain, while the others were processed without any addition of water. It has been reported previously (6-8) that wheat lipids oxidize very rapidly when water is added, due to lipoxygenase activity. The extent of this oxidation is dependent on the content of free fatty acids in the wheat, since free fatty acids are substrates to the lipoxygenase. The content of free fatty acids is in turn dependent on the storage time of the unprocessed wheat, since wheat lipases

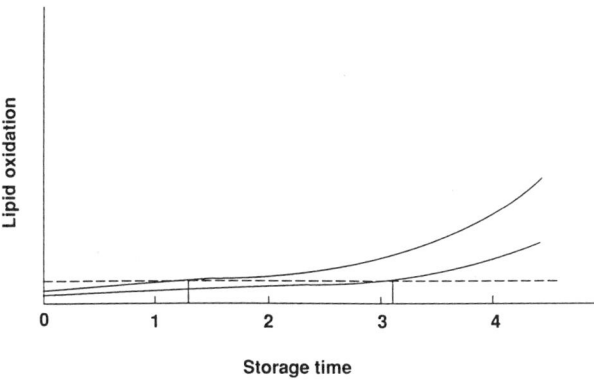

Figure 1a. Schematic illustration of the development of rancidity in food products during storage.

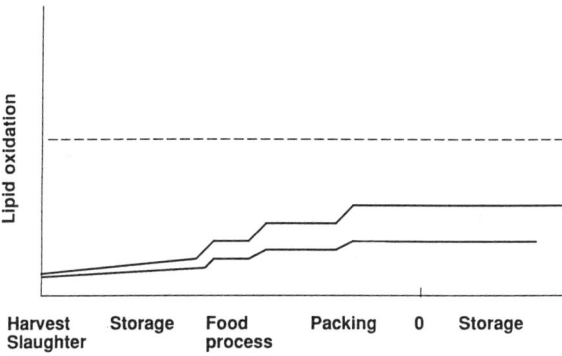

Figure 1b. Magnification of rancidity development from harvest/slaughter to start of storage.

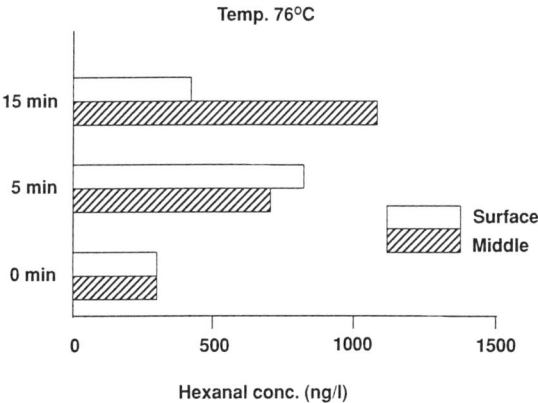

Figure 2. Hexanal formation during blanching of potato slices.

releasing the free fatty acids are active also at the relatively low water activities, prevailing during cereal storage.

Our conclusion is that the hydroperoxides that were enzymatically formed during processing of the roller-dried and extrusion-cooked samples decomposed during storage and stimulated further lipid oxidation. These experiments were examples of where it has been possible to measure differences in lipid oxidation during processing and where the extent of oxidation during processing seemed to have some relation to the oxidative stability during subsequent storage. (However, the hexanal concentration of the puffed sample was very similar to that of the roller-dried sample, while the puffed sample turned out to be far more stable.)

Sausage Production

In sausage production, the raw materials (beef, pork, water, starch, salt, etc.) are mixed and comminuted in a chopper. This is rather a rough process, during which lipids come in close contact with prooxidative hemoproteins and are exposed to the air atmosphere in the chopper. These conditions may promote lipid oxidation. We have studied the effect of using a vacuum or carbon dioxide atmosphere in the chopper when producing sausage.

From the chopper the sausage batter was transferred to a stuffer, where it was stuffed in casings. No protective gas was used when the batter was removed from the chopper or further on in the manufacturing process. The sausages were smoked, packed in plastic pouches under vacuum, frozen and stored at -20°C. During storage, samples were withdrawn and analyzed for rancidity by sensory evaluation and by gas chromatographic headspace analysis.

As seen in Figure 4, initial hexanal concentration was lower in samples chopped in a carbon dioxide or vacuum atmosphere than in the control sample, which had air atmosphere in the chopper. The vacuum and carbon dioxide samples were also considerably more stable during the 25-week storage time. These results are supported by the sensory analyses, Figure 5. From these data it was concluded that vacuum in the chopper improves flavor stability somewhat better than carbon dioxide.

These results indicate that sausages produced under vacuum or carbon dioxide oxidized to a lower extent during processing than those produced under air. An additional reason for the improved storage stability of the samples produced under vacuum or carbon dioxide is probably their lower content of remaining oxygen after packaging and during storage. We draw this conclusion from a parallel experiment using nitrogen in the chopper. The nitrogen-processed samples showed the same initial improvement in oxidation level as the carbon dioxide-processed samples. However, the sausages chopped under nitrogen were found to be inferior to the ones chopped under carbon dioxide with regard to their storage stability. According to our interpretation, this was due to carbon dioxide being far more soluble than nitrogen in the sausage batter. When the batter is removed from the chopper nitrogen is probably very quickly replaced by air, while carbon dioxide is bound to the batter. As a result of this, sausages prepared under carbon dioxide contain less oxygen during

16. LINGNERT *Influence of Food Processing* 297

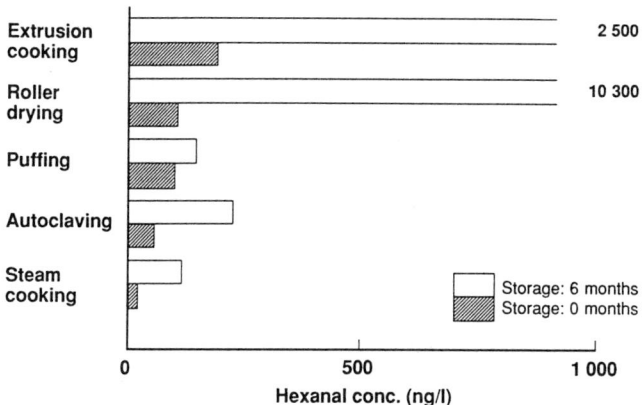

Figure 3. Hexanal concentration in whole wheat grain products.

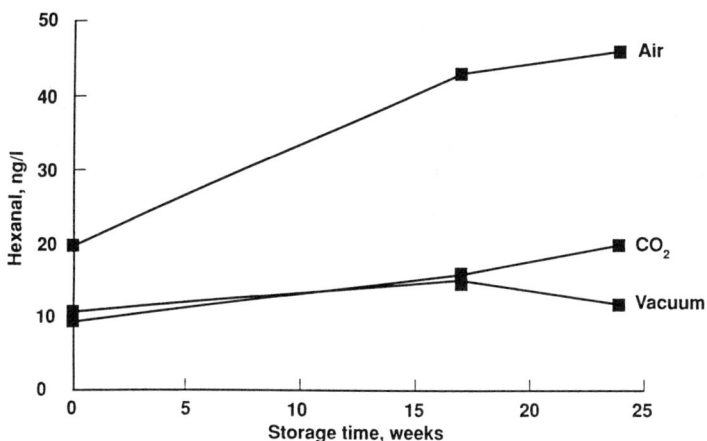

Figure 4. Hexanal formation in sausage during storage.

storage. In conclusion, the atmosphere used during processing may have a great influence on the stability of the final product.

Cookie Baking

In some cases there is a possibility that antioxidants, such as Maillard reaction products, are formed during heat processes. The baking process is one example of this. For instance, in the cookie industry it is well known that cookies baked to a high degree of browning are more stable than light-colored cookies. In previous experiments we have studied antioxidant formation during cookie-baking (9). Precursors to Maillard reaction products (histinine and glucose) were added to the cookie dough. As is shown in Figure 6 this addition resulted in a considerable improvement of the flavor stability during storage of the cookies. The main reason for this may be the formation of antioxidative Maillard reaction products during baking.

How is Lipid Oxidation affected by Food Processes?

The examples given above illustrate that the food process is of importance to flavor stability of the products during storage. Lipid oxidation occurs during food processing and this initial oxidation may influence further oxidation during subsequent storage. The process may also affect oxidation during storage by influencing prooxidants, antioxidants or the oxygen content of the final product.

Initiation or Stimulation of Lipid Oxidation during Processing. Food processes often involve conditions that are favorable to lipid oxidation, such as elevated temperatures, exposure of lipids to oxygen in the atmosphere or to lipid oxidation catalysts during mixing processes, exposure to light, etc. As has been indicated above, the process may also promote enzymatic oxidation (favorable temperature and water activity) prior to the inactivation of the enzymes. Although difficult to measure in food, such oxidation is very important to the oxidative stability of food during storage. A higher initial oxidation level will accelerate further oxidation during storage.

Influence on Prooxidants. Inactivation of oxidative enzymes is one important objective in many food processes. The blanching of vegetables prior to freezing, for instance, is performed exclusively to inactivate enzymes. However, during the heating process, the enzyme passes through its optimum temperature before it is heat-inactivated. Until the enzyme is inactivated, there is a risk that some oxidation will occur during processing. Therefore, good process control is important in order to minimize this risk.

Heat processes can also lead to activation of lipid oxidation catalysts. Eriksson et al. (10) showed that the catalytic effect of hemoproteins on lipid oxidation was increased by heat denaturation. Food processes may contaminate food with traces of prooxidative metals from the processing equipment.

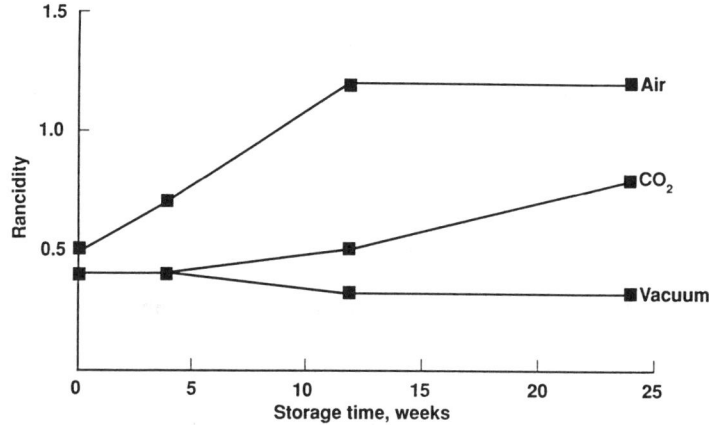

Figure 5. Development of rancid flavor in sausage during storage.

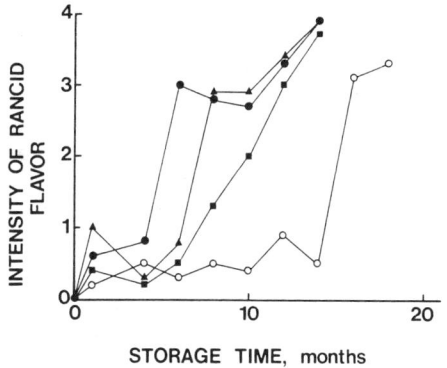

Figure 6. Development of rancid flavor in cookies during storage. Control (closed circles); 0.1% Histidine +1% Glucose added (open circles); Maillard reaction mixture 0.1%, added (triangles); 0.002% BHA/BHT added (squares). (Reproduced with permission from ref. 9. Copyright 1980 Food & Nutrition Press Inc.)

Influence on Antioxidants. Food processes may also involve the formation or loss of antioxidants. The formation of antioxidative Maillard reaction products during processes was reviewed by Lingnert (*11*). This is probably the best known example of process-induced antioxidant formation. However, it has also been reported that antioxidants may, for example, be formed during fermentation processes (*12*).

Loss of antioxidant effects can result from degradation of heat labile antioxidants or vaporization at high temperatures. For instance, partial loss of tocopherols and other phenolic antioxidants occurs during oil refining. Water soluble antioxidants, such as ascorbic acid, may also be lost by leakage to the process water in some food processes.

Influence on the Oxygen Level. Finally, the process may affect the conditions for further oxidation during storage by influencing the oxygen level in the final product. The product may be aerated or deaerated more or less during heating processes. One example is the production of UHT milk (milk sterilized by Ultra High Temperature treatment), where indirect heating processes normally result in higher contents of remaining oxygen than direct heating processes (*13*). The remaining oxygen content can also be affected by the use of vacuum or protective inert gases during processing (as in the sausage example above) and by controlling the final headspace volume in the packaging process.

Concluding Remarks

The food process is of great importance to the lipid oxidation and flavor stability of the food product. It is important to protect against oxidation at a very early stage and to choose processing conditions which ensure that lipid oxidation is avoided as much as possible. Early protection may be achieved by the use of inert gases in the process or by the early utilization of antioxidants. Special attention must be paid to avoiding enzymatic oxidation during processing. Appropriate conditions during food production is probably the best protection against oxidation during storage.

Literature Cited

1. Hall, G.; Andersson, J.; Lingnert, H.; Olofsson, B. *J.Food Qual.* **1985**, *7*, 153-190.
2. Lilja Hallberg, M. *Lipid Degradation during Production and Storage of Potato Granules.* Thesis, Chalmers University of Technology, Göteborg, Sweden 1990.
3. Lilja Hallberg, M.; Lingnert, H. *J. Am. Oil Chem. Soc.* **1991**, *68*, 167-170.
4. Siljeström, M.; Westerlund, E.; Björk, I.; Holm, J.; Asp, N.-G.; Theander, O. *J. Cereal Sci.* **1986**, *4*, 315-323.
5. Hall, G. *SIK Service-Serie* **1988**, No. 805.
6. Gaillard, T. *J. Cereal Sci.* **1986**, *4*, 33-50.
7. Gaillard, T. *J. Cereal Sci.* **1986**, *4*, 179-192.

8. Tait, S.P.C.; Gaillard, T. *J. Cereal Sci.* **1988**, *8*, 55-67.
9. Lingnert, H. *J. Food Proc. Preserv.* **1980**, *4*, 219-233.
10. Eriksson, C.E.; Olsson, P.A.; Svensson, S.G. *J. Am. Oil Chem. Soc.* **1971**, *48*, 442-447.
11. Lingnert, H.; Hall, G. In *Amino-Carbonyl Reactions in Food and Biological Systems.* Fujimaki, M.; Namiki, M.; Kato, H.,Eds; Developments in Food Science 13; Kodanska Ltd: Tokyo, Japan, 1986; pp. 273-279.
12. Murakami, H.; Asakawa, T.; Tereao, J.; Matsushita, S. *Agric. Biol. Chem.* **1984**, *48*, 2971-2975.
13. Thomas, E.L.; Burton, M.; Ford, J.E.; Perkin, A.G. *J. Dairy Res.* **1975**, *42*, 285-295.

RECEIVED February 19, 1992

Chapter 17

Factors Affecting Lipid Autoxidation of a Spray-Dried Milk Base for Baby Food

J. P. Roozen and J. P. H. Linssen

Department of Food Science, Wageningen Agricultural University, Netherlands

Dry milk base baby foods are sensitive to lipid autoxidation because of the relatively high amounts of polyunsaturated lipids and the addition of vitamins and minerals. Samples with and without additions were investigated on lipid autoxidation at water activities 0.11, 0.24 and 0.34 at 20 °C and 37°C. Autoxidation was followed by determining anisidine values in the free fat fractions after different periods of storage between 2 and 13 weeks. In a few cases headspace samples were analysed for the amount of hexanal by gas chromatography. Autoxidation was increased in samples containing added minerals and vitamins, while this reaction was suppressed and decreased in samples at water activity of 0.24. Storage at the higher temperature increased autoxidation slightly. Minerals added as carbonate-premixes caused more autoxidation than those added in chloride-premixes. The appearance of brown spots in the carbonate treated samples suggested much higher water activities resulting in increased browning and autoxidation reactions.

Manufacturers of dry baby foods enrich their humanised milk products (mother's milk analogs) with polyunsaturated fats, minerals (Fe, Cu, Zn) and vitamins (tocopherols, ascorbic acid) for nutritional and regulatory reasons. The shelf-life of these products is limited because of quality problems caused by peroxidized lipids (1). Important factors involved the method of drying and blending of the raw materials, packaging in reduced oxygen atmosphere and temperature during processing, transportation and distribution. Heating and/or shearing will redistribute the lipids in the product, while the activity of native antioxidants is reduced and the contact area with oxygen enlarged.

0097–6156/92/0500–0302$06.00/0
© 1992 American Chemical Society

Laakso (*2*) found that casein encapsulation of 1,4-pentadiene fatty acids can inhibit lipid peroxidation. Existing hydroperoxides, however, interact with milk proteins causing chemical changes like protein-protein crosslinks, protein scission, protein-lipid adducts and amino acid damage (*3*).

The main catalysts of non-enzymic lipid oxidation are metal ions added as a nutritional supplement or from contamination due to processing equipment (*4*). Fe^{2+} is a potential oxidation catalyst (*5*). Addition of synthetic antioxidants is inadvisable for both legal and marketing aspects. Most emphasis has been given to additions of trace metal complexes like citrates and lactates, which are considered to be less of a prooxidant (*6*). Usually, the protection achieved is not fully satisfactory, and the minerals are entrapped in microcapsules and denatured proteins. However, the best accepted protection of a susceptible product seems to be packaging under nitrogen gas.

Lipid Oxidation Measurement

Hydroperoxides, which are primary products of autoxidation, decompose into a series of secondary reaction products leading to oxidative rancidity. The fact that a number of different compounds are responsible for oxidative rancidity (*7, 8*), makes quantitative analysis very difficult (*9*). The International Union of Pure and Applied Chemistry (IUPAC) standardised an analytical method, "No. 2.504, the *p*-Anisidine Value", for measuring lipid oxidation (*10*). The anisidine value is defined as 100 times the absorbance of a solution resulting from the reaction of 1 g of lipids in 100 mL of a mixture of solvent and *p*-anisidine reagent in an iso-octane solution, measured spectrophotometrically at 350 nm in a 10 mm cell. Unsaturated aldehydes are primarily determined under the conditions of the test. This means that the method can give only an estimate of the content of carbonyl compounds in lipids.

Table I presents anisidine values of the total lipid (*11*) and free fat fraction (*12*) of rancid and bland smelling commercial baby foods. Correlation between off-taste and anisidine value of products varies considerably, because of the presence of different kinds of carbonyls with their own flavor characteristics and intensities (*13*), e.g. bland and rancid smelling samples can have similar anisidine values (Table I). The differences in anisidine values found between bland and rancid smelling samples are much bigger in the free fat fraction than in the total lipid fraction. This difference makes the free fat fraction more suitable for evaluation of lipid oxidation. Therefore, determination of anisidine value in free fat extracts was used throughout this study for analysing secondary products of lipid oxidation. Additionally, the effect of minerals on lipid oxidation was investigated by measuring the volatile compounds by gas chromatography. The major component was identified as hexanal. As seen in Figure 1, curves for hexanal concentration and for anisidine values were similar. Fischer and Grosch (*14*) reported that hexanal can be autoxidized to caproic acid at 38°C, which makes hexanal less suitable as an indicator for lipid oxidation.

Table I. Anisidine values of lipid extracts from commercial canned dry baby foods stored for three months at ambient conditions

Product description	Detected smell	Total lipid extract (11)	Free fat extract(12)
humanised milk	bland	2.9	7.4
humanised milk	rancid	6.8	50
humanised milk	rancid	9.6	64
acidified hum. milk	rancid	4.3	13
adapted lipids hum. milk	bland	3.0	2.7
free lactose hum. milk	rancid	3.4	6.1

Löliger (15) proposed a headspace procedure for the analysis of hydrocarbons, because that method is relatively simple when compared to the efforts required for analysis of other secondary oxidation products. However, lipid oxidation in dry milk products delivers only small amounts of hydrocarbons, i.e. mainly ethane from linolenic acid (15, 16). Consequently, the headspace procedure seems to be less suitable for analysing lipid oxidation in dry milk products.

Product Composition and Processing

Two different spray-dried milk bases for dry babyfood were taken from the ordinary commercial production of Dutch "Halffabricaat". One sample was the standard product (10 μg Zn, 2 μg Fe and 0.5 μg Cu per g dry matter) and the other, to which metal lactates were added before spray-drying, contained an enhanced mineral content (65 μg Zn, 16 μg Fe and 3 μg Cu per g dry matter). Both samples (25 g) were incubated at 20°C in glass petri dishes in stainless steel desiccators containing saturated lithium chloride solutions (Aw ≈ 0.11; moisture content of the samples were 2.1 ± 0.05 %). Lipid oxidation was determined by the anisidine value test and the hexanal content, as determined by gas chromatography. Figure 1 shows that no differences occured for all cases during seven weeks of incubation. Afterwards, the enhanced mineral sample curves for anisidine value and hexanal content increased exponentially. This result confirmed the practical experience that the milk base with enhanced mineral content had a much shorter shelf-life than the standard product. However, in order to meet nutritional requirements (17), the standard dry milk base (moisture content 2.4 ± 0.05 %) had to be supplemented with a mineral premix by adding carbonate or chloride salts. Carbonates are preferred because of their low ionic calcium content in the product, heat stability and slow release of metal ions in the digestive track (18). However, these salts accelerated lipid oxidation much more than chlorides did (Figure 2A). This "anion effect" may be related to the difference in hygroscopicity of the mineral premixes, as will be discussed in the next paragraph.

17. ROOZEN & LINSSEN *Lipid Autoxidation of Spray-Dried Milk Base* 305

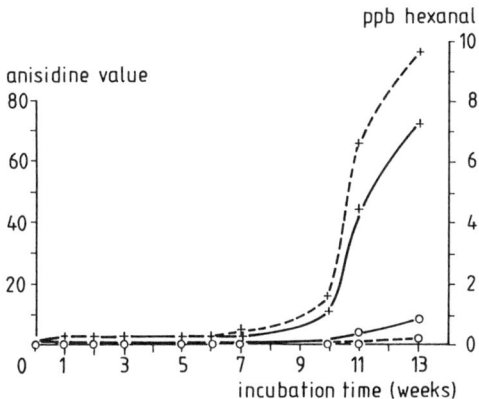

Figure 1. Lipid autoxidation in a standard spray-dried milkbase (O) and in a milkbase with enhanced mineral content (+); ——— anisidine values and ------ hexanal contents; Aw = 0.11 and temperature = 20 °C.

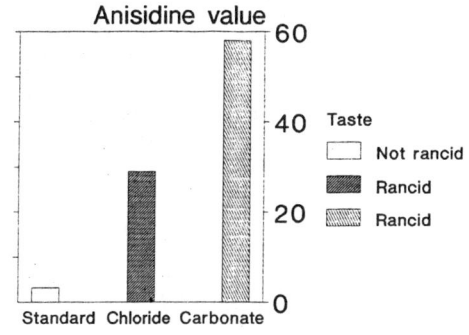

A. 4 weeks at ambient temp and rH

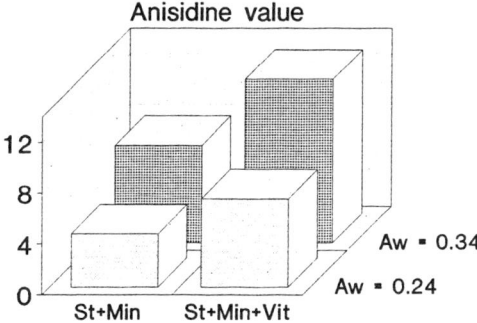

B. 8 weeks storage at temp. 20 °C

Figure 2. Lipid autoxidation in a standard spray-dried milkbase: A. after blending with a mineral premix and B. after blending with a mineral and a mineral/vitamin premix (both chlorides).

Mineral addition before or after spray-drying of the milk base appears to make no difference in the acceleration of lipid oxidation. Besides minerals, vitamins also have to be added to the dry milk base. Tocopherols and ascorbic acid are well known for their anti- and pro-oxidative action in the presence of iron (*19, 20*). Figure 2B demonstrates that the vitamins added to the milk base act as pro-oxidants when they are blended with the milk base as a vitamin/mineral premix. Similar effects were found in model systems with ascorbic acid and α-tocopherol, which reduce ferric ions (no effect) to ferrous ions producing (hydr)oxy- radicals (*21*).

Product Storage Conditions

In general, commercially produced dry baby food claims a shelf-life of three years, which requires special attention to the product storage conditions. The package should be kept gas-tight after oxygen has been reduced to about 2%. The usual practice for predicting shelf-life is storage of the final product at elevated temperatures up to 50°C. However, from time to time, products released for export to (sub)tropical countries become rancid upon arrival. Besides a rancid smell, brown spots appeared in some of the products.

Figure 3A presents the anisidine values determined in chloride minerals containing standard spray-dried milk bases, which were incubated at two temperature levels, 20°C and 37°C, but in desiccators equilibrated with a water activity of 0.34. The development of secondary lipid oxidation products appeared to be slightly little higher at the higher temperature, as was discussed by Karel (*22*). In this case, the water activity of the sample was regulated by the saturated magnesium chloride solution, while in closed packages an increase in temperature caused an increase of water activity of the sample (*23*).

Water activity had a much larger effect on lipid oxidation as shown in Figure 3B; its minimum effect was near Aw ≈ 0.24. In contrast the oxygen uptake of a thin lipid film exhibited a maximum rate of oxidation in this Aw range (relative humidity, 11-32%) (*24*).

As presented in Figure 2A, the carbonate metal premix sample had the highest anisidine value. In those samples, brown spots were present as pictured in Figure 4. These spots were absent, when the standard spray-dried milk base and the premixes containing metal carbonates and vitamins were equilibrated in desiccators (Aw =0.24) before blending. Probably, the carbonate minerals of the premixes are hygroscopic and able to create a micro-environment with enhanced water activity giving rise to lipid oxidation and browning reactions (*25*).

Conclusions and Recommendations

Different dry blends of a standard spray-dried milk base containing mineral and mineral/vitamin premixes were tested for lipid oxidation. The premixes accelerated lipid oxidation depending on composition, temperature and water activity. The latter is the most critical factor and should be kept near 0.24 to

17. ROOZEN & LINSSEN *Lipid Autoxidation of Spray-Dried Milk Base* 307

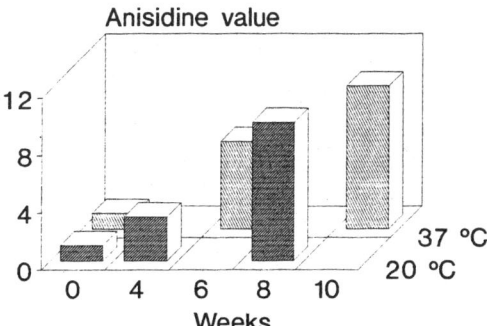

A. (water activity = 0.34)

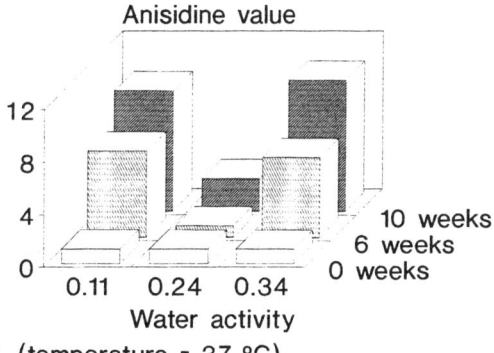

B. (temperature = 37 °C)

Figure 3. Effect of temperature (A) and water activity (B) on lipid autoxidation in a standard spray-dried milkbase blended with a mineral premix (chlorides).

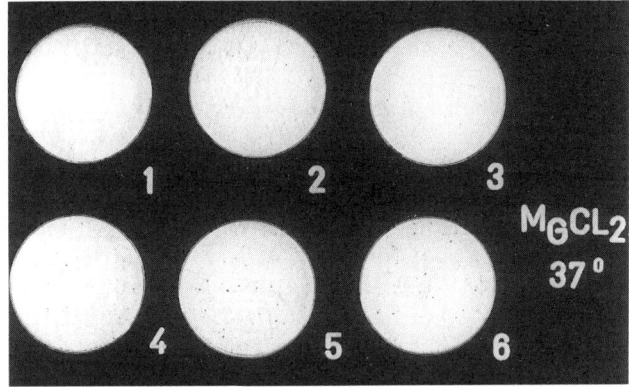

Figure 4. Picture of standard spray-dried milkbase [1] blended with a mineral premix [2, 3] and with a mineral/vitamin premix [4-6] (all carbonates). The samples were incubated for 13 weeks at Aw = 0.34.

minimize lipid oxidation. Temperature changes caused shifts in water activity of the packaged product, which made the product more liable for lipid autoxidation by moving toward Aw = 0.11 or Aw = 0.34. Accelerated storage tests at 50 °C for the prediction of shelf-life of a product were very limited in value for the same reason. Once minerals and vitamins have to be added, the best treatment for increasing shelf life of the product was to adjust the water activity of all components to 0.24 before blending.

Literature Cited

1. Ory, R.L.; St.Angelo, A.J. *J. Agric. Food Chem.* **1975**, *23*, 125.
2. Laakso, S. *Biochim. Biophys. Acta* **1984**, *792*, 11-15.
3. Gardner, H.W. *J. Agric. Food Chem.* **1979**, *27*, 220-229.
4. Pokorny, J. In *Autoxidation of Unsaturated Lipids*; Chan, H.W.S., Ed.; Food Science and Technology Series; Academic Press Inc.: London, UK, 1987; pp 141-206.
5. Grosch, W. *Z. Lebensm. Unters. Forsch.* **1976**, *160*, 371-375.
6. Wang, C.F. *A Chemical-Organoleptic Study on Iron Fortified Milk and the Biological Availability of the Iron*; Diss. Univ. of Maryland, College Park, MD, 1972; No. 72-20794.
7. Paquette, G.; Kupranycz, D.B.; Van de Voort, F.R. *Can. Inst. Food Sci. Technol. J.* **1985**, *18*, 197-206.
8. Ullrich, F.; Grosch, W. *Z. Lebensm. Unters. Forsch.* **1987**, *184*, 277-282.
9. Frankel, E.N.; Neff, W.E.; Selke, E. *Lipids* **1983**, *18*, 353-357.
10. Paquot, C. *Standard Methods for the Analysis of Oils, Fats and Derivatives*; 6th edition, IUPAC; Pergamon Press: Oxford, UK, 1979; pp. 143-144.
11. *Diätetische Lebensmittel. Gesamtfett-Bestimmung, Säureaufschluss-Methode*; In Schweizerisch Lebensmittelbuch, Zweiter Band; Eidg. Drucksachen und Materialzentrale: Bern, 1967, 22 (4.1); pp 31-32.
12. Paquot, C. *Standard Methods for the Analysis of Oils, Fats and Derivatives*; 6th edition, IUPAC; Pergamon Press: Oxford, UK, 1979; pp. 9-11.
13. Hall, G; Andersson, J. *J. Food Qual.* **1985**, *7*, 237-253.
14. Fischer, U.; Grosch, W. *Z. Lebensm. Unters. Forsch.* **1988**, *186*, 495-499.
15. Löliger, J. *J. Sci. Food Agric.* **1990**, 52, 119-128.
16. Frankel, E.N. *Prog. Lipid Res.* **1980**, *19*, 1-22.
17. Packard, V.S. *Human Milk and Infant Formula*; Food Science and Technology Series; Academic Press, Inc.: New York, NY, 1982.
18. McDermott, R.L. *Food Technol.* **1987**, *41(10)*, 91-103.
19. Niki, E. In *Selected Vitamins, Minerals, and Functional Consequences of Maternal Malnutrition*; Simopoulos, A.P., Ed; World Review of Nutrition and Dietetics; S. Karger AG: Basel, 1991, Vol 64; pp 1-30.

20. Niki, E.; Yamamoto, K; Takahashi, M. In *The Role of Oxygen in Chemistry and Biochemistry*; Ando, W; Moro-oka, Y., Eds.; Studies in Organic Chemistry; Elsevier Science Publishers: Amsterdam, 1988, Vol. 33; pp 509-514.
21. Mahoney, J.R.; Graf, E. *J. Food Sci.* **1986**, *51*, 1293-1296.
22. Karel, M. In *Concentration and Drying of Foods*; MacCarthy, D., Ed.; Elsevier Applied Science Publishers: London, UK, 1986; pp 37-51.
23. Fennema, O.R. In *Food Chemistry*; Fennema, O.R., Ed.; Food Science and Technology; Marcel Dekker, Inc.: New York, NY, 1985; pp 50-55.
24. Kahl, J.L.; Artz, W.E.; Schanus, E.G. *Lipids*, **1988**, *23*, 275-279.
25. Varshney, N.N.; Ojha, J.P. *J. Dairy Res.* **1977**, *44*, 93-101.

RECEIVED February 19, 1992

Chapter 18

Effect of Lipid Oxidation on Oil and Food Quality in Deep Frying

Edward G. Perkins

Department of Food Science, University of Illinois, Urbana, IL 61801

Deep frying is a popular method for food preparation, especially in fast food restaurants. However, lipid oxidation easily occurs at relatively high temperatures in the presence of air and produces a multiplicity of compounds and exerts both desirable and undesirable effects on food flavor and quality. A low level of oxidation actually improves the desirability of a food but higher levels decrease food quality. The presence of low molecular weight materials such as aldehydes, lactones and pyrazines influence the flavor of deep fried foods such as potatoes; while the presence of high molecular weight materials contribute to the overall quality deterioration of both the frying fat and the fried product. The application of GC-MS to the determination of flavor components in french fries and corresponding frying oil yielded information indicating a relationship between the presence of these materials and french fry quality. A larger scale study in which the amounts of high molecular weight products were determined showed the time course of oil deterioration and a relationship between total polar material formed and food quality.

Deep frying is a popular method for food preparation, especially in fast food restaurants. Although used primarily as a heat exchange medium to cook food, the oil used for deep frying contributes to the quality of the fried food. Lipid oxidation at higher temperatures in the presence of oxygen and steam, as in deep fat frying, produces a multiplicity of compounds which exert both desirable and undesirable effects on food flavor and quality. For instance (Figure 1), oxidative and chemical changes in frying fats during use are characterized by a decrease in the total unsaturation of the fat with increases in free fatty acid content, foaming, color, viscosity, polar materials, and polymeric material. All of these increase with time, at the expense of the original triglyceride via lipid oxidation through free radical mechanisms. Most of the free fatty acids found are from hydrolysis of the frying oil by steam generated from the food product.

Fats at any temperature appear to go through a free radical initiation period during oxidation. They then move into a free radical propagation or peroxide formation phase. The rate of formation then slows and a free radical termination phase is entered. In this phase peroxides are stabilized which, after a time period, begin to decompose. During this period volatile flavor and odor components and nonvolatile higher molecular weight products are formed as a result of oxidation (*1*). This process occurs at different rates depending upon the total unsaturation of the fat and many other associated factors (Figure 2).

Measurement of Lipid Oxidation

The measurement of lipid oxidation is essential to the determination of its effect on food and food oil quality. Lipid oxidation can be measured with physical methods such as molecular weight, refractive index, viscosity, specific gravity, and dielectric constant, all which increase with heating time. This type of data shows that polymerization is taking place and, in the case of the dielectric constant, indicates that there is oxygen incorporation into the fat. Spectrophotometry is usually used as a detector for certain chemical species. Ultraviolet absorption spectrophotometry is useful for determination of unsaturation particularly in the 232 and 369 nanometer region for conjugated dienoic and trienoic fatty acids respectively. Infrared absorption spectrophotometry may be used to measure hydroxyl, carboxyl and ester groups, and *cis* and *trans* unsaturation. Spectrophotofluorimetry is used to measure the presence of fluorescing materials in heated oils which are likely due to polymerization and decomposition of the original coloring materials such as carotenoids and tocopherols, remaining in the oil after processing.

Chemical methods for determination of lipid oxidation include the iodine value, which decreases as a result of destruction of unsaturation; saponification value, which will increase as the molecular weight of the sample increases. Peroxide value is less useful in the case of heated fats. At frying temperatures, peroxides do not exist since they would decompose as soon as they are formed. The thiobarbituric acid (TBA) test for the formation of malonaldehyde in fats and oils has also been used to measure lipid oxidation.

Determination of Volatile and Nonvolatile Oxidation Products

The low molecular weight components responsible for flavors and odors in many deep fried food products are formed via decomposition of lipid hydroperoxides by various mechanisms (*1*). For instance, hexanal, a very common aldehyde flavoring component, is formed by the destruction of a 13-hydroperoxide fatty acid derivative. Another very common off flavor component in fats and food products, 2,4-decadienal, is formed by cleavage of the 9-hydroperoxide. Further decompositions of hydroperoxides are responsible for the formation of complete homologous series of saturated and unsaturated hydrocarbons, saturated and unsaturated aldehydes, ketones, alcohols, and carboxylic acids. Other components termed pentyl furans and *trans, cis* and *trans, trans* dienals and pyrazines may be found in extracts of oils from food products such as french fries (*2*).

There are many non-volatile components present in a used fat. Unsaponifable materials such as aldehydes, hydrocarbons, and ketones are found. A large amount of nonoxidized material as well as cyclic monomeric products which contain cyclohexyl, cyclopentyl, and cyclohexenyl groups are found. Dimeric fatty acids containing both cyclic and noncyclic structures have been characterized. Several saturated and unsaturated keto- as well as mono- and di-hydroxy - and ketohydroxy-substituted fatty acids are formed. Trimeric triglycerides, and a mass of uncharacterized high molecular weight oligomeric triglycerides termed polymers are present. This polymeric material may be polar or nonpolar in nature (*3, 4*).

Monomers may result from the hydrolysis of triglycerides and may cyclize either in the triglyceride form or as a free fatty acid. They may also form by

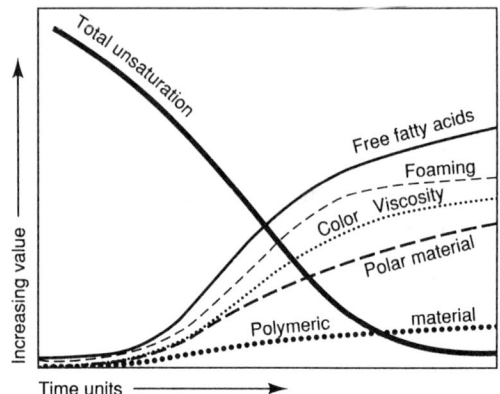

Figure 1. Changes in Frying Fats During Use.

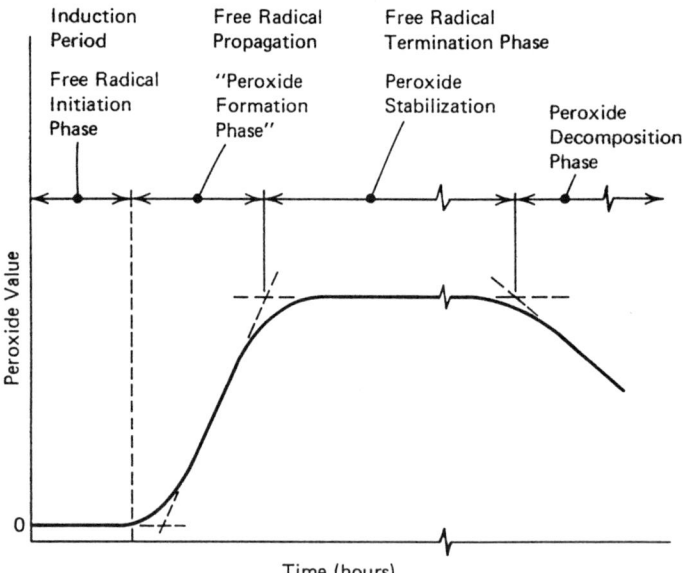

Figure 2. Effect of Time on Peroxide Formation and Decomposition.

decomposition of fatty acid hydroperoxides. Non thermal oxidation products may also be formed and consist of the cyclic monomers caused by initial cyclization of the fatty acid chain. Both aromatic rings and those containing a cyclohexenyl and cyclopentenyl group may be formed as part of the fatty acid chain (5).
Non volatile thermal oxidation products may be fractionated by using solvent partitioning. The polar material must solubilize into a relatively polar solvent. Non polar material must dissolve into a nonpolar solvent, such as hexane; this can be used to separate these compounds on a rough scale. Distillation will separate molecular weight species and urea adduction will separate compounds which differ in structure from stearate and linolenic acid. Gas liquid chromatography is used for analysis of the fatty acids but this could also be used for the determination of nonelutable species or polymers. Thin layer chromatography in many of its modes such as reverse phase, argentation, or adsorption on silica have been used to separate thermal oxidization products. High performance liquid chromatography (HPLC) has been successfully used to fractionate thermal oxidation products as well. One can collect component fractions from the HPLC and then perform gas chromatography on them. High performance size exclusion chromatography (HPSEC) will separate lipid oxidation products into their various molecular weight species.

Before discussion of the reactions of deep frying and absorption of oxidized components into food, analytical considerations for certain nonvolatile compounds should be addressed. Compounds such as cyclic monomers, dibasic acids, dimers, trimers, and higher molecular weight products as well, which may contain hydroxy and keto groups may need to be derivatized for increased volatility. This is usually done by preparing methyl esters or trimethyl silyl esters or ethers of hydroxyl and carboxyl groups prior to analysis by chromatographic methods. The cyclic fatty acids are formed in variable amounts up to 1500 to 2000 parts per million depending upon the unsaturated fatty acids. If linoleic or linolenic acid are the unsaturated fatty acids the structures shown in Figure 3 are formed. A cyclohexyl ring containing compound (as in the figure) occurs as a series of isomers which have been isolated, characterized and upon hydrogenation forms structure number 4 in Figure 3, a saturated cyclic monomer. There has also been evidence produced for the formation of cyclic monomer compounds 2 and 3. When an oil is high in linolenic acid one sees the formation of compound 5 containing a cyclic fatty acid. There is no direct evidence yet that the bicyclic fatty acid is formed during deep frying, but it is possible that it is formed and has not been determined. The analysis of cyclic monomers may be done in a preparatory mode by HPLC of the heated fat mixtures (6). Sebedio has shown that cyclic monomers are eluted after methyl palmitate and before methyl stearate on a reverse phase column with acetonitrile : acetone as mobile phase (6). If this fraction is collected, hydrogenated and then the mass spectra of the components collected by capillary gas chromatography - mass spectrometry evaluated, more than 10 different cyclic monomer structures in which the chain lengths range from ethyl to hexyl, as well as cis, trans isomers are found (6).

Dimeric fatty acids are formed which may or may not contain triglyceride. This depends upon whether or not intra or intermolecular dimerization has taken place. There is evidence for formation of noncyclic and cyclic dimers as well as bicyclic dimers in fats and oils which have been used for frying. Thermal dimers are simply those which are formed by heating and condensation. They are usually cyclic dimers which result from a Diels-Alder reaction. Others often form from condensation reactions of two fatty acid free radicals and do not have a cyclic structure. Seven different dimeric structures ranging from a thermal dimer to a tetrahydroxy dimer have been synthesized (7-9). All the synthetic dimers, when made into a model mixture, can be separated by HPLC (10). Now that mixtures of isomers of known dimeric structure are available, the dimeric fraction from an oil used for frying may be isolated by preparative size exclusion chromatography and then

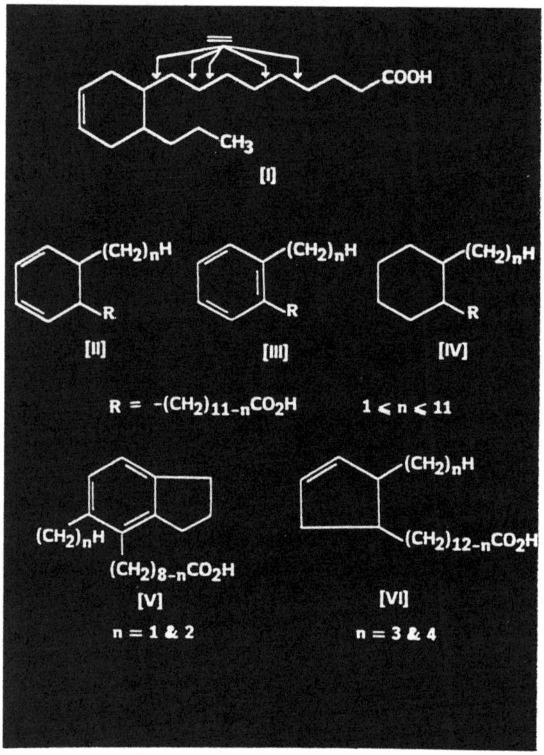

Figure 3. Structures of Cyclic Fatty Acids Formed During Deep Frying.

separated under the same conditions as those developed for model mixtures. All seven types of dimers were present in a used soybean oil used as a model. The same components were found in about the same concentrations relative to one another. It appears that the components in the used oils were absorbed by the french fries in concentrations approximating those present in the frying medium (*11*). There does not appear to be any preferential uptake by the french fry of any the non volatile components formed in the frying oils.

Deep Frying and Oil and Fry Quality

In a recent study of the effects of deep frying on the quality of frying oil and french fries, the fatty acid composition of the frying oil showed destruction of linolenic and linoleic acids with frying time (*11*). In addition the composition of frying oil and french fry components showed that there was a continuous increase in polar materials, polymers and in nonelutable components in both the used oil and in the french fries. There was no significant increase in oil absorbed by the fries during this study.

The volatile materials present in the frying oil were determined by placing 200 mg. of the used french frying oil on a small column of adsorbent and inserting this into an external closed inlet device (ECID) unit which is connected to a cryogenic gas chromatograph - mass spectrometer with a capillary column (*12*). Volatile components of the frying oil were then eluted from the ECID and their composition determined. To study the volatile components in the french fries a 200 gram portion of french fries were placed into a chamber such that the volatiles in the headspace within the chamber were carried with purified nitrogen into an ECID cartridge containing Tenax GC to adsorb the volatiles. After a period of time for collection of the volatiles, this cartridge was placed into the ECID unit and then desorbed onto the capillary gas chromatograph column. The components were identified by gas chromatography and mass spectrometry (*11*). The total volatiles in the frying oil and in the french fries reached the maximum at about 70 hours frying time. The hydrocarbons identified in the frying oil were hexene, hexane, heptane, decane, and nonane. All of these peaked at approximately 70 hours frying time. The major saturated aldehydes in frying oils, pentanal, hexanal, heptanal, octanal,and nonanal, peaked at around 60-70 hours frying time. The same major saturated hydrocarbons and aldehydes were present in the french fries and also peaked at about the same time. Alkenals in frying oil and in french fries were 2-heptenal, 2-octenal, 2-nonenal, 2-decenal and 2-undecenal. These and the alkyl dienals such as *trans, trans* 2,4-heptadienal, *trans, trans* 2,4-decadienal and *trans, cis* decadienals also had a peak formation time at about 70 hours. The content of α-pentylfuran in frying oil and french fries peaked at around 58 hours. Pyrazines in french fries cooked in this used oil were identified as methyl pyrazine and 2,5-dimethyl pyrazine. The maximum concentration of these compounds occurred at approximately 80 hours in the french fries, they are not present in the original frying oil because they arise from interaction between the hot oil plus french fries. In summary this study showed that the same components in approximately the same concentrations were present in both the frying oils and the corresponding fries at various times of usage of the oils. No evidence was found for the preferential uptake of oils or volatile material in the french fries with time and no evidence for any preferential uptake of free fatty acids or polymeric or polar components with time with these fries. The french fries produced during this study were all of high quality regarding texture, color, appearance and flavor.

Quality Evaluation of Fats

A number of factors influence fat stability and the formation of lipid oxidation products. Increased unsaturation, increased frying time, increased exposure of the oil to air and increased trace metal content will all result in decreased oxidative stability. The type of food fried influences fat stability. For instance, when frying chicken the fat from the chicken renders into the deep frying medium and changes the fatty acid composition of the oil. Fish does the same thing and replaces some of the shortening fat by fish oils which are highly unsaturated and influences fat stability. If both fish and chicken are batter-fried, then components of the batter will further degrade the fat more quickly and decrease stability. Nonfat foods such as french fries would degrade fat stability because of increased areation of the oil and the accumulation of small food particles that carbonize and, if left in the fat, cause it to deteriorate faster.

The presence of silicones in a frying oil will cause increased oil stability by yet unknown mechanisms. A more recent technique used to increase the frying life of an oil is treatment of the frying oil with "active" filtration media (*13,14*). Published data indicates that filtration of oils through certain active adsorbants will increase the useful frying life of an oil during actual fryer use by removal of colored materials, free fatty acids and other oxidation products.

Quality Evaluation of Fats. Quality evaluation of frying fats may be carried out in many ways. The first is sensory evaluation in which the flavor profile of the frying fat, as well as the food product, is determined using standardized sensory testing. The texture of the product fried in the fat will also indicate the quality of the used frying fat. The room odor, or the odor in the room where a deep fryer is used may give an indication of the quality of the frying fat (*15*). Peroxide values are an indication of frying fat quality if they are used in a very specific way. Usually peroxides decompose at about 150°C. Therefore at frying temperatures, the accumulation of peroxides does not occur. Peroxide values usually are a measure of lipid oxidation at lower temperatures such as those used for storage of fats or a product. The relationship between storage time and peroxide value can then be used to measure quality.

Volatile profile analysis has been used to evaluate frying fats as well as the food product fried in them. More recently, however, volatiles such as hexanal, pentane or hexane have been used to characterize the quality of a food product or its frying oil (*15*). The Schall oven test involves simply putting a small amount of the fat into a beaker and placing it into an oven under standardized conditions at 60°C to oxidize the sample. Samples are then taken and peroxide values determined on them.

Free fatty acids from frying fats can be determined by direct titration. Other components such as soaps may be formed from the reaction of free fatty acids with metal salts present in the oil but there does not seem to be any published evidence yet on the effect of soap formation upon quality of the frying fat or the food product fried in it. Industrially, frying oil quality is usually checked quickly by measurement of color and/or free fatty acid to tell an operator when a fat is ending its useful life. The foaming characteristics of the used fat will also lead one to the same conclusion. If one can't fry a food product because of undue high foaming levels, the oil has ended its useful life.

There are many other tests available to check frying oil quality, all which purport to tell the operator when to do something with the used fat - either filter it through active filters, discard it, or dilute it with a less degraded fat. Some tests which have been used to check frying oil quality are the saponification color index, 2,6-dichloroindole phenol color test, methylene blue color test, and iodine color scale. These tests allegedly determine when the fat has gone bad and can no longer produce a high quality food product. For instance, the Rau test from E. Merck is a colorometric test kit which contains redox indicators that react with total oxidized compounds in a sample. It has a four color scale and is used for diagnoses of fat quality. The fourth color scale indicates a bad oil and the oil should be discarded (*16*). The polar materials quick test kits from Libra laboratories purport to relate the color of a test solution with the percent polar materials present in the fat and the oil quality (*17*). Smith (*18*) surveyed 56 used shortenings and showed that the relationship of the food oil sensor (FOS) value, polar materials, free fatty acids and non-elutables by GLC can be used to find ranges for these measurements, which can indicate when frying fats should be discarded. This fat was used for frying chicken and french fries. The mean values from the 56 samples that were tested; 16.4% polar materials vs. 2.5 for fresh fat; 0.76% free fatty acids vs. 0.05 for fresh fat; nonelutable components 6.1% vs. 0.7 percent in the fresh fat.

COLOR. A quick colorometric test kit is available with a color scale to be used for measuring oil quality (*19*). The fat is discarded at the highest color scale value. Robern and Gray (*20*) developed a spot test to measure free fatty acids in which drops of used fat placed on glass covered with silica gel containing a pH indicator gave a three color test scale blue, green and yellow. This can indicate the amount of free fatty acids in a sample prior to its discard. Another test kit for measurement of alkaline contaminant materials in used frying fats has been described by Blumenthal and Stockler (*21*). A summary of results for 1800 used fats has been published by Begemann (*22*) in which the free acids values, soap color index, smoke point and any polar compounds present in the fats was estimated in the fresh fats vs. fats with no sensory change and a distinct sensory change vs. a bad fat. The bad fat had an acid value of 9.8 vs. 0.2 for the fresh fat, soap color index 32.5 vs. 0.1, for the fresh fat, smoke point 110 degrees centigrade vs. 205 for the fresh fat and 64% polar components vs. less than 10% of the fresh fat.

POLAR COMPONENTS. The presence of polar components in a frying fat may be misleading. However, the determination of polar materials in frying fats is an official method used in Europe. This test involves separation of used fats into a polar and a non-polar fraction via silica gel chromatography (*23*). However, the polar fraction contains not only polymeric and monomeric material but also free fatty acids, and mono and diglycerides. These are the major portion of such fractions and their effects on frying oil and food product quality are not understood. In Figure 4 the top most HPSEC curve represents a used oil at its mid-life point. The oil was obtained from a mixed menu frying operation. Monoglycerides, diglycerides, triglycerides, dimeric triglycerides, trimeric triglycerides and oligomeric triglycerides are present (*24*). If this fat is analyzed for its nonpolar fraction, using the method described above, one peak is present which is pure triglyceride. This is expected since the silica gel column was eluted with a 13% solution of ether in hexane and that should only remove nonpolar or triglyceride material. The next chromatogram, Figure 4B, shows the polar fraction from the mixed menu frying fat resulting from the column separation. Larger concentrations of monoglycerides, diglycerides, and dimeric diglyceride, triglyceride dimer, trimeric and some oligomeric triacylglycerols are present.

Consideration of days of frying vs. weight percent of the component formed during the mixed menu frying study indicates that only total polar material increases uniformly in concentration. The relationship between percent acceptability for flavor, texture, appearance, odor of the product and weight percent of the component of trimer, percent polymer, percent dimer, percent non elutables by gas chromatography and the percent of polar components vs. days of frying in this mixed menu experiment is quite interesting (Figure 5)(*24*). Only the polar components increased linearly with days of frying. All of the parameters for flavor, texture, appearance and

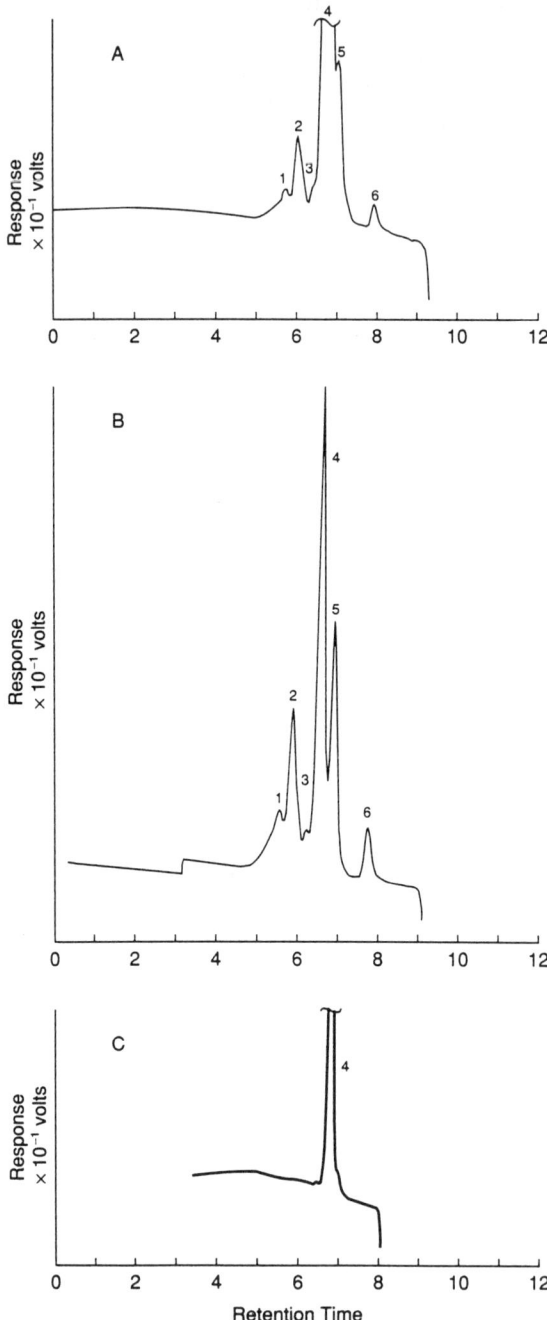

Figure 4. High Performance Size Exclusion Chromatography of Used Frying Fat. (A) Used Fat, (B) Polar Fraction (C) Non-Polar Fraction.

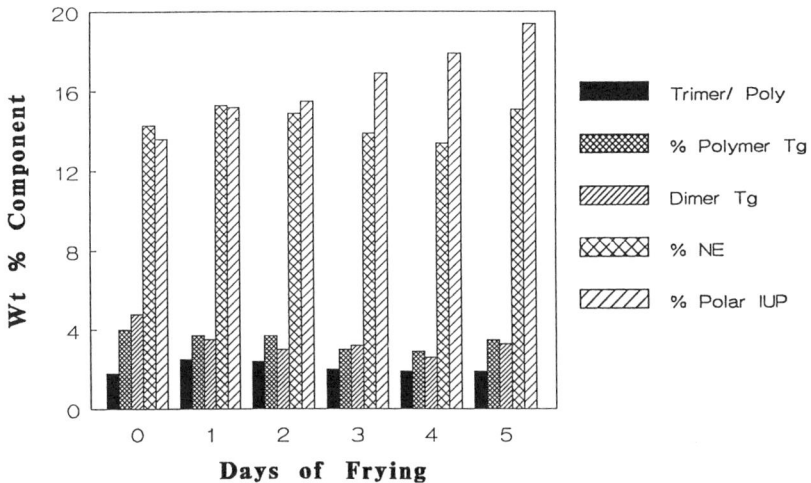

Figure 5. Weight Percent of Compounds Formed vs. Days of Frying.

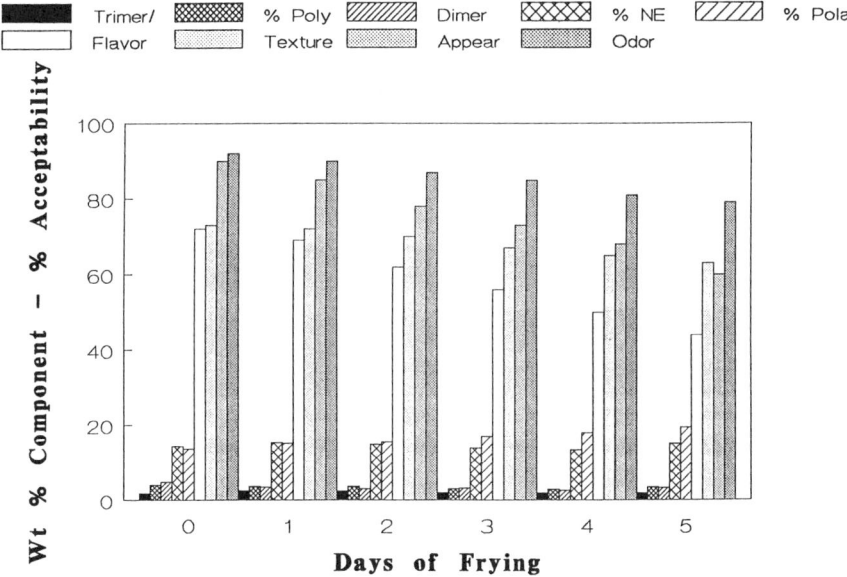

Figure 6. Weight Percent of Component Formed and Acceptability vs. Days of Frying.

odor of product decreased uniformly. Based upon the limited data that we have available, however, the only parameter that correlates well with the sensory evaluation of the product is the percent of total polars formed a shown in Figure 6 (24).

The direct relationship of free fatty acids, percent of polar material and food oil sensor values is shown in Figure 7 using data of Smith et al.(18). The smoke point vs. acid value and polymers relationship indicates that, as smoke point decreases, the acidic material and the polymeric material increases (Figure 8). The

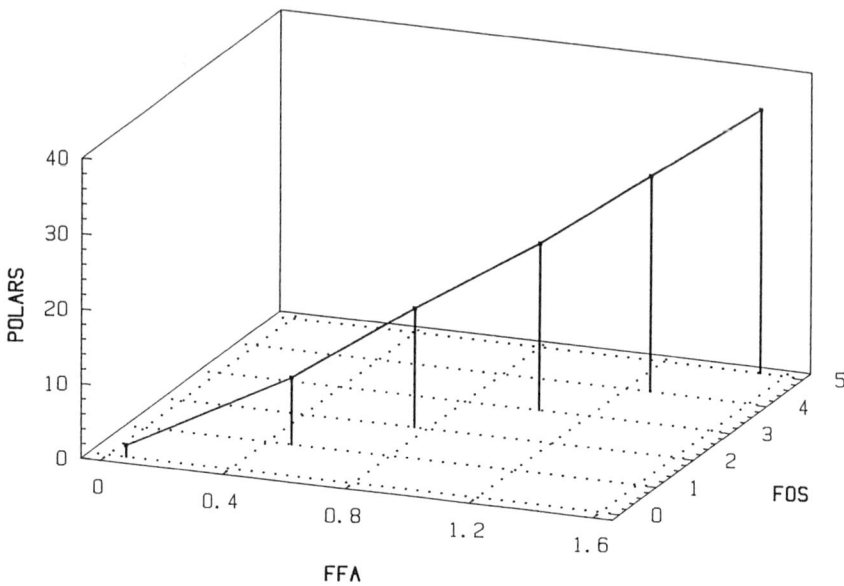

Figure 7. Relationship between amount of Polar Material, Free Fatty Acid content (FFA), and Food Oil Sensor Readings (FOS).

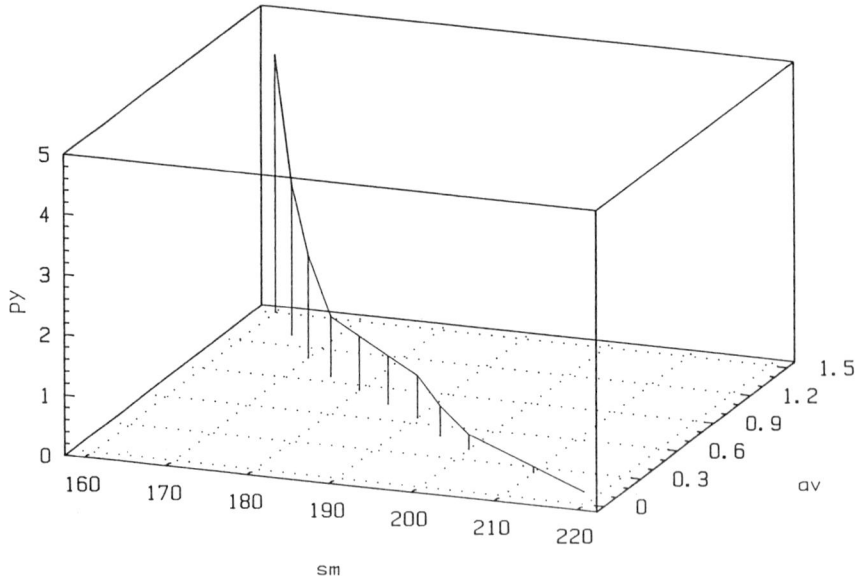

Figure 8. Relationship between polymeric materials formed (py), Smoke Point (sm), and amount of Free Fatty Acids (av) formed in Deep Frying.

relationships of lipid oxidation products to each other when plotted as shown in figures 7 and 8 allows one to predict values which assist in making a decision regarding when to discard a fat.

In summary, the effect of lipid oxidation on food and oil quality during deep frying is wide ranging. The type of food fried in the oil as well as the age of the oil and the conditions under which they are used will influence the quality of the food product. In order to obtain the maximum quality of food product, the oil should be maintained at its maximum quality so that the build up of polymeric materials, free fatty acids and color is held at a minimum.

Literature Cited
1. Frankel, E.F., In *Flavor Chemistry of Fats and Oils*; Min, D.B. and Smouse, T.H., Eds.; American Oil Chemists Society: Champaign, IL, 1985, pp. 1-37.
2. Chang, S.S.; Vallese, F.M., Hwang, L.S., Hseih, O.A.L., and Min, D.B. *J. Agr. Food Chem.* **1977**, *25*, 450.
3. Artman, N.R.; In *Adv. Lipid Res.*; Paoletti, R., and Kritchevsky, D., Eds.; Academic Press: New York, NY, 1969, 7, 245.
4. Nawar, W.W.; In *Flavor Chemistry of Fats and Oils*; Min, D.B. and Smouse, T.H., Eds.; American Oil Chemists Society: Champaign, IL, 1985, 39.
5. Pokorny, J.; In *Flavor Chemistry of Lipid Foods*; Min, D.B.and Smouse, T.H., Eds.; American Oil Chemists Society: Champaign, IL, 1985, 113.
6. Sebedio, J.L.; Prevost, J. and Grandgirard, A. *J. Amer. Oil Chemists Soc.* **1987**, *67*, 1026.
7. Christopoulou, C.N.; Perkins, E.G. *J. Amer. Oil Chemists Soc.* **1989**, *66*, 1344.
8. IBID, pg 1353.
9. IBID, pg 1338.
10. IBID, pg 1360.
11. Qian, C.; Perkins, E.G. *Inform* **1991**, *2*, 323.
12. Legendre, M.; Fisher, G.S., Shuller, W.H., Dupuy, H.P.,and Rayner, E.T. *J. Amer. Oil Chemists Soc.* **1977**, *54*, 445.
13. Yates, R.A.; Caldwell, J.D. *Inform* **1991**, *2*, 323.
14. McNeill.J.; Kakuda, Y., and Kamal, B. *J. Amer. Oil Chemists Soc.* **1989**, *66*, 1017.
15. Warner, K., In *Flavor Chemistry of Fats and Oils*, Min D.B. and Smouse, T.H., Eds.; American Oil Chemists Society: Champaign, IL, 1985, 207.
16. Meyer, H.; *Fette Seifen Anstr.* **1979**, *81*, 524.
17. Blumenthal, M.; *J. Amer.Oil Chemists Soc.* **1988**, *65*, 482.
18. Smith, L.M.; Cliford, A.J., Hamblin, C.L., and Creveling, R.K. *J. Amer. Oil Chemists Soc.* **1986**, *63*, 1017.
19. Blumenthal, M.; *J. Amer.Oil Chemists Soc.* **1985**, *62*, 1319.
20. Robern, H.; Grey, L. *Can. Inst. Food Sci. Technol.* **1981**, *14*, 159.
21. Blumenthal, M.; Stockler, J. *J. Amer. Oil Chemists Soc.* **1986**, *63*, 687.
22. Begemann, O.; *Fette Seifen Anstr.* **1986**, *88*, 124.
23. Paquot,C.; Hautfenne, A. *Standard Methods for the Analysis of Oils Fats and Derivatives*; Blackwell Scientific Publications: Palo Alto, CA, 1987, pp. 216-219.
24. Perkins E.G.; Qian, C., Caldwell, J.D., and Yates, R.A. *J. Amer Oil Chemists Soc.* **1989**, *66*, 483.

RECEIVED April 23, 1992

Chapter 19

Effect of Storage on Roasted Peanut Quality
Descriptive Sensory Analysis and Gas Chromatographic Techniques

Karen L. Bett and T. D. Boylston[1]

Agricultural Research Service, U.S. Department of Agriculture, Southern Regional Research Center, 1100 Robert E. Lee Boulevard, New Orleans, LA 70124

>Flavor of roasted peanuts, as well as the profile of volatile compounds change during storage. Florunner peanuts from two crop years were sorted by commercial grade sizes and roasted. After roasting, peanuts were stored in open containers at 37°C until sampling for a maximum of 12 weeks. Six flavor descriptors (roasted peanutty, sweet aromatic, cardboardy, painty, fermented/fruity, and woody/hulls/skins) were evaluated for intensity. Volatile compounds were analyzed during the 12 weeks using headspace analysis by gas chromatograph, with trapping on Tenax adsorbent. Lipid oxidation descriptors, such as painty and cardboardy flavor intensities, as well as lipid oxidation products, such as hexanal, octanal and 2-octanone increased during storage. In contrast, roasted peanutty flavor intensity decreased during storage along with the alkylpyrazines. Commercial seed size affected the concentration of carbonyl compounds, but not the alkylpyrazines.

Roasted peanut flavor is composed of a complex blend of heterocyclic and other volatile compounds formed during roasting through thermal degradation reactions, including Maillard reactions between carbohydrate, free amino acid and protein (*1,2*). The alkylpyrazines, which have nutty flavor characteristics, are predominant compounds in the volatile profile of roasted peanuts.

Lipid oxidation during the storage of peanuts and other lipid-containing foods has long been recognized as contributing to the development of undesirable flavors in these foods (*3*). These reactions lead indirectly to the formation of numerous aliphatic aldehydes, ketones, and alcohols with "cardboardy" or "painty" flavor characteristics (*4*). The free-radicals and hydroperoxides formed during the autoxidation of lipids may also interact with other components of the food system,

[1]Current address: Washington State University, Department of Food Science and Nutrition, Pullman, WA 99164

including amino acids, proteins, and other nitrogen-containing compounds, and further affect overall flavor quality (5). Differences in the seed size and maturity of peanuts have contributed to variability in the content and composition of free amino acids, sugars (6,7), lipids (8), roasting characteristics and flavor quality (7,9). Seed size and maturity are related based on work of Sanders et al. (9) that showed a decrease in the percentage of immature peanuts in a commercial size class as the size of the peanuts increases. The potential for desirable roasted peanutty flavor increases and the formation of undesirable off-flavors decreases as seed size increases and/or peanuts mature (9).

The objectives of this study were to determine the effect of elevated storage temperature on the flavor and volatile compounds, determined through sensory evaluation and instrumental analysis, of roasted peanuts. This study was designed as a model to further understand the relationships between the sensory and instrumental analysis of flavor quality of roasted peanuts.

Methods for Evaluating Flavor and Flavor Compounds

Preparation of Samples. Florunner peanuts from two crop years (1988 and 1989) were sorted by commercial grade sizes into #1's which rode a 5.95mm screen and fell through a 7.14mm screen and medium/jumbos which rode a 7.15mm diameter screen. The jumbos and mediums were not separated because Sanders et al. (9) did not find significant differences between the two commercial seed sizes, and combining the two resulted in a larger sample to work with. The four different treatment combinations (two crop years and two commercial grade sizes) of peanuts were placed uncovered in separate glass 190 x 100 evaporation dishes and placed in a 37°C oven for storage. Peanuts were stirred every two weeks. Samples were taken at 0, 4, 8, and 12 weeks for sensory analysis and 0, 2, 4, 6, 8, and 12 weeks for gas chromatographic (GC) analysis. The peanuts were reduced to a paste according to Sanders et al. (10) to assure a uniform sample for GC and sensory purposes.

Volatile Flavor Compound Analysis. The volatile flavor compounds were isolated from the peanut butter samples using headspace analysis techniques with trapping on Tenax-GC adsorbent (Boylston, T.D. and Vinyard, B.T. *J. Food Sci.*, in press.). The purge (water bath) temperature was 90°C and volatiles were collected for 4 hr. Tetradecane (25 µg) was included as quantification standards with the sample prior to purging and pentadecane (5 µg) was placed on the top of the adsorbent prior to elution of volatiles. The volatiles were eluted from the adsorbent with 15 ml HPLC grade hexane and concentrated to 100 µl.

The volatile flavor compounds were separated on a cross-linked, 5% phenylmethyl silicone fused silica capillary column (HP-5, 50 m, 0.32 mm od, 0.52 µ film thickness, Hewlett-Packard) installed in a gas chromatograph equipped with a flame ionization detector (Model 5890A, Hewlett-Packard). The GC oven temperature was initially held at 35°C for 15 min, then increased at a rate of 2°C/min to a final temperature of 250°C and held for 45 min. Injector and detector temperatures were set at 200°C and 250°C, respectively. The extracts (2.5 µl) were injected using a splitless injection, with a column flow rate of 1.1 ml/min and a purge flow rate of 2.0 ml/min. Contents of the volatile compounds were quantified based on standard curves for the quantification standard, pentadecane (20 to 500 ng).

Identification of the volatile compounds in the samples was based on the comparison to retention times and mass spectra of standards and mass spectra published in the literature (Boylston, T.D. and Vinyard, B.T. *J. Food Sci.*, in press.). A gas chromatograph-quadrupole mass spectrometer (Model 4500, Finnigan-MAT) interfaced with an Incos data system was used for confirmation of the identity of the volatile compounds in the extracts. GC conditions were as for the chromatographic analysis. The conditions for the mass spectrometer were set as follows: ionizing voltage, 70 eV; emission current, 0.3 mA; electron multiplier voltage, 1800 kV; ion source temperature, 150°C; ionization chamber pressure, 6.0×10^{-6} atm; and scan range 33 to 250 m/z in 0.95 sec with a 0.05 sec hold.

Descriptive Sensory Analysis. Descriptive flavor analysis was accomplished using the Spectrum Method described by Meilgaard et al. (*11*). The descriptive panel consisted of 12 persons selected for availability and normal abilities to taste and smell. The six descriptors monitored were roasted peanutty (RPT), sweet aromatic (SAC), cardboardy (CBD), painty (PTY), woody/hulls/skins (WHS), and fruity/fermented (FFY) (*10,12*). Each descriptor was evaluated for intensity using a 15-point scale (*11*). Since there were 16 samples and only eight samples were presented at each session, year effect was confounded with session effect. Year effect was considered replication by the authors and was not as important a comparison as storage time or peanut size. Samples were served in 1 oz. plastic cups identified with 3-digit random numbers under red lights to mask color differences. Each sample was presented twice to the panel as a repeated measure. The data was checked for outliers and faulty panelists using the methods of Crippen et al. (Crippen, K.L., Shaffer, G.P., Vercellotti, J.R., Sanders, T.H., Blankenship, P.D. *J. Sensory Stud.*, in press.). The means of panelists were calculated for each treatment combination/repeated measure sample, and used as the data set for the statistical analysis.

Statistical Analysis. Peanuts from the two crop years provided the replication for the experiment. Duplicate analyses for each storage time, seed size, and crop year treatment were conducted for the volatile analyses. The data for each sensory attribute and GC peak was subjected to a repeated measures analysis of variance (*13*) to identify significant size, crop year, storage, and/or interaction effects. All sensory data exhibited a spherical error structure (*14*) required for the further use of a repeated measures analysis to identify storage trends. Storage trends respective to each sensory attribute were identified for each level of a significant effect by using polynomial, profile, and 0-day to each subsequent day, contrasts. Some of the data collected from volatile compounds did not possess this required structure. Hence, storage trends for the GC data were identified by fitting individual regressions to data from each level of a significant effect (*13*).

Results of Peanut Storage

Peanut flavor descriptors were significantly affected by the experimental treatments. Means across storage were significantly different for cardboardy, painty, roasted peanutty, sweet aromatic, and woody/hulls/skins (Table I). Roasted peanutty, sweet aromatic, and woody/hulls/skins had significant interaction effects as well as significant main effect.

Volatile flavor compounds representative of desirable and undesirable flavor characteristics of roasted peanuts were selected to evaluate the influence of storage time and seed size on flavor composition. These compounds were the major peaks on the GC trace of each treatment.

The effect of the storage time and peanut seed size treatments on the content of the volatile flavor compounds varied depending on the nature of the flavor compounds (Tables II-IV). Storage time had a significant effect ($p<0.05$) on the content of 9 of the 12 heterocyclic compounds (Table II). Peanut seed size and crop year and the interactions between these treatments and storage time were not significant ($p>0.05$) for the heterocyclic compounds. For the lipid oxidation products, storage time had a significant effect on 15 of the 18 compounds analyzed in this study. However, peanut seed size, crop year, and interactions between storage time and seed size, and storage time, seed size, and crop year had a significant effect ($p<0.05$) on the content of the lipid oxidation products (Table III) in contrast to the heterocyclic compounds. The significant effect of seed size on the content of several of the lipid oxidation products has been attributed to differences in the lipid content and composition of the peanuts, with respect to maturity, and thus seed size (8). The effect of storage time and seed size on the content of the selected volatile flavor compounds will be discussed further. Benzaldehyde and limonene are other compounds affected by storage time (Table IV).

Heterocyclic Compounds. Heterocyclic compounds are formed through the Maillard and other thermal degradation reactions between amino acids and sugars. These compounds have roasted, browned, and nutty flavor notes. In particular, the nutty, earthy flavor characteristics of roasted peanuts have been attributed to the alkylpyrazines (1,15,16). The content of the heterocyclic compounds decreased significantly ($p<0.05$) with storage. As shown in Table V, the decrease in concentration was generally greatest during the initial weeks of storage. With continued storage, the rate of decrease in the content of the alkylpyrazines concentration diminished. The decreases in heterocyclic compound concentrations followed linear, quadratic, and cubic trends, with no apparent relationship between the content and nature of the heterocyclic compounds and observed trend. The actual mechanism for the observed decrease of the pyrazines and 2-methylfurfural during the 12 week storage period is unknown. The decrease in the content of the heterocyclic compounds may be attributed to either the degradation by lipid radicals and peroxides or flavor entrainment by complexes between proteins and lipid hydroperoxides or its secondary products (5).

The higher content of free amino acids and sucrose in #1 peanuts, in comparison to that of jumbo/medium peanuts, contributed to the significantly different roasting characteristics and less desirable flavor quality of the smaller peanuts (7). In this study, the content of the heterocyclic compounds were not significantly different ($p>0.05$) with respect to peanut seed size, although trends indicated slightly higher contents of several pyrazines, including 2,5-dimethylpyrazine, in the jumbo/medium peanuts (data not shown). Thus, perceived differences in the intensity of the roasted

Table I. Significance of treatment effects on sensory attributes, as determined by repeated measures analysis of variance

Flavor Descriptor	Main Effects			Interactions			
	Time	Size	Year	Time × Size	Time × Year	Size × Year	Time × Size × Year
Roasted Peanutty	0.0001	0.0136	NS	NS	0.0073	NS	NS
Sweet Aromatic	0.0001	NS	NS	NS	0.0103	NS	NS
Woody/Hulls/Skins	0.0100	NS	NS	NS	NS	NS	0.0384
Painty	0.0001	0.0119	NS	0.0584	0.0527	NS	NS
Cardboard	0.0135	NS	NS	NS	NS	NS	NS
Fruity/Fermented	NS	0.0001	0.0004	0.0435	NS	0.0049	NS

NS = intensity means were not significantly different (p>0.05)

Table II. Significance of treatment effects on heterocyclic compounds, as determined by repeated measures analysis of variance

Compound	Main Effects			Interactions		
	Time	Size	Year	Time × Size	Time × Year	Time × Size × Year
2,5-Dimethylpyrazine	0.0001	NS	NS	NS	NS	NS
Ethylpyrazine	0.0001	NS	NS	NS	NS	NS
2,3-Dimethylpyrazine	0.0001	NS	NS	NS	NS	NS
2-Ethyl-6-methylpyrazine	NS	NS	NS	NS	NS	NS
2-Ethyl-5-methylpyrazine	0.0071	NS	NS	NS	NS	NS
Trimethylpyrazine	0.0001	NS	NS	NS	NS	NS
2-Ethyl-3,6-dimethylpyrazine	0.0002	NS	NS	NS	NS	NS
2-Ethyl-3,5-dimethylpyrazine	NS	NS	NS	NS	NS	NS
2,5-Diethylpyrazine	0.0320	NS	NS	NS	NS	NS
2-Isobutyl-3-methylpyrazine	0.0095	NS	NS	NS	NS	NS
2,3-Diethyl-5-methylpyrazine	NS	NS	NS	NS	NS	NS
2-Methylfurfural	0.0097	NS	NS	NS	NS	NS

NS = mean peak contents were not significantly different ($p > 0.05$)

Table III. Significance of treatment effects on lipid oxidation products, as determined by repeated measures analysis of variance

	Main Effects			Interactions		
Compound	Time	Size	Year	Time × Size	Time × Year	Time × Size × Year
Hexanal	0.0001	NS	NS	0.0257	NS	NS
Heptanal	0.0010	0.0015	0.0206	0.0056	NS	NS
2-Heptenal	0.0013	0.0031	NS	NS	NS	NS
Octanal	0.0001	0.0106	NS	0.0001	NS	NS
2-Octenal	0.0001	0.0009	NS	0.0003	0.0194	NS
Nonanal	0.0007	0.0084	NS	0.0120	0.0389	0.0476
Decanal	0.0001	0.0114	NS	0.0001	0.0059	0.0048
2-Decenal	0.0001	0.0048	NS	0.0002	NS	NS
t,c-2,4-Decadienal	0.0065	NS	NS	NS	NS	NS
t,t-2,4-Decadienal	0.0001	NS	NS	NS	NS	NS
2-Hexen-1-ol	0.0012	NS	NS	NS	NS	NS
1-Octen-3-ol	NS	0.0337	NS	NS	NS	NS
2-Heptanone	0.0005	NS	0.0342	NS	NS	NS
2-Octanone	0.0002	NS	NS	NS	NS	NS
2,3-Octanedione	NS	0.0075	NS	NS	NS	NS
3-Octen-2-one	0.0015	NS	NS	NS	NS	NS
2-Nonanone	0.0001	0.0005	NS	0.0001	0.0010	0.0025
2-Pentylfuran	0.0001	NS	NS	0.0036	NS	NS

NS = Mean peak contents were not significantly differently ($p > 0.05$)

Table IV. Significance of treatment effects on contents of miscellaneous compounds, as determined by repeated measures analysis of variance

Compound	Main Effects			Interactions			
	Time	Size	Year	Time × Size	Time × Year	Time × Size	Time × Size × Year
Benzaldehyde	0.0395	0.0090	0.0348	0.0261	NS	NS	NS
Limonene	0.0001	NS	NS	NS	NS	NS	NS
Phenylacetaldehyde	NS	NS	NS	NS	NS	NS	NS
Naphthalene	NS	NS	0.0400	NS	NS	NS	NS
Vinylphenol	NS	NS	NS	NS	NS	NS	0.0974

NS = Mean peak contents were not significantly different ($p > 0.05$)

Table V. Influence of storage time on the content of heterocyclic compounds (ng/g)[a]

Compound	RI	Storage Time (weeks)					Model
		0	2	4	6	8	12	
2,5-DiMethylpyrazine[b]	909	300.86	135.98	84.80	49.79	42.18	26.09	Cubic
Ethylpyrazine[b]	914	12.76	5.73	3.22	1.33	1.25	0.11	Cubic
2,3-Dimethylpyrazine[b]	917	8.86	3.91	2.08	0.73	0.59	0.00	Cubic
2-Ethyl-6-methylpyrazine	997	26.71	16.55	22.27	12.44	27.08	19.30	NS
2-Ethyl-5-methylpyrazine[b]	1000	129.50	88.62	66.85	54.88	48.19	58.88	Quadratic
Trimethylpyrazine[b]	1001	115.76	76.70	57.56	44.44	39.85	21.25	Quadratic
2-Ethyl-3,6-Dimethylpyrazine[b]	1080	102.53	84.29	72.70	60.94	55.43	33.51	Linear
2-Ethyl-3,5-Dimethylpyrazine	1085	9.15	7.64	6.82	5.26	5.02	7.33	NS
2,5-Diethylpyrazine[b]	1087	17.11	15.57	15.10	12.98	11.99	8.42	Linear
2-Isobutyl-3-methylpyrazine[b]	1140	7.31	4.39	3.76	3.82	3.52	4.24	Quadratic
2,3-Diethyl-5-methylpyrazine	1157	4.76	4.48	4.08	3.38	3.17	1.56	NS
2-Methylfurfural[b]	965	4.27	3.53	2.25	1.86	1.77	7.26	Cubic

[a] Data for 2 seed sizes (#1, Jumbo/medium (JM)), 2 crop years (1988, 1989), and 2 replications have been pooled. Interactions between storage time, seed size, and crop year were not significant ($p>0.05$).
[b] Storage time had a significant effect ($p<0.05$) on the content of compounds as determined by repeated measures analysis of variance.
RI = Retention index

peanutty flavor would be attributed to factors other than the content of these heterocyclic compounds.

Flavor Intensity

Roasted Peanutty. Storage time significantly affected the roasted peanutty flavor, but there was a significant crop year by storage interaction, which means the change over storage time is not consistent across lots of peanuts. For both crop years roasted peanutty intensity decreased during storage, although the 1989 crop year had a greater initial drop than the 1988 crop year. The 1988 crop year experienced a linear decrease during storage while the 1989 crop year decreased with a quadratic trend (Table VI), which accounts for the interaction effect between storage time and crop year. The change in means between 0 and 4 weeks was significant for both crop years (p=0.01), but between 4 and 8 weeks, and 8 and 12 weeks there were no significant differences. When data for both crop years are combined there is a significant increase between 4 and 8 weeks (p=0.05) and 8 and 12 weeks (p=0.04), individually these increases were not significant. Each storage time was significantly different than the 0 week for both crop years with p<0.02 for each comparison. It is not known if roasted peanutty flavor decreases or if it is masked by the increase in painty flavor. Since the pyrazines decreased during storage, the authors feel that roasted peanutty flavor actually decreases and masking is minimal. Size had a significant effect on roasted peanutty flavor intensity. Overall the #1 size peanuts had a lower intensity than the jumbo/medium size. The general trend was that roasted peanutty flavor decreased with storage and the trend of the decrease depends on crop year and possibly seed size or maturity.

Sweet Aromatic. Sweet aromatic displayed a significant crop year by storage interaction and a significant storage time main effect. Sweet aromatic increased slightly during initial storage for 1988, but the 1989 crop year decreased continuously. Intensity decreased between 8 and 12 weeks for the samples from both crop years (p=0.01) and across storage time, ignoring crop year (p=0.001). Observations of the sweet aromatic intensity data indicate no consistent trends during storage. Variations caused by crop year and seed size or maturity tend to over shadow any slight trends that may be present.

Woody/Hulls/Skins. Woody/hulls/skins had a significant storage by size by crop year interaction effect as well as a significant storage effect. Intensity of woody/hulls/skins had a decreasing trend across storage time for the 1989 #1's and the 1988 jumbo/mediums. The 1988 #1's and the 1989 jumbo/mediums exhibited similar quadratic trend with the lowest intensities occurring at 4 and 8 weeks. Storage effect on woody/hulls/skins intensity was not obvious or easily interpreted.

Lipid Oxidation Products. Lipids compose approximately 52% of the dry weight of peanuts. Over 75% of the fatty acids present are unsaturated, with oleic and linoleic acids making up 48% and 31% of the total fatty acid composition, respectively. Linolenic acid is not present in significant amounts (*17*). Roasting initiates lipid oxidation and the formation of carbonyl compounds in peanuts (*2*).

Table VI. Influence of storage time, commercial seed size and the interactions on the mean intensities of descriptive flavor attributes

Descriptor	Storage Time (weeks)			
	0	4	8	12
Storage time effect				
Cardboardy[ah]	0.5	0.6	0.8	0.8
Painty[ahi]	0.4	0.8	0.9	1.4
Roasted Peanutty[chi]	5.0	4.3	4.0	3.7
Sweet Aromatic[ch]	2.7	2.7	2.8	2.4
Woody/Hulls/Skins[ch]	2.1	2.0	1.8	2.0
Storage time by seed size effect				
Fruity/Fermented[bfij]				
JM	0.2	0.5	0.5	0.4
#1	1.1	1.0	1.0	0.7
Storage time by crop year effect				
Roasted Peanutty[dhi]				
1988	4.5	4.3	4.1	3.6
1989	5.5	4.4	4.0	3.8
Sweet Aromatic[dh]				
1988	2.5	2.8	2.9	2.3
1989	2.9	2.6	2.6	2.4
Storage time by seed size by crop year				
Woody/Hulls/Skins[ehi]				
1988				
JM	2.2	2.2	1.9	1.9
#1	2.1	1.8	1.7	2.0
1989				
JM	2.0	1.8	1.9	2.0
#1	2.0	2.3	1.9	2.0

Seed size effect	Commercial Seed Size	
	Jumbo/Medium	No. 1's
Painty[gi]	0.6	1.1
Roasted Peanutty[gi]	4.6	4.0

[a] Data for 2 seed sizes (#1, Jumbo\medium (JM)), 2 crop years (1988, 1989), and 2 replications have been pooled. Interactions between storage time, seed size, and/or crop year were not significant ($p > 0.05$).
[b] Data for 2 crop years (1988, 1989) and 2 replications have been pooled.

(Table VI, continued)

c Data for 2 seed sizes (#1, JM), 2 crop years (1988, 1989), and 2 replications have been pooled. Interactions that were significant were considered separately.
d Data for 2 seed sizes (#1, JM) and 2 replications have been pooled. Significant interactions ($p<0.05$) between storage time and crop year dictate that each factor be considered separately.
e Data reported is for 2 replications. Significant interactions ($p<0.05$) between storage time, seed size, and crop year dictate that each factor be considered separately.
f Data for 2 crop years (1988, 1989) and 2 replications have been pooled. Since Fruity/fermented flavor was not significantly affected by storage, other significant main effects and interactions were ignored.
g Data for 4 storage times, 2 crop years, and 2 replications have been pooled.
h Storage time had a significant effect ($p<0.05$) on the flavor intensity as determined by repeated measures analysis of variance.
i Seed size had a significant effect ($p<0.05$) on the flavor intensity as determined by repeated measures analysis of variance.
j Crop year had a significant effect ($p<0.05$) on the flavor intensity of compounds as determined by repeated measures analysis of variance.

Lipid oxidation continues during storage at elevated temperatures, resulting in further increases in the content of the aliphatic aldehydes, ketones, alcohols, and other products of lipid oxidation. Hexanal, heptanal, 2-heptenal, octanal, 2-octenal, 2-decenal, 2,4-decadienal, 1-octen-3-ol, and 2-pentylfuran have been identified as products of linoleic acid oxidation. Heptanal, octanal, and 2-decenal, in addition to nonanal and decanal are also products of oleic acid oxidation. These compounds typically have cardboardy, painty, rancid, and oxidized fat flavor notes (*4,18*).

In this study, the rate of increase in the content of lipid oxidation products, as a function of storage time, varied depending on the specific compound and peanut seed size (Table VII). During the 12-week storage period, the contents of hexanal, heptanal, 2-heptenal, octanal, 2-octenal, nonanal, decanal, 2-decenal, 2-hexen-1-ol, 2-heptanone, 2-octanone, 3-octen-2-one, 2-nonanone, and 2-pentylfuran increased significantly ($p<0.05$) following linear, quadratic, or cubic trends. In contrast to the heterocyclic compounds, the greatest increases in the content of these compounds occurred at the end of the storage time, with the concentrations of the lipid oxidation compounds being less susceptible to change during the initial storage period.

The content of the 2,4-decadienal isomers decreased, rather than increased during the storage period. These compounds may undergo further oxidation and degradation to more stable, shorter chain compounds, such as hexanal and 2-octenal (*19-21*).

As peanuts mature, there is a decrease in the mole percent of linoleic acid and an increase in the mole percent of oleic acid (*8*). The differences in the distribution of fatty acids with maturity would contribute to the observed significant effects of seed size and storage time by seed size interactions on the content of a majority of the lipid oxidation products in the roasted peanuts. The contents of heptanal, 2-heptenal, octanal, 2-octenal, nonanal, 2-decenal, 1-octen-3-ol, 2,3-octanedione, and 2-nonanone were significantly ($p<0.05$) higher in the #1 peanuts than in the jumbos/medium. The content of these compounds was not significantly different ($p<0.05$) with respect to seed size initially. However, during storage, the content of these compounds increased at a greater rate in the #1 peanuts than in the jumbo/medium peanuts. The higher percentage of linoleic acid in the total fatty acids of the #1 peanuts would contribute to a greater susceptibility of these peanuts to lipid oxidation and thus a greater rate of production of off-flavor compounds. These results concur with those of Sanders et al. (*8*), which reported an increase in the oxidative oven stability of pressed peanut oils with an increase in peanut maturity. The differences in fatty acid distribution, in addition to possible variability in the content of antioxidants and prooxidants, may contribute to the increased susceptibility of the #1 peanuts to oxidation.

Regression analysis of the increase in the content of the lipid oxidation products with storage time indicated that, in general, the power on the 'storage time' term was higher for the jumbo/medium peanuts in comparison to the #1 peanuts. Whether this observation bears any significant relationship to the difference in fatty acid content of the jumbo/medium and #1 peanuts would require further research.

Flavors Related to Lipid Oxidation

Painty. Painty intensity means were significantly different across storage times with no significant interactions. The change over time exhibited an overall linear

Table VII. Influence of storage time on the content of lipid oxidation products (ng/g)

Compound	RI	Storage Time (weeks)						Model
		0	2	4	6	8	12	
Hexanal[a,f]	809							
#1		43.39	50.15	43.10	60.14	91.76	147.78	Linear
JM		46.42	32.12	20.99	46.52	44.34	72.73	Quadratic
Heptanal[a,f,g,h]	902							
#1		10.38	9.09	11.59	10.07	16.03	34.71	Quadratic
JM		8.75	5.93	3.20	7.28	8.64	11.06	Cubic
2-Heptenal[b,f,g]	958	44.06	41.24	36.06	40.05	51.18	61.47	Quadratic
Octanal[a,f,g]	1004							
#1		21.94	32.52	38.16	42.75	55.20	121.66	Cubic
JM		21.98	22.46	19.70	33.71	33.84	27.61	No trend
2-Octenal[f,g]	1059							
#1[a]		29.15	25.25	34.52	44.11	54.89	126.95	Quadratic
JM		25.10	13.89	10.74	21.19	22.40	23.64	Cubic
1988[c]		22.18	18.70	22.51	30.57	39.60	81.60	Linear
1989		32.07	20.44	22.76	34.73	37.68	62.56	Linear
Nonanal[d,f,g]	1105							
#1 - 1988		58.91	92.46	105.46	111.67	179.01	278.00	Quadratic
JM - 1988		52.59	60.66	110.62	86.16	99.54	121.74	Linear
#1 - 1989		52.34	125.03	112.98	112.15	124.78	202.13	Cubic
JM - 1989		85.32	75.56	61.47	118.42	121.59	115.56	Linear

Continued on next page

(Table VII, Continued)

Compound	RI	Storage Time (weeks)					Model
		0	2	4	6	8	12	
Decanal[d,g]	1208							
#1 - 1988		2.08	3.48	6.54	6.55	11.66	27.22	Quadratic
JM - 1988		1.32	4.48	11.26	4.32	5.18	8.90	Cubic
#1 - 1989		1.64	5.11	8.08	6.78	9.44	18.17	Linear
JM - 1989		2.41	2.72	2.74	7.44	8.60	7.84	Linear
2-Decenal[a,f,g]	1260							
#1		4.36	6.88	11.17	9.22	13.56	34.04	Linear
JM		3.46	4.21	5.36	3.80	5.96	6.02	Quadratic
t,c-2,4-Decadienal[b,f]	1296	21.92	4.90	5.60	4.37	3.75	6.28	Cubic
t,t-2,4-Decadienal[b,f]	1320	154.30	81.79	94.10	69.04	59.94	55.15	Quadratic
2-Hexen-1-ol[b,f]	872	4.71	3.84	1.19	1.01	4.27	6.56	Quadratic
1-Octen-3-ol[e,g]	981							
#1		19.65	22.05	25.55	21.45	26.06	31.89	Linear
JM		18.17	15.35	10.82	18.81	24.52	19.50	No trend
2-Heptanone[b,f,h]	891	4.89	3.76	3.77	4.95	6.77	12.38	Quadratic
2-Octanone[b,f]	993	7.18	11.55	19.88	15.27	26.38	33.46	Linear
2,3-Octanedione[e,g]	984							
#1		12.64	15.39	16.20	13.32	11.88	15.79	No trend
JM		10.29	11.84	8.89	12.42	12.75	12.91	Linear
3-Octen-2-one[b,f]	1043	11.10	85.02	120.52	168.31	177.28	161.26	Quadratic

	RI							Regression
2-Nonanone[d,f,g]	1094							
#1 - 1988		1.38	8.46	25.68	39.42	62.24	197.02	Cubic
JM - 1988		3.54	3.34	7.63	9.98	13.30	23.63	Quadratic
#1 - 1989		2.48	14.04	33.12	38.73	54.48	122.50	Cubic
JM - 1989		3.16	7.10	8.70	23.34	26.30	23.75	Linear
2-Pentylfuran[a,f,g]	991							
#1		10.02	26.94	47.39	43.79	44.62	42.37	Quadratic
JM		6.79	11.88	9.96	30.69	46.38	48.63	Cubic

[a] Data for 2 crop years (1988, 1989) and 2 replications have been pooled. Significant interactions (p<0.05) between storage time and seed size dictate that each factor be considered separately.

[b] Data for 2 seed sizes (#1, Jumbo/Medium (JM)), 2 crop years (1988, 1989), and 2 replications have been pooled. Interactions between storage time, seed size, and crop year were not significant (p>0.05).

[c] Data for 2 seed sizes (#1, Jumbo/Medium (JM)) and 2 replications have been pooled. Significant interactions (p<0.05) between storage time and crop year dictate that each factor be considered separately.

[d] Data reported is for 2 replications. Significant interactions (p<0.05) between storage time, seed size, and crop year dictate that each factor be considered separately.

[e] Data for 2 crop years (1988, 1989) and 2 replications have been pooled. Interactions between storage time, seed size, and crop year were not significant (p>0.05).

[f] Storage time had a significant effect (p<0.05) on the content of compounds as determined by repeated measures analysis of variance.

[g] Seed size had a significant effect (p<0.05) on the content of compounds as determined by repeated measures analysis of variance.

[h] Crop year had a significant effect (p<0.05) on the content of compounds as determined by repeated measures analysis of variance.

RI = Retention Index

increase. Painty flavor continued to increase across all storage times and was still increasing at 12 weeks. There was a significant difference between means at 0 and 4 weeks (p=0.01), but between 4 and 8 weeks there was no significant difference (p=0.31). The difference between means at 8 and 12 weeks was significant (p=0.01). In this data set painty flavor intensity increased initially, leveled off between 4 and 8 weeks, and then increased again. Crop year did not have a significant effect, but during years with adverse growing conditions, such as drought, peanuts are more susceptible to off-flavor development. Mean intensity of painty was significantly different between sizes, which is related to maturity (9). The #1's had a more intense painty flavor than the jumbo/medium's. The #1's seem to be more susceptible to the development of painty flavor than the jumbo/medium's. As mentioned earlier oleic/linoleic acid (O/L) ratio plays an important role in susceptibility of peanuts to storage off-flavor and O/L ratio is directly related to maturity.

Cardboardy. All peanut samples became more cardboardy as storage time progressed, regardless of crop year or seed size. Between 0 and 4 weeks (p=0.22) and 0 and 8 weeks (p=0.07) means were not different, but between 0 and 12 weeks they were significantly different (p=0.01). The differences between 4 and 8 weeks (p=0.25) and 8 and 12 weeks (p=0.46) were not significant either. Even though there was a steady increase, the change was not significant until 12 weeks.

Miscellaneous Compounds and Flavors

Gas Chromatograph. In addition to the heterocyclic compounds and lipid oxidation products, the effects of storage and seed size on the content of several other volatile compounds were determined (Table VIII). Phenylacetaldehyde is formed during roasting through the Strecker degradation of free and peptide-bound phenylalanine present in raw peanuts (22). Phenylacetaldehyde contributes to the sweet background (or bouquet) of roasted peanut aroma (23). Storage time and seed size had no significant effect on the content of phenylacetaldehyde.

Storage time, seed size, and the interaction between these two factors had a significant effect on the content of benzaldehyde (Table IV). Several mechanisms have been suggested for the formation of benzaldehyde, although which mechanism occurs in roasted peanuts is questionable based on the results of this study. Benzaldehyde is thought to be a Strecker degradation product of phenylglycine (24) formed during roasting. However, the presence of phenylglycine in peanuts has not been reported (6,25). An alternative mechanism involves the thermal oxidation of linoleic acid (26). Although this pathway is feasible, it would not explain why the content of benzaldehyde is higher in the jumbo/medium peanuts, rather than in the #1 peanuts, as would be expected due to the higher percentage of linoleic acid in the more immature #1 peanuts (8).

Limonene content increased significantly (p<0.05) during the 12 week storage period. Limonene is a terpene biosynthesized through the isoprenoid pathway. However, the mechanism for the increased limonene content during storage is unknown.

Table VIII. Influence of storage time on the content of miscellaneous volatile compounds (ng/g)

Compound	RI	Storage Time (weeks)					Model
		0	2	4	6	8	12	
Benzaldehyde[acde]	962							
#1		24.98	28.82	30.34	19.10	20.71	24.83	Cubic
JM		48.74	39.17	19.35	31.15	38.46	28.78	Cubic
Limonene[ac]	1030	6.81	10.08	9.98	15.91	16.06	29.34	Quadratic
Phenylacetaldehyde[b]	1047	14.24	15.68	18.08	27.11	22.96	34.10	NS
Naphthalene[be]	1177	5.31	5.54	8.90	4.74	5.16	3.82	NS
Vinylphenol[b]	1222	403.18	440.95	549.72	455.53	374.11	322.21	NS

[a] Data for 2 crop years (1988, 1989) and 2 replications have been pooled. Significant interactions ($p<0.05$) between storage time and seed size dictate that each factor be considered separately.
[b] Data for 2 seed sizes (#1, Jumbo/Medium (JM)), 2 crop years (1988, 1989), and 2 replications have been pooled. Interactions between storage time, seed size, and crop year were not significant ($p>0.05$).
[c] Storage time had a significant effect ($p<0.05$) on the content of compounds as determined by repeated measures analysis of variance.
[d] Seed size had a significant effect ($p<0.05$) on the content of compounds as determined by repeated measures analysis of variance.
[e] Crop year had a significant effect ($p<0.05$) on the content of compounds as determined by repeated measures analysis of variance.
RI = Retention Index

The thermal degradation of 4-hydroxycinnamic acid esters results in the formation of vinylphenol (27). Vinylphenol was identified as a major component in the volatile profile of roasted peanuts, however, storage time and seed size did not have a significant effect (p>0.05) on its content.

Fruity/Fermented. Although there was a significant size by storage time interaction, the main effect of storage time did not significantly affect fruity/fermented flavor. The mean intensities within size varied significantly over storage time (Table VI). It is documented in the literature (9) that size has an effect on intensity of fruity/fermented flavor. Based on the seed size by storage time interaction, the less mature #1 size peanuts decreased in intensity over time, while medium peanuts showed little change in fruity/fermented. It is uncertain if fruity/fermented was really decreasing in intensity or if it was being masked by the increase in painty and cardboardy flavors.

As was anticipated from the results of Sanders et al. (9) size had an effect on fruity/fermented. The #1's had a more intense fruity/fermented. There was a significant size by crop year interaction (Table I). The #1's from the 1988 crop year were more intense than those from the 1989 crop year or the Jumbo/mediums from either year. The development of fruity/fermented flavor in peanuts can be attributed to improper curing or drying, freeze damage, or drought condition (10,28).

Comparison of flavor attributes and chemical compounds

Many of the heterocyclic compounds decreased during storage as did roasted peanutty and sweet aromatic. No heterocyclic compounds that had regression patterns comparable to these two flavors, though. When the data for the two crop years were combined the trends for roasted peanutty and sweet aromatic were linear which was similar to 2-ethyl-3,6-dimethylpyrazine and 2,5-dimethylpyrazine. 2-Ethyl-3,6-dimethylpyrazine has a roasted/nutty, coffee, cocoa, woody or burnt aroma. 2,5-Dimethylpyrazine has a similar aroma, but may tend to have more roast beef or grilled chicken aroma. These compounds likely add to the peanutty bouquet but are not the sole contributing compounds to roasted peanutty or sweet aromatic flavor (29).

Woody/hulls/skins displayed a quadratic trend similar to 2-ethyl-5-methylpyrazine and 2-isobutyl-3-methylpyrazine. 2-Ethyl-5-methylpyrazine has a roasted, nutty, grassy aroma. 2-Isobutyl-3-methylpyrazine has a green, bell pepper aroma, or a roasted hazel nut flavor (29). According to Crippen et al. (30) woody/hulls/skins correlates highly with $\underline{N}$-methylpyrrole, which was not observed in this experiment.

Many of the lipid oxidation products increased during storage along with painty and cardboardy flavor. 2-Octanone expressed a linear increase like painty and cardboardy. 2-Decenal was linear in the #1 size peanuts, but not in the jumbo/medium's. Nonanal was linear for the jumbo/medium's but not the #1's. Decanal was linear for the 1989 crop year, but not the 1988 crop year. 2-Nonanone increased over storage time, but sometimes it displayed a quadratic or cubic curve. All of these compounds are related to lipid oxidation reactions, but their contribution to painty and cardboardy flavors is uncertain. 2-Octanone has a herbaceous, spicy, butter, fruity, musty aroma; 2-decenal has an orange, floral, green, fatty, tallow aroma; nonanal has a citrus, floral, waxy, tallow, green aroma; decanal has a waxy, floral, citrus, tallowy aroma; and 2-nonanone has a fruity, floral, musty aroma *(29,31-34)*. These compounds do not necessarily have cardboardy or painty characteristics. The other lipid oxidation products displayed trends that were less obviously related to painty and cardboardy flavors. 2-Hexen-1-ol and 2-heptanone displayed trends similar to woody/hulls/skins. 2-hexen-1-ol has a green grass aroma and 2-heptanone has a fruity, spicy, cinnamon aroma *(34)*.

Summary of Findings

Storage of roasted peanuts at elevated temperatures resulted in significant changes in the volatile profile, and five of the six flavor attributes. Only fruity/fermented was not affected. The pyrazines, which may contribute to the desirable roasted peanut flavor characteristics, decreased over the 12 week storage period. In contrast, aliphatic aldehyde, alcohol, and ketone contents increased. These compounds are products of lipid oxidation reactions.

Commercial seed size had a significant effect on the formation of the lipid oxidation products during storage. Aliphatic aldehyde, alcohol, and ketone contents were significantly higher in the #1 peanuts than in the jumbo/medium peanuts. Immature peanuts have less potential for the development of desirable roasted peanut flavor and a greater potential for off-flavor formation *(9)*. Because the percentage of immature peanuts in a given size class decreases as the size of the peanuts increases, the larger peanuts have less potential for undesirable peanut flavor development *(9)*. In comparing the effects of size on the contents of the pyrazines and lipid oxidation products, it was noted that size did not have a significant effect on the content of the pyrazines, but size did have a significant effect on the content of the lipid degradation compounds or the pyrazines. Since size does not affect the pyrazine content, but it does roasted peanutty flavor, this indicates that roasted peanutty flavor is more complicated than this combination of pyrazines alone.

Acknowledgements. The authors wish to acknowledge Bryan Vinyard for his statistical assistance with the data analysis and interpretation. Carolyn Vinnett, the sensory panel leader and the other sensory laboratory personnel, as well as the sensory panel members were instrumental in making the sensory data possible.

Literature Cited

1. Johnson, B.R.; Waller, G.R.; Burlingame, A.L. *J. Agric. Food Chem.* **1971**, *19*, 1020-1024.
2. Buckholz, L.L., Jr.; Daun, H. In *Quality of Selected Fruits and Vegetables of North America*; Teranishi, R., Barrera-Benitez, H., Eds.; ACS Symp. Ser. 170, American Chemical Society: Washington, DC, **1981**; pp 163-181.
3. Litman, I.; Numrych, S. In *Lipids as a Source of Flavor*; M.K. Supran, M.K., Ed., ACS Symp. Ser. 75, American Chemical Society: Washington, DC, **1978**; pp 1-17.
4. Forss, D.A. *Prog. Chem. Fats Other Lipids* **1972**, *13(4)*, 177-258.

5. Gardner, H.W. In *Xenobiotics in Foods and Feeds*; Finley, J.W., Schwass, D.E., Eds.; ACS Symp. Ser. 234; American Chemical Society; Washington, DC, **1983**; pp 63-84.
6. Pattee, H.E.; Young, C.T.; Giesbrecht, F.G. *Peanut Sci.* **1981**, *8*, 113-116.
7. Pattee, H.E.; Pearson, J.L.; Young, C.T.; Giesbrecht, F.G. *J.Food Sci.* **1982**, *47*, 455-460.
8. Sanders, T.H.; Lansden, J.A.; Greene, R.L.; Drexler, J.S.; Williams, E.J. *Peanut Sci.* **1982**, *9*, 20-23.
9. Sanders, T.H.; Blankenship, P.D.; Vercellotti, J.R.; Crippen, K.L. *Peanut Sci.* **1990**, *17*, 85-89.
10. Sanders, T.H.; Vercellotti, J.R.; Crippen, K.L.; Civille, G.V. *J. Food Sci.* **1989**. *54*, 475-477.
11. Meilgaard, M.C.; Civille, G.V. and Carr, B.T. *Sensory Evaluation Techniques*, Vol. I, CRC Press, Inc. Boca Raton, FL. **1987**; pp 1-23.
12. Johnsen, P.B.; Civille, G.V.; Vercellotti, J.R.; Sanders, T.H.; Dus, C.A. *J. Sensory Stud.* **1988**, *3*, 9-18.
13. SAS Institute Inc. *SAS/STAT User's Guide*, Release 6.03 Ed., Cary, NC, **1988**; pp 576-578, 602-609.
14. Milliken, G.A. and Johnson, D.E. *Analysis of Messy Data - Volume I: Designed Experiments*, Van Nostrand Reinhold Company, New York, **1984**; pp 323-326.
15. Mason, M.E.; Johnson, B.; Hamming, M. *J. Agric. Food Chem.* **1966**, *14*, 454-460.
16. Buckholz, L.L., Jr.; Daun, H.; Trout, R. *J. Food Sci.* **1980**, *45*, 547-554.
17. Conkerton, E.J. and St. Angelo, A.J. In *CRC Handbook of Processing and Utilization in Agriculture. Vol. II: Part 2 Plant Products*; Wolff, I.A., Ed.; CRC Press, Inc., Boca Raton, FL., **1982**; pp 157-185.
18. Badings, H.T. *Ned. Melk Zuiveltijdschr.* **1970**, *24*, 147. Cited in: Grosch, W. In *Food Flavours*; Morton, I.D., MacLeod, A.J., Eds.; Elsevier Scientific Publishing: New York, **1982**; pp 325-398.
19. Grosch, W. In *Food Flavours*; Morton, I.D., MacLeod, A.J., Eds.; Elsevier Scientific Publishing: New York, **1982**; pp 325-398. 20. Frankel, E.N. In *Recent Advances in the Chemistry of Meat*; Bailey, A.J., Ed.; The Royal Society of Chemistry: London, **1984**; pp 87-118.
21. Josephson, D.B.; Lindsay, R.C. *J. Food Sci.* **1987**, *52*, 1186-1190, 1218.
22. Mason, M.E.; Newell, J.A.; Johnson, B.R.; Koehler, P.E.; Waller, G.R. *J. Agric. Food Chem.* **1969**, *17*, 728-732.
23. Mason, M.E.; Johnson, B.; Hamming, M.C. *J. Agric. Food Chem.* **1967**, *15*, 66-73.
24. Heath, H.B. *Source Book of Flavors*, AVI Publishing Company, Inc.: Westport, CT, **1981**; pp 78-147.
25. Oupadissakoon, C.; Young, C.T.; Giesbrecht, F.G.; Perry, A. *Peanut Sci.* **1980**, *7*, 61-67.
26. Kawada, T.; Krishnamurthy, R.G.; Mookherjee, B.D.; Chang, S.S. *J. Am. Oil Chem. Soc.* **1967**, *44*, 131-135.
27. Walradt, J.P.; Pittet, A.O.; Kinlin, T.E.; Muralidhara, R.; Sanderson, A. *J. Agric. Food Chem.* **1971**, *19*, 972-979.

28. Crippen, K.L.; Vercellotti, J.R.; Butler, J.L.; Williams, E.J.; Clary, B.L.; Wright, F.S. and Porter, D.M. *Preceedings of American Peanut Research and Education Society, Inc.* **1989**, *21*, 36.
29. Fors, S. In *The Maillard Reaction in Foods and Nutrition.* G.R. . Waller, G.R. and Feather, M.S., Eds.; American chemical Society, Symposium Series 215, Washington, DC. **1983**; pp 185-286.
30. Crippen, K.L.; Vercellotti, J.R.; Lovegren, N.V.; Sanders. T.H. *Proceedings of the 7th International Flavor Conference*, Pythagorion, Samos, Greece, June 26-28, **1992**. Ed. George Charalambous. In Press.
31. Grosch, W. In *Autoxidation of Unsaturated Lipids.* Ed., H.W.-S. Chan. Academic Press, London, U.K. **1987**; pp 95-140.
32. Gasser, U.; Grosch, W. *Z. Lebensm. Unters Forsch* **1988**, *186*, 489-494.
33. Gasser, U.; Grosch, W. *Z. Lebensm. Unters Forsch* **1990**, *190*, 3-8.
34. Anonymous. *Flavors and Fragrances Catalogue.* Aldrich Chemical Co., Milwaukee, WI. **1991**; pp 1-51.

RECEIVED February 19, 1992

Chapter 20

Changes in Lipid Oxidation During Cooking of Refrigerated Minced Channel Catfish Muscle

M. C. Erickson

Department of Food Science and Technology, Food Safety and Quality Enhancement Laboratory, University of Georgia, Georgia Agricultural Experiment Station, Griffin, GA 30223

> Refrigerated storage of minced channel catfish affected oxidative changes induced by cooking. Following refrigeration for 0, 2, 5 or 7 days, samples were analyzed before and after baking for thiobarbituric acid-reactive substances (TBA-RS), fluorescent pigment and tocopherol content. Larger increases in TBA-RS were found after cooking 2-day refrigerated samples than fresh samples, whereas smaller increases were found in 5 and 7-day samples. In contrast, the largest increases in fluorescent pigment content after cooking were found in the 5-day refrigerated samples. Loss in gamma-tocopherol upon cooking remained fairly constant (15%) whereas losses of alpha-tocopherol were greater in the 2- and 5- day refrigerated samples (40%) than in the 7-day refrigerated sample (14%).

Oxidative decomposition of meat lipids during cooking is a recognized event (*1-11*) and is responsible for many desirable flavor characteristics. Unfortunately, due to the autocatalytic nature of lipid oxidation, cooking also accentuates undesirable flavors in products which have experienced lipid oxidation during storage in the raw state. While storage of raw product prior to cooking alters the oxidative response during cooking, both promotion and inhibition of lipid oxidation during cooking has been noted (*1-3*). To further explore the lipid oxidative response to cooking, lipid oxidation was monitored by a variety of methods in cooked, minced channel catfish which had previously been refrigerated for various periods of time.

Materials and Methods

Samples. Live channel catfish (*Ictalurus punctatus*) obtained locally were stunned, deheaded, gutted, skinned and filleted. The fillets from each fish were pooled and minced to obtain a homogeneous sample. The minced samples were distributed into cups or tubes and heated for five minutes at 177°C.

TBA-RS. Minced channel catfish muscle tissue (10.0 gm) was placed in a shallow cup and spread in a thin layer. Both cooked and uncooked meat were scraped from the cup and analyzed for thiobarbituric acid reactive substances (TBA-RS) by the distillation procedure described by Rhee (*12*). Results were expressed in terms of mg malonaldehyde (MA)/g tissue.

Lipid Extraction. Minced channel catfish muscle tissue (1.0 gm) was placed in 16 x 125 mm screw-cap test tubes and spread in a thin layer. Lipids from both cooked and uncooked samples were extracted by chloroform:methanol (2:1) as described by Erickson (*13*).

Fluorescent Pigments. Fluorescent pigments were determined on 10.0 ml of a lipid extract (25 ml total) which had been washed with 2.50 ml of 0.88% KCl. Diluted samples of both the aqueous and organic layers were measured in a Turner fluorometer, Model 112, using a quinine sulfate standard (1 x 10^{-8} M) set equal to 100 fluorescence units (FU).

Fatty Acid Esterification. Total lipid extracts were esterified in the presence of an internal standard, heptadecanoic acid, using 4% H_2SO_4 in methanol as described by Erickson and Selivonchick (*14*). In addition, both cooked and uncooked tissue samples (1.0 gm), distributed into 16 x 125 mm screw-cap test tubes, were freeze-dried, followed by direct esterification using 4% H_2SO_4 in methanol. A Hewlett-Packard 5790A Series gas chromatograph equipped with a glass capillary column (J & W DB-225; 30 m x 0.25 mm, 0.15 μm film) was used for separation of the esterified fatty acids.

Tocopherol. Under saponification conditions described by Erickson (*13*), extracted tocopherols were analyzed using the reverse-phase high performance liquid chromatographic conditions designated by Vatassery and Smith (*15*). The chromatographic system consisted of a Micromeritics 752 Gradient Programmer, Micromeritics 750 Solvent Delivery System, and a Brinkmann 656 Electrochemical detector.

Statistical Analyses. Analysis of variance and Fisher's least significant difference test were used to analyze the data for statistical differences.

Results and Discussion

Response to Cooking of Fresh Product. Level of TBA-RS increased by over 500% after cooking the fresh channel catfish samples (Table I). Previously, the

Table I. Oxidative Response to Cooking of Fresh Minced Channel Catfish[a]

	TBA-RS (mg MA/g)	Extraction/ Esterification (% PUFA)	Direct Esterification (% PUFA)
Raw Fish	0.54 ± 0.32	22.0 ± 0.3	22.2 ± 0.3
Cooked Fish	2.91 ± 0.84	21.9 ± 0.2	22.1 ± 0.2

[a] Reported values are expressed as mean ± standard deviation based on five observations.

TBA number increased 406% for baked minced carp (*4*), 190% for cooked chicken (*5*) and 405% for cooked beef (*6*). Fatty acid composition of the tissue, TBA analytical methodology, adjustments for drip loss, as well as temperature and time of heating must all be considered to account for these variations exhibited by different products in response to cooking. Other measurements of oxidation which have similarly been noted to increase with cooking are the fluorescent pigments (*5,7*), total carbonyls (*8*) and oxysterols (*9*). The increase in lipid oxidation that occurs during cooking may be attributed to increases in nonheme iron content (*16*), disruption of muscle membranes, as well as to the increased catalytic activity encountered at higher temperatures (*17*).

In the present study, the polyunsaturated fatty acids (PUFA) were also monitored before and after cooking to detect if changes occurred in the levels of the oxidative substrate. Two procedures for esterification of catfish samples were employed to determine if differences exist in their sensitivities for detecting losses of PUFA in cooked tissues. In one procedure, the lipids were extracted from tissue by chloroform:methanol (2:1) prior to esterification. In the second procedure, freeze-dried tissue was esterified directly to ensure that all unoxidized fatty acids were released. While direct esterification recovered 52.1 ± 3.1 and 50.0 ± 2.6 mg total FAME/g in raw and cooked tissue, extraction followed by esterification recovered 48.2 ± 2.9 mg total FAME/g in raw tissue but only 32.9 ± 2.2 mg total FAME/g in cooked tissue. When PUFA content was expressed as a percentage of the total fatty acids, however, no preferential losses of PUFA could be detected with either procedure (Table I). The inability to detect losses of PUFA due to cooking, though, is not uncommon in muscle foods with large triglyceride contents (*4,18*). Similar to the situation in chicken breast (*10*), preferential oxidation and loss of phospholipid PUFA may have occurred in catfish, however, the large triglyceride content (43 mg triglyceride/g tissue) would have obscured these losses.

Response to Cooking of Stored Product. Due to the propagative nature of lipid oxidation, the degree of oxidation incurred during cooking should theoretically be dependent on the initial oxidative rate in the raw product. Under such an assumption, an increase in the initial levels of oxidation during refrigerated or frozen storage would lead to much larger quantities of oxidative products being generated during cooking. Support for this line of reasoning was reported in studies examining the sequential treatments of frozen storage and cooking of chicken breast and leg meat (*3*). The results reported by Igene *et al.* (*2*), however, contradict this expectation that frozen raw meat is more susceptible to oxidation during cooking than fresh unfrozen meat. Similarly, when minced channel catfish was stored under refrigeration for varying periods of time prior to cooking, cooking did not lead to significant increases in TBA-RS for those samples stored more than 3 days, whereas the positive response to cooking of samples stored 3 days or less was statistically significant (Table II). Since a similar trend in TBA numbers has been observed for broiler tissues when stored under refrigeration for different times prior to cooking (*1*), the results suggest that additional factors other than the initial oxidation levels must be accounted for when estimating the potential effects of cooking.

To aid in identifying the mechanism for this modification in the response to cooking, another batch of minced catfish muscle tissue was stored for 2, 5 and 7 days of refrigeration. For these samples, fluorescent pigments were also analyzed in addition to TBA-RS in the raw and cooked product (Table III). The levels of TBA-RS identified in these raw and cooked samples were greater than those found previously (Table II) which may be attributed to different lipid and/or antioxidant compositions in the catfish. Similar to the data presented in Table II, though, the data in Table III shows that cooking of 5- and 7-day samples resulted in much smaller increases in TBA-RS than cooking of 2-day fish. Increases in fluorescent

Table II. Response to Cooking of Refrigerated Minced
Channel Catfish as Measured by TBA-RS[a]

Days Stored	mg MA/g tissue		
	Raw Fish	Cooked Fish	Change[b]
1	0.14 ± 0.06 [a]	0.55 ± 0.06 [cd]	0.41
2	0.40 ± 0.11 [bc]	0.89 ± 0.15 [e]	0.49
3	0.24 ± 0.06 [ab]	0.98 ± 0.18 [e]	0.74
4	0.46 ± 0.16 [cd]	0.60 ± 0.18 [d]	0.14
5	0.41 ± 0.10 [bc]	0.48 ± 0.09 [cd]	0.07
6	0.27 ± 0.06 [ab]	0.36 ± 0.07 [bc]	0.09

[a] Reported values are expressed as mean ± standard deviation based on four observations. Raw and cooked values followed by the same superscript letter are not significantly different ($P \leq 0.05$).
[b] TBA-RS of cooked fish minus TBA-RS of raw fish.

pigment content with cooking, on the other hand, were maximal for 5- day refrigerated samples. The decreased levels of TBA-RS found in cooked 5-day samples could therefore be ascribed to the further reaction of the TBA-RS to form fluorescent pigments. The decreased quantities of both TBA-RS and fluorescent pigments found in cooked 7-day samples compared to 2- and 5-day samples, would suggest, however, that an overall decrease in the oxidative reactions had occurred with cooking of 7-day samples.

The tocopherol data for the samples would also support the explanation that the amount of oxidation which had occurred upon cooking 7-day tissue was less than 2-day or 5-day tissue (Table III). While losses of gamma-tocopherol upon cooking were seen to be fairly constant for all refrigerated samples, losses of alpha-tocopherol upon cooking were greater in the 2- and 5-day refrigerated samples than in the 7-day refrigerated samples. Therefore, the decreased loss of tocopherol together with the decreased increase in oxidative products indicates that, overall, the oxidative reactions occurring with cooking have been reduced in 7-day refrigerated samples compared to 2- and 5-day refrigerated samples.

Proposed Mechanisms for Alteration in Response to Cooking. Several possibilities exist to explain the inhibition of lipid oxidation on cooking of samples refrigerated for extended periods of time. These possibilities may be divided into 2 groups: those from microbiological effects and those from chemical/biochemical effects.

Consideration must first be given that with refrigerated storage of muscle tissue, microbial loads will increase. This increase in microbial load in turn leads to increased concentrations of polyamines, compounds which act as antioxidants (19,20). The mechanism of polyamine's antioxidant action may involve a complexation between the polyamine, phospholipid polar head and the iron catalyst, or alternatively, the increased concentrations of polyamines and other amines may affect the catalytic activity of nonheme iron by altering the pH of the muscle system (21). Inhibition of lipid oxidation by microorganisms may also occur through their ability to remove hydroperoxides and carbonyl oxidative products. In stored adipose tissue, for example, several Pseudomonas species have been shown to decrease the concentration of peroxides (22).

Table III. Oxidative Response to Cooking of Refrigerated Minced Channel Catfish[a]

Days Stored	Product	TBA-RS (mg MA/g)	Aqueous FP (FU/g)	Organic FP (FU/g)	Alpha Tocopherol (μg/g)	Gamma Tocopherol (μg/g)
2	Raw	0.66 ± 0.07 [a]	37.73 ± 1.75 [a]	147.74 ± 3.68 [a]	1.78 ± 0.03 [c]	1.58 ± 0.05 [c]
	Cooked	3.54 ± 0.40 [c]	50.53 ± 5.78 [bc]	198.05 ± 5.78 [ac]	1.08 ± 0.05 [a]	1.31 ± 0.04 [a]
	Change[b]	+ 2.88	+ 12.80	+ 50.31	- 0.70	- 0.27
5	Raw	0.56 ± 0.11 [a]	40.43 ± 0.63 [ab]	207.01 ± 32.68 [c]	1.82 ± 0.03 [cd]	1.49 ± 0.06 [b]
	Cooked	1.64 ± 0.21 [b]	98.77 ± 14.29 [d]	280.82 ± 26.53 [d]	1.07 ± 0.07 [a]	1.28 ± 0.03 [a]
	Change	+ 1.08	+ 58.34	+ 73.81	- 0.75	- 0.21
7	Raw	0.64 ± 0.26 [a]	44.49 ± 1.10 [abc]	150.03 ± 9.78 [ab]	1.86 ± 0.03 [d]	1.86 ± 0.02 [d]
	Cooked	1.70 ± 0.28 [b]	55.27 ± 3.12 [c]	187.13 ± 9.20 [abc]	1.60 ± 0.03 [b]	1.62 ± 0.04 [c]
	Change	+ 1.06	+ 10.78	+ 37.10	- 0.26	- 0.24

[a] Reported values are expressed as mean ± standard deviation. All measurements with the exception of TBA/cooked were conducted on triplicate samples. TBA/cooked measurements were conducted on quadruplicate samples. Values within a column followed by the same superscript letter are not significantly different ($P \leq 0.05$).
[b] TBA-RS of cooked tissue minus TBA-RS of raw tissue.

The role of microbiological factors, on the other hand, has been discounted as a factor in the oxidative stability of broiler meat during cooking (1). Observing no differences between inoculated and control samples for TBA and major volatiles, Ang et al. (1) postulated that their decrease in TBA numbers of cooked meat as a function of raw storage time was due to enzymic degradation of raw meat during storage followed by an increased formation of antioxidative Maillard reaction products during cooking.

Phospholipid hydrolysis, observed in both refrigerated (23,24) and frozen storage (25), could also play a biochemical/chemical role in the altered response displayed here in cooking refrigerated catfish and in Igene et al's (2) altered response in cooking frozen chicken. Free fatty acids released from phospholipids during storage may serve to decrease the lipid oxidation reactions induced during cooking by altering the membrane's integrity and ability to propagate the lipid oxidation process (26).

Conclusions. The degree of lipid oxidation induced in minced channel catfish muscle tissue by cooking has been shown to be dependent on the length of refrigeration that the product has undergone. Cooking of product stored for 5 days or less generated greater amounts of either TBA-RS or fluorescent pigments than cooking of product stored for 7 days. The decreased loss of alpha-tocopherol observed in cooked 7-day product compared to cooked 2- or 5-day product lends further support that an inhibition of lipid oxidation during cooking had occurred in the 7-day sample. Additional studies will be required to identify that factor or factors responsible for the inhibition but may involve pH, polyamines, Maillard reaction products, removal of hydroperoxides by microorganisms and/or phospholipase activity. Elucidation of this factor or factors could prove beneficial in providing the knowledge necessary to develop new and unique treatments or modification of existing treatments to optimally extend the shelflife of stored muscle foods.

Acknowledgments. This research was supported in part by State and Hatch funds allocated to the Georgia Agricultural Experiment Stations. The author wishes to thank Sandra O'Pry for technical assistance.

Literature Cited

1. Ang, C.Y.W., Lillard, H.S. and Searcy, G.K. *Poultry Sci.* **1989**, *68*, 1470-1477.
2. Igene, J.O., Pearson, A.M., Merkel, R.A. and Coleman, T.H. *J. Animal Sci.* **1979**, *49*, 701-707.
3. Pikul, J., Leszczynski, D.E., Niewiarowicz, A. and Kummerow, F.A. *J. Food Technol.* **1984**, *19*, 575-584.
4. Mai, J. and Kinsella, J.E. *J. Sci. Food Agric.* **1981**, *32*, 293-299.
5. Pikul, J., Leszczynski, D.E. and Kummerow, F.A. *Poultry Sci.* **1985**, *64*, 93-100.
6. St. Angelo, A.J., Vercellotti, J.R., Legendre, M.G., Vinnett, C.H., Kuan, J.W., James, C. and Dupuy, H.P. *J. Food Sci.* **1987**, *52*, 1163-1168.
7. Pikul, J. and Kummerow, F.A. *J. Food Sci.* **1990**, *55*, 30-37.
8. Keller, J.D. and Kinsella, J.E. *J. Food Sci.* **1973**, *38*, 1200-1204.
9. Pie, J.E., Spahis, K. and Seillan, C. *J. Agric. Food Chem.* **1991**, *39*, 250-254.
10. Igene, J.O., Pearson, A.M. and Gray, J.I. *Food Chem.* **1981**, *7*, 289-303.

11. Salih, A.M., Price, J.F., Smith, D.M. and Dawson, L.E. *Poultry Sci.* **1989**, *68*, 754-761.
12. Rhee, K.S. *J. Food Sci.* **1978**, *43*, 1776-1781.
13. Erickson, M.C. *Comp. Biochem. Physiol.* Part A., in press.
14. Erickson, M.C. and Selivonchick, D.P. *Lipids* **1988**, *23*, 22-27.
15. Vatassery, G.T. and Smith, W.E. *Anal. Biochem.* **1987**, *167*, 411-417.
16. Chen, C.C., Pearson, A.M., Gray, J.I., Fooladi, M.H. and Ku, P.K. *J. Food Sci* . **1984**, *49*, 581-584.
17. Nawar, W.W. In *Lipids*; Fennema, O.R., Ed.; Food Chemistry, 2nd edition; Marcel Dekker, Inc.: New York, NY, 1985; pp. 139-244.
18. Hearn, T.L., Sgoutas, S.A., Sgoutas, D.S. and Hearn, J.A. *J. Food Sci.* **1987**, *52*, 1430-1431.
19. Tadolini, B. *Biochem. J.* **1988**, *249*, 33-36.
20. Lovaas, E. *J. Am. Oil Chem. Soc.* **1991**, *68*, 353-358.
21. Tichivangana, J.Z. and Morrissey, P.A. *Irish J. Food Sci. Technol.* **1985**, *9*, 99-106.
22. Moerck, K.E. and Ball, H.R. Jr. *J. Agric. Food Chem.* **1979**, *27*, 854-859.
23. Ohshima, T., Wada, S. and Koizumi, C. *Bull. Jap. Soc. Sci. Fish.* **1983**, *49*, 1213-1219.
24. Sklan, D., Tenne, Z. and Budowski, P. *J. Sci. Food Agric.* **1983**, *34*, 93-99.
25. Shewfelt, R.L. *J. Food Biochem.* **1981**, *5*, 79-100.
26. Shewfelt, R.L. and Hultin, H.O. *Biochim. Biophys. Acta* **1983**, *751*, 432-438.

RECEIVED February 19, 1992

Author Index

Ackman, R. G., 208
Bailey, Milton E., 122
Berger, Ralf, 74
Bett, Karen L., 140,232,322
Bland, John M., 104
Boylston, T. D., 322
Civille, G. V., 279
Decker, Eric A., 33
Dus, C. A., 279
Erickson, M. C., 344
Flick, George J., Jr., 183
German, J. Bruce, 74
Grosch, Werner, 266
Gunnlaugsdottir, H., 208
Guth, Helmut, 266
Hong, Gi-Pyo, 183
Hultin, Herbert O., 33
Jovanovic, Slobodan V., 14
Kanner, Joseph, 55
Knobl, Geoffrey M., 183
Konopka, Ute Christine, 266
Labuza, Theodore P., 93
Lingnert, Hans, 292
Linssen, J. P. H., 302
Miller, James A., 104
Mills, O. E., 232
Nelson, Katherine A., 93
Niki, Etsuo, 14
Perkins, Edward G., 310
Roozen, J. P., 302
Shahidi, Fereidoon, 161
Simic, Michael G., 14
Spanier, Arthur M., 1,104,140
St. Angelo, Allen J., 1,140
Sullen, D. L., 232
Um, Ki Won, 122
Vercellotti, J. R., 1,232
Zhang, Hongjian, 74

Affiliation Index

A.R.O., The Volcani Center, 55
Agricultural Research Service, 1,104,140,232,322
Deutsche Forschungsanstalt für Lebensmittelchemie, 266
Institute for Nuclear Science Vinca, 14
Memorial University of Newfoundland, 161
Sensory Spectrum, Inc., 279
SIK—The Swedish Institute for Food Research, 292
Technischen Universität München, 74
Technology University of Nova Scotia, 208
U.S. Department of Agriculture, 1,104,140,232,322
University of California—Davis, 74
University of Georgia, 344
University of Illinois, 310
University of Kentucky, 33
University of Maryland—Baltimore, 14
University of Massachusetts—Amherst, 33
University of Minnesota, 93
University of Missouri, 122
University of Tokyo, 14
Virginia Polytechnic Institute and State University, 183
The Volcani Center, A.R.O., 55
Wageningen Agricultural University, 302

Subject Index

A

N-Acetyl-β-glucosaminidase, free radicals vs. activity, 111,113f
Add-back process, description, 293
Additives
  lipid oxidation in seafood, effect during storage, 203t,204f
  use for control of oxidation in foods, 122
Adenosine phosphates, role in lipid oxidation, 43
Albumin, biological role, 61
Alkanes, biomarkers of lipid oxidation, 29,30f
American Society for Testing Materials, development of descriptive terminology for sensory evaluation of lipid oxidation off-flavors, 283
Aminoreductones, inhibition of lipid oxidation, 126
Anisidine value
  commercial baby foods, 303,304t
  definition, 303
Antioxidant(s)
  depletion via lipoxygenases, 84,86f,87
  inhibition of lipid peroxidation, 67,68t
  lipid oxidation
    food processing, 300
    storage of seafood, 201,202t
  lipid oxidation inhibition
    kinetics, 25,26f
    mechanisms, 25,26f
    occurrence in foods, 23
    one-electron oxidation potentials, 27t
    redox potential, 25,27t,28,30f
  Maillard reaction
    inhibition of lipid oxidation, 125–128
    oil preservation, 125
Antioxidant-treated meat, chemical and sensory evaluation of flavor, 140–158
Arachidonic acid, peroxidation, 75,76f
Aroma compounds, localization in volatile fraction, 266
Aroma extract dilution analysis
  lipid oxidation product, effect on food quality, 6
  off-flavors, characterization, 266–277
Ascorbate
  catalysis of lipid oxidation, 39–41
  reduction of iron, 39

Ascorbic acid, role in curing, 171,173t
Atlantic redfish
  fat distribution, 213
  frozen storage life, 213t
Atmosphere, lipid oxidation in seafood, effect during storage, 202
Autoxidation
  linoleic acid 18:2, 75
  seafoods, 184

B

Baby foods, dry milk base, See Dry milk-base baby foods
Beef
  flavor descriptions, 283,284t
  flavor intensity ratings vs. storage, 284,285t
Beef diffusate, antioxidant properties, 129,131,132f
Beef peptide, free-radical effect, 111,114–117
Biochemical oxidation processes in situ, fish, 217,219,220f
Biomarkers of lipid oxidation
  alkanes, 29,30f
  hydroperoxides, 28
  4-hydroxyl-2-nonenal, 29
  thiobarbituric acid test, 29
Bligh and Dyer procedure, 222
Butter
  furanoid fatty acid content, 276,277t
  3-methylnonane-2,4-dione formation, 275–277t
Butter oil
  flavor source, 275
  fluorescent light vs. flavor, 275
  furanoid fatty acid content, 276,277t
  light vs. flavor dilution factor, 275,276t
  3-methylnonane-2,4-dione formation, 275–277t
tert-Butyl hydroperoxide, lipoxygenase of sturgeon, 87,88f

C

C-18 fatty acids, relative oxidation rates, 161,163
Carbonyl compounds, cellular signal transduction effect, 14

INDEX

Cardboardy flavor intensity in peanuts
  source compounds, 340–341
  storage effects, 338
Carp, lipoxygenase activity vs. storage
  time, 87,89f,90
Catfish muscle, changes in lipid oxidation
  during cooking, 344–349
Cathepsin D, free radicals vs. activity,
  111,115f
Cereal
  free fatty acid content vs. lipid
    oxidation, 294,296
  hexanal concentration after processing
    and after storage, 294,297f
  processing methods, 294
Ceruloplasmin
  biological role, 61
  ionic iron, role in lipid oxidation, 38
  lipid oxidation, prevention by redox
    iron, 50
  lipid peroxidation, inhibition, 68,69t
Chelators
  ionic iron, role in lipid oxidation, 38
  lipid oxidation, prevention by redox
    iron, 50
Chemical evaluation of flavor, untreated
  and antioxidant-treated meat, 140–158
Chicken, flavor intensity ratings vs.
  storage, 285t
Color, quality evaluation of fats, 317
Cooked meats
  beefy flavor notes, decrease on
    storage, 140–141
  lipid oxidation products, decrease on
    storage, 140
  warmed-over flavor, inhibition by Maillard
    reaction products, 129,130f
Cookie, baking vs. rancid flavor
  development during storage, 298,299f
Cooking
  lipid oxidation, 188–190t
  refrigerated minced channel catfish muscle,
    changes in lipid oxidation, 344–349
Creatine phosphokinase, free radicals vs.
  activity, 111,113f
Curing adjuncts, lipid oxidation,
  171,173–176
Cytochrome b_5 reductase, role in lipid
  oxidation, 42
Cytosol, Fe-mediated lipid peroxidation,
  68–70f

D

2,4-Decadienal, formation, 311
Deep frying, oil vs. fried food quality, 310
  fats, quality evaluation, 316–321
  fatty acid composition vs. frying time, 315
  lipid oxidation measurement, 311
  oxidative product determination,
    311,313–315
  volatile components, 315
Deep skinning, economic effect, 213
Deheated mustard flavor, lipid oxidation
  inhibition, 175,176f
Descriptive analytical methods, sensory
  evaluation of lipid oxidation
  off-flavors, 282
Descriptive sensory analysis
  procedure, 324
  storage vs. roasted peanut quality, 324–341
Desirable flavor quadrant, description, 158
Discoloration of fish muscle, 184–185
Disodium salt of ethylenediaminetetraacetic
  acid, lipid oxidation inhibition, 166,167f
Docosahexenoic acid, measurement of lipid
  oxidation in fish, 46–48f
Drum temperature, effect on liquid
  oxidation in fish, 194t
Dry milk-base baby foods
  anisidine values, 303–305f
  composition, 304
  hexanal concentration, 303–305f
  lipid oxidation measurement, 303–305
  metal carbonate vs. lipid oxidation and
    browning, 306,307f
  metal ions as lipid oxidation catalysts, 303
  mineral premix addition vs. lipid
    oxidation, 304–306
  processing, 304
  shelf life, 302
  temperature of storage vs. lipid
    autoxidation, 306,307f
  vitamin addition vs. lipid oxidation, 305f,306
  water activity of storage vs. lipid
    autoxidation, 306,307f
Drying, lipid oxidation in fish, 191,193–196

E

Edible oil
  descriptors of oxidized oil, 282t
  flavor quality scale, 281,282t

Endogenous agents, initiation of lipid
  oxidation, 15–16,17t
Enzymatic oxidation, lipoxygenases, 77
Enzymes
  lipid oxidation effect, 189–190,192f
  meat flavor, 106–107
$trans$-4,5-Epoxy-(E)-2-decenal,
  warmed-over flavor of meat, 6
Erythorbic acid, role in curing, 171,173t
Estrogen antioxidants, examples, 28,30f
Ethylenediaminetetraacetic acid (EDTA)
  inhibition of lipid oxidation, 41
  lipid oxidation effect, 200–201
  prevention of lipid oxidation by redox
    iron, 49
Exogenous agents, initiation of lipid
  oxidation, 17–19t,21f

F

Fats
  distribution in fish
    cod, 209
    mackerel, 209,210f
    microwave pretreatment vs. free fatty
      acid changes, 209,212f
    microwave pretreatment vs. peroxide
      values, 209,211f
    removal by deep skinning, 213
  quality evaluation, 316–321
  stability, influencing factors, 316
Fatty acids, number of double bonds vs.
  number of carbons susceptible to
  hydrogen abstraction, 82
Fenton reaction, example of ionic iron
  catalysis of lipid oxidation, 33–34
Ferritins
  iron source in muscle, 37–38
  lipid oxidation effect, 200
  loss of iron, 61t
  source of free iron in cells, 59
Ferrous iron, lipid oxidation effect, 200–201
Ferryl iron, generation, 58–60f
Fish
  biochemical oxidation processes in situ,
    217,219,220f
  fat distribution, 209–213
  frozen storage life, 213t
  muscle types, 213–214

Fish—$Continued$
  nomenclature, 208
  prevention of flavor degradation caused
    by lipid oxidation, 5
Fish fillets
  antioxidant effect on lipid oxidation,
    216–218f
  fatty acids, 216t
  lipid content vs. muscle type, 214t
  origin of free fatty acids, 214,215t
  total lipid analysis, 214,215t
Fish lipids
  chemical composition, 183
  degree of unsaturation, 183
  oxidation as cause of quality loss
    during storage, 184–185
Fish lipoxygenases
  sequential reactions, 82,84,85f
  specificity, 77–83
Fish meal
  antioxidant additive effect, 224
  fat content, 221
  fatty acid composition of lipids vs.
    storage conditions, 222–223,224–225t
  free fatty acid generation, 221
  growth of salmon vs. fish meal
    processing temperatures, 222t
  lipid content, 221
  lipid extraction procedures, 222,223t
  lipid extraction using Bligh and Dyer
    procedure, 222,223t
  market, 221
  production process, 221
Fish muscle, preservation from lipid
  oxidation, 9
Fish muscle sarcoplasmic reticulum, lipid
  peroxidative activity, 46
Fish oils
  mechanisms of volatile generation, 75,77
  oxidation processes, 217,218f
Fish silage
  animal feed, use, 225
  content, 225
Fishy aromas, source, 217
Flavor
  chemical and sensory evaluation in
    untreated and antioxidant-treated
    meat, 140–158
  lipid oxidation effect, 169
Flavor dilution factor, definition, 6,266

INDEX

Flavor evaluation in untreated and
antioxidant-treated meat
  antioxidant effect on off-flavor, 157–158
  chelator effects, 157–158
  chemical analytical procedure, 142
  chemical and flavor attributes, correlation, 143,144f
  chemical and instrumental data statistical evaluation, 151,154t
  desirable flavor quadrant, 158
  hexanal measurement, 156
  instrumental methodology, 142–143
  loss of beefy/brothy aroma, 156
  principal component analysis of chemical and flavor treatments, 154–156
  sample preparation, 141–142
  sensory, chemical, and instrumental data statistical evaluation, 154t
  sensory analytical procedure, 142
  sensory data statistical evaluation, 151,153t
  statistical analytical procedure, 143
  vacuum and additives
    vs. chemical attributes, 143,145f,147
    vs. instrumental attributes, 143,145–146f,147
    vs. off-flavor attributes, 147,148–149f
  vacuum vs. desirable flavor attributes, 147,150f–152t
Flavor of meat, relationship among cooking, curing, storage, and meat flavor deterioration, 171,172f
Flavor production, properties of lipoxygenases relevant to role, 87,88–89f,90
Flavor retention, role of nitrite-free curing systems, 175,177,178t
Flavor stability, food processing effect, 292–300
Flounder, lipid oxidation, 47,48f
Food(s)
  lipid oxidation, 1–9
  lipoxygenases in lipid oxidation, 74–90
  Maillard reaction products as antioxidants, 128–129
  prevention of autoxidation, 4–5
  sensory evaluation of lipid oxidation, 279–288
Food flavor(s)
  mechanisms of deterioration, 141
  volatiles, analytical difficulties, 232
Food preservation, control of oxidation, 122

Food processing
  effect on lipid oxidation and flavor stability
    antioxidant effect, 300
    cereal, 294,296,297f
    cookies, 298,299f
    headspace analytical procedure for volatiles, 293
    initiation and stimulation of lipid oxidation, 298
    oxygen level effect, 300
    potato granules, 293–295f
    prooxidant effect, 298
    sausage, 296,297f,298,299f
    sensory evaluation procedure, 293
  lipid oxidation effect, 7–9
Food quality, oxidative process effect, 14
Free copper, role in lipid peroxidation, 61–62
Free iron, See Low molecular weight iron in muscle
Free metal ions
  content in turkey red muscles, 59,61t
  generation, 59,61t,62
  source in cells, 59
Free radicals
  beef enzyme activity effect, 111,113f,115f
  damage to cellular membranes, 109,110f
  generation, induction of lipid oxidation, 109,111,112f
  induction, 109
  oxidation generating system
    beef enzyme activity effect, 111,113f,115f
    description, 107,109
    specific beef peptide effect, 111,114–117
  scavengers, preservation of meat flavor by inhibition of lipid oxidation, 4–5
  specific beef peptide effect, 111,114–117
Free volume, definition, 98
Freezing, lipid oxidation effect, 186–188t
Fresh flavor, components, 74–75
Freshness, description, 75
FROG, See Free radicals, oxidation generating system
D-Fructosamine, inhibition of lipid oxidation, 126,127f
Fruity/fermented flavor intensity in peanuts, storage effects, 340
Frying fats
  changes during use, 310,312f
  time vs. peroxide formation and decomposition, 310–312f

Furan fatty acids, role in off-flavors, 6–7
Furanoid fatty acid, 3-methylnonane-2,4-
  dione formation during photooxidation,
  273t,274f

G

Gas chromatographic analyses
  lipid oxidation volatiles in foods
  advances affecting food flavor
    measurements, 232–234
  correlation to sensory data, 234
  instrumentation, 234
  peanut flavor volatile analysis by
    concentration on solid supports, 235–236
  peanut flavor volatiles
  adsorbent material vs. recovery, 243,245f
  adsorbent trap preparation, 238
  compound affinity for supports vs.
    recovery, 247–248
  compounds used for purge and trap
    evaluation, 240t
  direct GC procedure, 236–237
  flame ionization detector calibration
    procedure, 238
  geometry of sample vessel vs. recovery,
    241,243,244f
  olfactory attributes, comparison,
    251,252–257f,258
  packed column vs. wide-bore capillary,
    258,260,261–262f
  purge and trap system
    experimental procedure, 238–239
    factors affecting performance, 240
    problems in quantitation, 247
  recovery
    calculation procedure, 239
    studies, test solution preparation, 238
  study samples, GC procedure, 239
  volatiles from oil, 243,246f
  reproducibility, 243,247t
  response standard preparation, 248–250
  retention times of lipid oxidation
    products, 258,259t
  sample tube preparation, 236
  sparging vs. recovery, 243,245f
  time vs. recovery, 240–241,242f
  vacuum in adsorbent tube vs. recovery,
    243,244f

Gas chromatographic analyses—*Continued*
  peanut flavor volatiles—*Continued*
    vapor pressure vs. recovery, 247,248t
    wide-bore capillary GC procedure, 237–238
Gas chromatography and sniffer port,
  assessment of lipid oxidation product effect
  on food quality, 5–6
Glass transition temperature, definition, 98
Glass transition theory
  applications, 98
  polymer, *See* Polymer glass transition theory

H

Heme proteins, catalysis of lipid
  peroxidation, 62
Hemoglobin
  iron source in muscle, 36–37
  lipid oxidation effect, 200
Hemoproteins, role as prooxidant in lipid
  oxidation of meat, 166
Hemosiderin, iron source in muscle, 38
Herbs, lipid oxidation inhibition, 175
Herring, lipid oxidation, 46–48f
Hexanal
  formation, 311
  indicator of lipid oxidation, 179
  measurement in cooked meat, 156
Histidine, role in lipid oxidation, 43,44–45t
Homogenized roasted peanuts, commercial
  importance of flavor, 236
Hydrogen peroxide, generation, 57
Hydroperoxides
  biomarkers of lipid oxidation, 28
  DNA effect, 14
  formation of secondary oxidation
    products, 161
4-Hydroxy-2-nonenal, biomarker of
  lipid oxidation, 29
Hydroxyl radicals
  formation of C-central, S-centered, and
    heterocyclic radicals, 16–17
  generation, 57,58t
  scavengers, inhibition of
    lipid peroxidation, 67,68t

I

Ingredients, lipid oxidation effect,
  191,192f

INDEX

Inhibition of lipid oxidation
 antioxidants, 23,25–28,30
 definition, 177
 food, 4–5
Initiation of lipid oxidation
 endogenous agents, 15–17t
 exogenous agents, 17–19t
Insoluble proteins, iron sources in muscle, 38
Ionic iron, importance of chelators and
 ceruloplasmin in establishing function, 38
Ionizing radiation, initiation of lipid
 oxidation, 18
Iron
 contact vs. lipid oxidation in fish,
  194,197t,198f
 function for life, 33
 lipid oxidation
  initiation, 15
  prevention, by reducing concentration, 50
  role, 2
 muscle foods, content, 33
 tissue, distribution, 34,35t,36
Iron, lipid oxidation dependence
 ascorbate, 39–41
 enzymes involved, 42–49
 inhibition by EDTA, 41
 NADH and NADPH, 41
 prevention strategies, 39
 sodium chloride, 41–42
 superoxide anion, 39
Iron, lipid peroxidation dependence on
 redox cycle
 control and prevention, 69
 cytosol effect, 68–70f
 Fe supplementation effect, 69,70f
 inhibition
  by ceruloplasmin, 68,69t
  by hydroxyl radical scavengers and
   antioxidants, 67,68t
  by superoxide dismutase, 66–67t
 initiation by ferrous ions, 64–65
 production of superoxide, hydrogen
  peroxide, and hydroxyl radical by
  ferrous ions, 64
Iron-catalyzed lipid oxidation,
 influencing factors, 34
Iron sources, protein bound, *See*
 Protein-bound iron sources in muscle
Iron supplementation, lipid peroxidation
 effect, 69,70f

L

Lactotransferrin, iron source in muscle, 38
Linoleic acid
 peroxidation, 75,76f
 peroxidation mechanism, 20,21f
Linoleic acid 18:2, autoxidation, 75
Linolenic acid, peroxidation, 75,79f
Lipid(s)
 cell metabolism, 1
 oxidation, 1–9
 polyunsaturated fatty acids, flavor
  deterioration in cooked meat, 106,108f
Lipid oxidation
 biomarkers, 28–30f
 cereal, 294,296,297f
 cookies, 298,299f
 flavor effect, 169
 food processing effect, 292–300
 induction, 109
 inhibition by antioxidants, 23,25–28,30
 lipoxygenases in foods, 74–90
 measurement in dry milk-base baby foods,
  303,304t,305f
 measurement methods, 311
 mechanisms, 15–30,93
 muscle tissue, relationship to pigment
  oxidation, 47
 off-flavors, development, 184
 potato granules, 293–295f
 prevention, Maillard reaction products,
  122–136
 prevention, nitrite, 164–168
  antioxidant and flavor effects, 164–168
  assessment, 177,179
  curing adjuncts effect, 171,173–176
  curing of meat, 163–164
  nitrite-free curing systems, 175,177,178t
  volatiles of nitrite-cured meats, 168–172
 product determination, 311,313–315
 quality deterioration, cause in muscle
  foods, 161
 rates, influencing factors, 93
 reaction rate vs. water activity, 94,95f
 redox iron
  enzymes involved, 42–49
  prevention strategies, 49–50
 roasted peanut flavor effect, 322
 sausage, 296–299f
 sensory evaluation in foods, 279–288

Lipid oxidation—*Continued*
  volatiles, concentration and
    separation, 233–234
Lipid oxidation, changes during cooking of
    refrigerated catfish muscle
  chemical and biochemical effects, 349
  cooking of fresh product, response, 345t,346
  cooking of stored product, response,
    346,347–348t
  fatty acid esterification procedure, 345
  fluorescent pigment determination, 345
  lipid extraction procedure, 345
  mechanisms, 347,349
  microbiological effects, 347,349
  sample preparation, 344
  statistical analyses, 345
  thiobarbituric acid reactive substance,
    analytical procedure, 344–345
  tocopherol, analytical procedure, 345
Lipid oxidation, foods
  catalytic systems, 1
  flavor degradation in fish, 5
  food processing effect, 7–9
  food quality, methods for assessment of
    effect, 5–7
  free-radical mechanisms, 2
  inhibition by free-radical scavengers, 4–5
  iron catalysis, 2
  lipoxygenases, 3
  muscle foods, problem during
    frozen storage, 46
  pathways of nonenzymic lipid
    peroxidation after cellular injury, 2–3
  polypeptide effect, 3–4
  research importance, 9
  seafood, during storage
    additive effect, 203t,204f
    antioxidant effect, 201,202t
    atmosphere effect, 202
    cooking effect, 188,189–190t
    enzyme effect, 189–190,192f
    fish part effect, 191
    freezing effect, 186–187,188t
    influencing factors, 185–199
    ingredient effect, 191,192–193f
    metal catalysis, 200–201
    packaging material effect, 202,204f
    processing technology effect, 191,193–199
    relative humidity effect, 203
    season effect, 190–191

Lipid oxidation, foods—*Continued*
  seafood, during storage—*Continued*
    species effect, 190
    temperature effect, 186
    water content effect, 185
  storage
    development, 292,295f
    influencing factors, 292
    water activity vs. oxidation, 3
Lipid oxidation, meat proteins
  background, 106–107,108f
  free radicals vs. beef enzyme activity,
    111,113f,115f
  free radicals vs. specific beef peptide,
    111,114–117
  induction of lipid oxidation and free
    radicals, 109,110f
  lipid oxidation induction by free-
    radical generation, 109,111,112f
  liposome preparation, 107,108f,109
Lipid peroxidation
  free-radical mechanism, 184
  influencing factors, 184
  initiators, 163
  quality deterioration in stored foods, cause, 55
Lipoxygenases, role in lipid oxidation, 3
Lipoxygenases in lipid oxidation in foods
  autoxidation, 75–77,79f
  depletion of antioxidants and
    peroxidases, 84,86f,87
  enzymatic oxidation, 77
  flavor production, 87,88–89f,90
  sequential reaction of fish
    lipoxygenases, 82,84,85f
  specificity of fish lipoxygenases, 77–83
Low molecular weight iron in muscle
  description, 34
  muscle tissue type vs. concentration, 34,35t
  processing conditions vs. concentration, 35
  sources, 35–36
  species vs. concentration, 34
  storage condition vs. concentration, 35

M

Macadamia nuts, moisture content vs.
  oxidative stability, 96,98
Maillard reaction
  definition, 123
  description, 123,124f,125

INDEX

Maillard reaction—*Continued*
  lipid oxidation prevention
    antioxidant(s), 125–128
    antioxidant properties of beef
      diffusate, 129,131,132f
    antioxidant properties of melanoidins,
      128,130f
    application as antioxidants in food,
      128–129
    inhibition of warmed-over flavor in
      cooked meat, 129,130f
    synthetic meat flavor, effect on meat
      flavor, 131,133–136
  pathways to intermediate compound
    formation in flavor, 123–125
Malondialdehyde, inhibition of formation,
  177,178t
Meat
  beef flavor descriptions, 283,284t
  beef flavor intensity ratings vs.
    storage, 284,285t
  descriptors for warmed-over flavor, 280t,281
  nitrite curing, 163–164
Meat flavor
  deterioration
    cause, 140
    degree of unsaturation of fatty acid
      constituents vs. rate, 161–163
    identification of two mechanisms, 157
    influencing factors, 141
    lipids, 106,108f
    nitrite-free curing systems, role in
      prevention, 175,177,178t
    rate, factors affecting, 163
    See also Warmed-over flavor
  Maillard reaction products vs. sensory
    scores, 131,133f,134
  NaCl vs. sensory scores, 134,136f
  preservation by inhibition of lipid
    oxidation with free-radical
    scavengers, 4–5
  quality
    influencing factors, 104–105
    warmed-over flavor, 105
  synthetic meat flavor mixture as
    preservative, 131,133–136
  synthetic meat flavor vs. sensory
    scores, 134,135f
Meat proteins, lipid oxidation effect, 106–117
Meat tenderization, mechanism, 106–107

Meaty flavor of muscle foods, nonlipid
  portion as contributing factor, 105
Mechanisms of lipid oxidation
  inhibition by antioxidants, 23,25–28,30
  initiation
    by endogenous agents, 15–17t
    by exogenous agents, 17–19t,21f
  propagation, 19–24,26
Melanoidins
  antioxidant properties, 128,130f
  inhibition of lipid oxidation, 126,128
Metal carbonates, lipid extraction of dry
  milk-base baby food effect, 306,307f
Metal catalysis, lipid oxidation of seafoods
  EDTA effect, 200–201
  ferritin effect, 200
  ferrous iron effect, 200–201
  hemoglobin effect, 200
  hemoprotein and nonheme iron, relative
    importance as catalysts, 200
  metal ions, relative activity, 200
  myoglobin effect, 200–201
  nitrite effect, 200–201
Methemoglobin, initiation of lipid
  peroxidation, 58–60f
Methyl linoleate, oxygen absorption, 96,97f
3-Methylnonane-2,4-dione
  formation during photooxidation of butter
    and butter oil, 276,277t
  formation during photooxidation of
    soybean oil, 273,274f
Metmyoglobin, initiation of lipid
  peroxidation, 58–60f
Milk-base dried powdered infant
  formulations, preservation from lipid
  oxidation, 8
Milk fat, 3-methylnonane-2,4-dione
  formation, 275–277t
Mineral premix, addition vs. lipid oxidation
  of dry milk-base baby foods, 304–306
Moisture content, relationship to water
  activity, 94,95f
Muscle, protein-bound iron sources, 36t–38
Muscle foods
  iron content, 33
  lipid oxidation via redox iron, 33–50
  nonenzymic lipid peroxidation, mechanism,
    55–70
  prevention of lipid oxidation by nitrite
    and nitrite-free compositions, 161–179

Muscle in fish, types, 213–214
Muscle type, lipid oxidation in fish
  effect, 197t
Myoglobin
  initiation of lipid oxidation, 16
  iron source in muscle, 36–37
  lipid oxidation effect, 200–201

N

Nicotinamide adenine dinucleotide
  (phosphate), reduced NAD(P)H, role
  in enzyme-catalyzed lipid oxidation,
  42–43
Nitrite
  antioxidant effect in cooked meat, 164–168
  Fe^{2+}, reaction, 166t
  health hazard, 175
  inhibition of lipid oxidation for
    preservation of meats, 4
  lipid oxidation effect, 200–201
  lipid stabilization, mechanism, 166
Nitrite curing of meat
  history, 163
  lipid oxidation, assessment, 177,179
  NaCl, use, 163–164
  nitrite, role, 164
  scientific investigations, 163
  volatiles in cured meats, 168–172
Nitrite-free curing systems
  importance, 175
  inhibition of malondialdehyde formation,
    177,178t
  lipid oxidation, assessment, 177,179
S-Nitrosocysteine, antioxidant effect, 167
Nitrosylheme compounds, antioxidant
  effect, 166–167
Nonenzymic browning, reaction rate vs.
  water activity, 94,95f
Nonenzymic lipid oxidation via redox iron
  ascorbate, 39–41
  inhibition by EDTA, 41
  NADH and NADPH, 41
  sodium chloride, 41–42
  superoxide anion, 39
Nonenzymic lipid peroxidation
  alternate pathways after cellular injury, 2–3
  muscle foods
    activation of oxygen species and metal
      compounds, 56–65

Nonenzymic lipid peroxidation—*Continued*
  muscle foods—*Continued*
    cytosolic fraction effect, 62–64
    ferryl ion reduction by carnosine, 64,65f
    glucose oxidase system effect, 62,63f
    heme proteins as catalysts, 62
    initiation pathways, 56,60f
    iron redox cycle dependent peroxidation,
      64–70
    model systems vs. in situ systems, 62–65f
Nonvolatile oxidation products,
  determination, 311,313–315
North Atlantic cod, fat distribution, 209
North Atlantic mackerel, fat distribution,
  209,210f

O

Off-flavors
  characterization by aroma extract dilution
    analysis
    butter and butter oil, 275–277t
    soybean oil, 269–275
    warmed-over flavor, 267,268–269t
  sensory evaluation of lipid oxidation,
    279–288
  soybean oil
    capillary GC, 270,271f
    concentration and flavor threshold values
      of odorants, 269,270f
    formation of 3-methylnonane-2,4-dione,
      273t,274f
    light exposure vs. odorant content,
      269,270t
    quantification of fatty acids, 273,275t
    structures of fatty acids, 270,272f
Olfactory response analysis of roasted
  peanut volatiles
  comparison to GC chromatograms,
    251,252–258
  procedure, 251
Omega-3 fatty acids, seafoods and fishery
  byproducts, 208–225
One-electron oxidation potentials,
  antioxidants, 27t
Ovotransferrin, iron source in muscle, 38
Oxidation in food products, development
  during storage, 292,295f
Oxidation products, determination,
  311,313–315

INDEX

Oxidative processes, changes caused in food quality, 14
Oxygen level, lipid oxidation effect during food processing, 300
Oxyl radicals, reaction rate constants with biologically related substrates, 17t
Ozone, initiation rate constants of fatty acid oxidation, 18,19t

P

Packaging, lipid oxidation in seafood, effect during storage, 202,204f
Painty flavor intensity in peanuts
  source compounds, 340–341
  storage effects, 334,338
Parts of fish, lipid oxidation effect, 191
Peanut(s)
  comparison of fresh to oxidized paste flavor intensity ratings, 287t
  descriptors of flavor, 281
  flavor descriptors, 286t,287
  roasted, See Roasted peanuts, quality
Peanut flavor volatiles
  analysis by concentration on solid supports, 235–236
  GC analysis, 237–262
Pearson correlation coefficients, statistical evaluation of flavor, 151,153–156
Perhydroxyl radical
  generation, 56
  initiation of lipid peroxidation, 56
Peroxidases, depletion via lipoxygenases, 84,86f,87
Phenoxyl radical, mesomeric forms, 25,26f
Phosphates, role in curing, 173t
Phospholipases, inhibition of lipid oxidation
  A_2, 45
  C, 45–46
Phytate, prevention of lipid oxidation by redox iron, 49
Polar components, use for quality evaluation of fats, 317,318–321
Polymer glass transition theory
  collapse of polymer system, 99
  crystallization vs. transition temperature, 99,101
  description, 98
  free volume, 98
  oxidation of limonene in orange oil, 99,100f

Polymer glass transition theory—*Continued*
  theoretical state diagram, 98,100f
  water as plasticizing agent, 98
Polypeptides, lipid oxidation effect, 3–4
Polyphosphates
  lipid oxidation inhibition, 173–175
  role in curing, 173t
Polyunsaturated fatty acids modification
  catalysis of secondary sites, 75–77,79f
  depletion of antioxidants and peroxidases, 84,86f,87
  enzymatic oxidation via lipoxygenases, 77
  oxidation as source of off-flavors in foods, 74
  peroxidation, 55
Porous polymer trapping methods, analysis of food flavor volatiles, 233
Postharvest storage, lipid oxidation, 7–9
Potato chips, measurement of oxidation, 96,97f
Potato granule
  add-back process for production, 293
  fatty acid release during production, 293
  hexanal formation during blanching, 294,295f
  lipid oxidation, mechanism, 294
Poultry, descriptors for warmed-over flavor, 280t,281
Precooked beef patties, storage vs. flavor, 106,108f
Processing, lipid oxidation effect, 7–9
Processing technology, lipid oxidation effect, 191,193–199
Prooxidants, lipid oxidation effect during food processing, 298
Propagation of lipid oxidation
  chain branching, 23
  conditions, 22
  elements, 19
  energetics of peroxidation, 19–21f
  initiation rate, 22–24f
  kinetics, 20
  location of H atom abstraction, 20,21t
  mechanisms, 20
  reaction kinetics of free radicals with oxygen, 19
  resonance, 20,21f
  structure of aggregates, 23,24f
  temperature, 23,26f

Protein, role in meat flavor, 106–107
Protein-bound iron sources in muscle
   insoluble proteins, 38
   soluble proteins, 36t,37–38
Protein extenders, lipid oxidation
   inhibition, 175

Q

Quality evaluation of fats
   color, 317
   polar components, 317,318–321
   procedure, 316–317

R

Rancidity in foods, evidence for lipid
   oxidation as source, 234
Redox iron, 39–49
   See also Iron, lipid oxidation
     dependence *and* Iron, lipid
     peroxidation dependence on redox cycle
Redox potentials, antioxidants, 27t
Reductone, Maillard reaction products,
   125–126
Refrigerated minced channel catfish
   muscle, changes in lipid oxidation
   during cooking, 344–349
Roasted peanuts
   flavor
     composition, 322
     influencing factors, 322–323
     lipid oxidation effect, 322
   flavor intensity
     source compounds, 340
     storage effects, 331,332–333t
   lipid oxidation effect, 8–9
   quality, storage effects, 323–341

S

Salmon, sensitivity to oxidation products,
   224–225
Salt, role in curing, 171,173t
Sausage
   CO_2 and vacuum vs. air for storage
     stability, 296,298
   hexanal formation during storage, 296,297f
   production method, 296

Sausage—*Continued*
   rancid flavor development during
     storage, 296,299f
Seafood, lipid oxidation during storage,
   183–204
Season, lipid oxidation effect, 190–191
Sensory characteristics of lipid-based
   food, terminology, 279
Sensory evaluation
   flavor, untreated and antioxidant-
     treated meat, 140–158
   food
     importance, 7
     measurements, 7
   lipid oxidation off-flavors
     descriptive analytical methods, 282
     past terminology and methodology,
       280–282t,283
     recent development in descriptive
       terminology, 283–288
Sequential reactions of fish lipoxygenases, enzymatic catalysis of
   polyunsaturated fatty acids, 82,84,85f
Sexual maturation, lipid oxidation
   effect, 219
Shellfish, recommended frozen storage
   life, 213t
Silicones, frying oil stability effect, 316
Singlet oxygen
   generation, 56
   initiation of lipid oxidation, 17–18
   initiation of lipid peroxidation, 56
   initiation rate constants of fatty acid
     oxidation, 18,19t
Smoke, use as antioxidant, 177
Smoking
   lipid oxidation effect, 219
   lipid oxidation in fish effect, 197,199t
Sodium chloride
   meat flavor effect, 134,136f
   nonenzymic lipid oxidation effect, 41–42
Sodium pyrophosphate, role in curing, 173
Sodium tripolyphosphate
   curing, role, 173t
   lipid oxidation inhibition, 166,167t
Soluble proteins, iron sources in muscle,
   36t,37–38
Soybean oil, light-induced off-flavors,
   269–275
Species, lipid oxidation effect, 190

Specificity of fish lipoxygenases
  arachidonic acid
    oxidation products, 78,79f
    primary metabolite, 78,80f
  concentrations of discriminating volatiles in trout gill homogenates, 78,81t
  elution of trout gill lipoxygenases from hydroxylapatite column, 81,83f
  esculetin effect on release of hydroxy fatty acids, 82,83f
  hydroxylated polyunsaturated fatty acids from endogenous fatty acids, 78,80f
  specificity vs. final flavor profile, 77–78
  substrate specificity and products generated by $N-6$ lipoxygenase, 81,82t
Spices, lipid oxidation inhibition, 175
Storage condition, lipid extraction of dry milk-base baby food effect, 306,307f
Storage effects
  roasted peanut quality
    cardboardy flavor, 338
    descriptive sensor analytical procedure, 324
    flavor attributes vs. chemical compounds, 340–341
    fruity/fermented flavor, 340
    lipid oxidation products, 331,334,335–337t
    painty flavor, 334,338
    peanut seed size vs. flavor, 325,327–329t
    roasted peanutty flavor intensity, 331,332–333t
    sample preparation, 323
    statistical analysis of data, 324
    storage time vs. flavor, 325,327–329t
    storage time vs. heterocyclic compound content, 325,330t,331
    sweet aromatic flavor intensity, 331,332–333t
    treatment vs. sensory attributes, 324–325,326
    volatile compounds, 338–340
    volatile flavor compound analytical procedure, 323–324
    woody/hulls/skins flavor intensity, 331,332–333t
  seafood, lipid oxidation, 183–204
Strecker degradation, role in intermediate compound formation in flavor, 123–125
Sulfanilamide, indicator of lipid oxidation, 179

Sulfite, initiation rate constants of fatty acid oxidation, 18,19t
Superoxide
  generation, 56
  initiation of lipid peroxidation, 56
Superoxide anion
  preparation, 39
  reduction of iron, 39
Superoxide dismutase, inhibition of lipid peroxidation, 66–67t
Superoxide radicals, initiation of lipid oxidation, 16
Sweet aromatic flavor intensity in peanuts
  source compounds, 341
  storage effects, 331,332–333t
Synthetic meat flavor
  preparation, 131
  sensory score effect, 134,135f

T

Temperature
  initiation of lipid oxidation, 18,21f
  lipid extraction of dry milk-base baby food effect, 306,307f
  lipid oxidation effect, 186
Termination reactions, water effects, 96
2-Thiobarbituric acid test
  biomarker of lipid oxidation, 29
  lipid oxidation in nitrite and nitrite-free cured meats, assessment, 177,179
  oxidation in fish tissue, measurement, 216
Thiyl radicals, initiation rate constants of fatty acid oxidation, 18,19t
α-Tocopherol
  inhibition of hydroperoxide formation, 25,26f
  seasonal effects, 219,220f
Transferrin, iron source in muscle, 38
Trimethylamine oxide, quality indicator, 217
Triose reductone, formation, 126
Trout
  lipid oxidation, 47,49
  lipoxygenase activity vs. storage time, 87,89f,90
  sensitivity to oxidation products, 224–225
Trout gill homogenates, concentrations of discriminating volatiles, 78,81t

U

Ultraviolet light, initiation of lipid oxidation, 18
Untreated meat, chemical and sensory evaluation of flavor, 140–158

V

Vegetable oils
  flavor intensity ratings vs. storage, 287,288t
  lipid oxidation effect, 8
Vitamins, addition vs. lipid oxidation of dry milk-base baby foods, 305f,306
Volatile compounds of nitrite-cured meats
  chemical composition, 168t
  curing vs. aldehyde concentration, 169t
  hexanal content vs. linoleic acid content, 168–170f
Volatile flavor compound analysis
  procedure, 323–324
  storage vs. roasted peanut quality, 324–341
Volatile lipid oxidation products, odor thresholds, 169,171t
Volatile oxidation products, determination, 311,313–315

W

Warmed-over flavor
  beef flavor
    descriptions, 283,284t
    intensity ratings vs. storage, 284,285t
  causes, 267
  compound identification, 267
  definition, 105
  description, 280

Warmed-over flavor—*Continued*
  inhibition in cooked meat by Maillard reaction products, 129,130f
  storage vs. flavor dilution factors, 267,268t
  storage vs. odorant levels, 268,269t
  terminology, 280t
  See also Meat flavor, deterioration
Washing, lipid oxidation in fish effect, 194,196f,197
Water
  content, lipid oxidation effect, 185
  food system function, 93
  plasticizing agent function, 99
Water activity
  control of lipid oxidation, 93
  definition, 94
  lipid extraction of dry milk-base baby food effect, 306,307
  lipid oxidation effect, 3
  moisture content effect, 94,95f
Water–lipid oxidation rate relationship
  influencing factors, 98
  interfacial bonding, 94,96
  mechanism, 96
  moisture sorption isotherm, 94,95f
  polymer glass transition theory, 98–99,101
  reaction rate vs. water activity, 94,95f
  studies, 96–98
  water activity vs. termination reaction, 96
Woody/hulls/skins flavor intensity in peanuts
  source compounds, 341
  storage effects, 331,332–333t

X

Xanthine oxidase, initiation of lipid oxidation, 16

Production: Paula M. Bérard
Indexing: Deborah H. Steiner
Acquisition: Barbara C. Tansill

Printed and bound by Maple Press, York, PA